DEVELOPMENTS IN DAIRY CHEMISTRY—3

Lactose and Minor Constituents

CONTENTS OF VOLUMES 1 and 2

Volume 1: Proteins

Volume 2: Lipids

DEVELOPMENTS IN DAIRY CHEMISTRY—3

Lactose and Minor Constituents

Edited by

P. F. FOX

Department of Dairy and Food Chemistry,
University College, Cork, Republic of Ireland

ELSEVIER APPLIED SCIENCE PUBLISHERS
LONDON and NEW YORK

ELSEVIER APPLIED SCIENCE PUBLISHERS LTD
Crown House, Linton Road, Barking, Essex IG11 8JU, England

Sole Distributor in the USA and Canada
ELSEVIER SCIENCE PUBLISHING CO., INC.
52 Vanderbilt Avenue, New York, NY 10017, USA

WITH 97 ILLUSTRATIONS AND 59 TABLES

British Library Cataloguing in Publication Data

Developments in dairy chemistry.—(The Developments
   series)
   3 : Lactose and minor constituents
   1. Dairy produce—Analysis
   I. Fox, P. F.
   637′.01′543   SF253

The Library of Congress has cataloged this work as
   follows:
Developments in dairy chemistry. — 1 – — London; New York:
   Applied Science Publishers, c1982–
       v.: ill.; 23 cm.—(Developments series)
   ISSN 0264-8407 = Developments in dairy chemistry.

   1. Dairy products—Analysis—Collected works.   I. Series.
   [DNAL: 1. Dairy products—Analysis]
   SF253.D44   637′.028′7—dc19   84-644649

ISBN   0-85334-370-5

Photoset in Malta by Interprint Ltd
Printed in Great Britain by Page Bros. (Norwich) Limited

PREFACE

This volume is the third in the series on the chemistry and physical chemistry of milk constituents. Volumes 1 and 2 dealt with the commercially more important constituents, proteins and lipids, respectively. Although the constituents covered in this volume are of less direct commercial importance than the former two, they are nevertheless of major significance in the chemical, physical, technological, nutritional and physiological properties of milk.

Lactose, the principal component of the milks of most species, is a rather unique sugar in many respects—it has been referred to as one of Nature's paradoxes. It is also the principal component in concentrated and dehydrated dairy products, many of the properties of which reflect those of lactose. The chemistry and principal properties of lactose have been thoroughly researched over the years and relatively little new information is available on these aspects; this new knowledge, as well as some of the older literature, is reviewed in Chapter 1.

Although lactose has many applications in the food, pharmaceutical and chemical industries, not more than 10% of the potentially available lactose is actually recovered as such. Like other sugars, lactose may be modified by a multitude of chemical reagents; some of these are reviewed in Chapter 2 and some applications of the derivatives discussed. The enzymatic hydrolysis of lactose to glucose and galactose has considerable technological as well as nutritional significance, and the recent literature on this subject is reviewed in Chapter 3. Lactose is not digestible by the majority of the world's population, and the current views on this nutritionally important problem are discussed in Chapter 4. A deficiency of either of two enzymes involved in the Leloir pathway for galactose metabolism leads to the inability to metabolize galactose produced from lactose (or other galactose-containing sugars) and causes two relatively rare congenital diseases referred to as glactosaemia, the literature on which is reviewed in Chapter 5.

Quantitatively, the salts of milk are minor constituents but they play a disproportionately important role in many of the technologically important properties of milk, some of which have been discussed in Volume 1 of this series. Recent literature on the rather complex chemistry of the milk salts *per se* is reviewed in Chapter 6. Many of the inorganic constituents of milk, some of which are present only at trace levels, are also of very considerable nutritional significance. Since a variety of minerals are required for proper growth and development, and milk is the sole source of these requirements at a critical stage of infant growth, the significance of milk as a source of dietary minerals is discussed in Chapter 7.

The flavour/off-flavour of milk and dairy products is undoubtedly technologically important and extremely complex. This topic could easily occupy a full volume in this series but a comprehensive summary is presented in Chapter 8.

Many people may regard milk simply as a source of lipids, proteins, carbohydrates and minerals, with very little biological activity as such. This, in fact, is not the case; milk contains a great variety of biologically active species, some of which, e.g. enzymes, may cause undesirable changes in milk and dairy products during storage, while others, e.g. vitamins, immunoglobulins, are of very considerable nutritional and biological significance. Chapters 9, 10 and 11 review the recent literature on the indigenous enzymes in milk, indigenous antibacterial systems and vitamins, respectively. The importance of at least some of the indigenous enzymes and vitamins is well established but the indigenous antibacterial systems may be of much greater significance than considered heretofore, and it is hoped that Chapter 10 will stimulate further research in this area.

I wish to thank sincerely the 13 authors who have contributed to this volume; their cooperation and effort made my task as editor rather simple.

P. F. Fox

CONTENTS

LIST OF CONTRIBUTORS

F. M. Cremin
Department of Nutrition, University College, Cork, Republic of Ireland

Lenora R. Davis
Department of Maternal and Child Health, School of Hygiene and Public Health, Johns Hopkins University, 615 North Wolfe Street, Baltimore, Maryland 21205, USA

Albert Flynn
Department of Nutrition, University College, Cork, Republic of Ireland

C. Holt
Hannah Research Institute, Ayr, Scotland KA6 5HL, UK

Barry J. Kitchen
Gilbert Chandler Institute of Dairy Technology, Werribee, Victoria 3030, Australia. Present address: Cadbury Schweppes Pty Ltd, PO Box 200, Ringwood, Victoria 3134, Australia

R. R. Mahoney
Department of Food Science and Nutrition, University of Massachusetts, Amherst, Massachusetts 01003, USA

D. J. MANNING
Food Research Institute, University of Reading, Shinfield, Reading, Berkshire RG2 9AT, UK

P. A. MORRISSEY
Department of Food Chemistry, University College, Cork, Republic of Ireland

H. E. NURSTEN
Department of Food Science, University of Reading, Whiteknights, Reading, Berkshire RG6 2AP, UK

DAVID M. PAIGE
Department of Maternal and Child Health, School of Hygiene and Public Health, Johns Hopkins University, 615 North Wolfe Street, Baltimore, Maryland 21205, USA

PAUL POWER
Department of Nutrition, University College, Cork, Republic of Ireland

BRUNO REITER
Hon. Research Fellow, Department of Paediatrics, University of Oxford, Oxford OX1 3PD, UK. Present address: 23 Brompton Court, Ray Park Avenue, Maidenhead, Berkshire SL6 8EA, UK

L. A. W. THELWALL
Tate and Lyle Group Research and Development, Philip Lyle Memorial Research Laboratory, PO Box 68, Whiteknights, Reading, Berkshire RG6 2BX, UK

Chapter 1

LACTOSE: CHEMICAL AND PHYSICOCHEMICAL PROPERTIES

P. A. MORRISSEY

*Department of Food Chemistry,
University College, Cork, Republic of Ireland*

1. INTRODUCTION

Lactose is the major carbohydrate in the milk of the most mammals and it is generally accepted that non-mammalian sources of lactose are very rare. Lactose usually occurs free but small amounts occur in the form of lactose-containing oligosaccharides while galactose is present in a number of oligosaccharides and glycoproteins. Lactose constitutes only a small proportion of the free carbohydrate in the milk of monotremes: 8% in the echidna, 1% in the platypus,[1] values corresponding to milk lactose concentrations of 0·1% or less, compared with 7% for human milk. The main carbohydrates in the milk of the echidna are the trisaccharide fucosyl-lactose (29%) and the tetrasaccharide difucosyl-lactose (13%), which are, incidentally, minor components of human milk[2] and absent from the milk of marsupials.[3] Marsupial milk usually contains less than 3% of lactose and about 5% of oligosaccharides that yield galactose on hydrolysis.[4] Lactose is the principal carbohydrate in the milk of most eutherians (placentals). The concentration of free lactose may vary from 2% to ∼10%;[5] however, lactose is present at very low levels in milks of several species of seals[6] and sea lion[7] which appear to contain no free carbohydrate. The contrast between lactose contents of the milks of wild bears (0·2–0·7%) and those of bears in zoos (1·9–2·8%) is notable.[5] The lactose content in the milk of a particular species is dependent on such factors as breed, individuality of the animal, and mammary gland infection. The lactose content of normal cow's milk is generally in the range 4·4–5·2%, averaging 4·8% of anhydrous lactose.[8]

The literature on lactose is voluminous and consequently no attempt is made in this chapter to review it in a comprehensive manner. Most of the discussion will be from literature dealing with technological aspects of lactose which now appear important in a rapidly changing food industry.

2. BIOSYNTHESIS OF LACTOSE

The biosynthetic pathway of lactose was finally established with the discovery that free glucose is the true galactosyl acceptor in the reaction catalysed by lactose synthetase.[9] The amount of glucose absorbed by the mammary gland is quantitatively more than adequate to meet the requirements for lactose synthesis. The pathway involves the initial conversion of one molecule of glucose, via glucose-6-P and glucose-1-P, into UDP-glucose which is epimerized to UDP-galactose. In the presence of α-lactalbumin and the enzyme galactosyl transferase (EC 2.4.1.22), UDP-galactose condenses with a further molecule of glucose to form lactose.[9] Lactose synthetase is unique to mammary tissue and consists of two dissociable units, A and B.[10] The A unit is a non-specific galactosyl transferase, common to many tissues, which normally catalyses the transfer of galactosyl residues to the carbohydrate side-chains of glycoproteins. The acceptor specificity of the A component is modified by the presence of the B component resulting in the formation of a bimolecular complex capable of transferring galactose from UDP-galactose to glucose to form lactose.[11] The B component is identical with α-lactalbumin,[12] the principal whey protein in the milk of human and many other monogastric species and the second most numerous protein in the milks of ruminants. It has been postulated that secretion of α-lactalbumin in the mammary gland controls the level of lactose synthesis,[13] and a high correlation has been found between the lactose and α-lactalbumin contents of milks from a number of species.[14] Since lactose is the principal constituent in milk affecting osmotic pressure, its synthesis must be rigidly controlled, and this is probably the physiological role of α-lactalbumin. Each molecule of α-lactalbumin regulates lactose synthesis for a short period and is then discarded and replaced. The disposable nature of the α-lactalbumin component may be the mechanism by which the cell can shut off lactose synthesis quickly and thus protect the cell against osmotic damage.[15] However, changes in galactosyl transferase activity, Ca^{2+} concentration and proton generation may also be involved in the rising yield of lactose during early lactation.[16]

3. OSMOTIC REGULATION

The osmotic pressure of milk varies within narrow limits and is close to that of blood and other body fluids; a high correlation between the freezing points of mammary venous blood and milk has been reported.[17] Over 50% of the freezing point depression of milk, and hence its osmolality, is due to lactose; the other principal contributors are K^+, Na^+ and Cl^-.[18] It is now generally considered that, since lactose is formed in the Golgi apparatus of the mammary secretory cell and carried in secretory vesicles to the lumen of the alveolus, water is drawn osmotically to dilute the newly synthesized sugar.[19] The mechanisms of secretion of lactose, water and the major monovalent ions are closely coupled, and the composition of the aqueous phase of milk, at least as far as the major components are concerned, is constant at 'normal' rates of lactose synthesis.[20] Several investigators[21–23] have confirmed the complementary relationship between lactose and chloride: a decrease in the lactose content of milk is accompanied by a corresponding increase in chloride content.[24] A variation of 0·1% chloride is equivalent to 1·75% lactose.[24] In some physiological conditions, e.g. mastitis, the integrity of the mammary secretory epithelium, normally impermeable to the main constituents of milk,[19] is disrupted, thereby allowing small molecules and ions to pass directly between extracellular fluid and milk down their respective concentration gradients. Thus, since the concentrations of sodium and chloride are higher in extracellular fluid than in milk, the concentrations of these species in milk increase, while those of lactose and potassium, which pass in the reverse direction, decrease. It is for this reason that the chloride/lactose ratio in milk (Koestler number) is used to diagnose udder infection.

It has been proposed[5] that a disaccharide (lactose) has an evolutionarily selective advantage over a monosaccharide because a given weight of lactose exerts only about half the osmotic pressure of the same weight of a monosaccharide. Thus, twice the calorific value can be accommodated for a given osmotic increment.

4. CHEMISTRY OF LACTOSE

Lactose is a disaccharide which yields D-glucose and D-galactose on hydrolysis. It may be designated 4-*O*-β-D-galactopyranosyl-D-gluco-pyranose. Lactose occurs in both α and β forms, which differ in the steric configuration of the H and OH around C-1 of glucose. The chemistry of lactose and evidence of structure have been reviewed extensively[8,25] and will not be discussed here.

Lactose is not as sweet as sucrose, glucose or fructose. For example, 1·08%, 2·2% and 4·91% sucrose are equivalent in sweetness to 3·75%, 7·5% and 15·0% equilibrium lactose solutions.[26] β-Lactose is 1·05–1·22 times as sweet as α-lactose[26] but this small difference in sweetness is of no practical value because equilibration of the anomers occurs rapidly in solution.

5. PHYSICAL PROPERTIES OF LACTOSE

Lactose is present in milk in two anomeric forms which are in equilibrium. In manufactured products, lactose is present in either of two crystalline forms—α (hydrate) and β (anhydrous)—or it may be present as a 'glass' mixture of α- and β-lactose. The designations α and β refer to the configuration of substituents on the asymmetric number one carbon of glucose. This configuration is not stable and can readily pass from the α to the β form or vice versa when a solution of either form is prepared. The transition from one form to the other may be followed by measuring the changes in optical rotation using a polarimeter, which furnishes a useful method for determining the amounts of α- and β-lactose in dairy products.[27] The α and β forms have specific rotations in water at 20°C of $+89\cdot4°$ and $+35°$, respectively (anhydrous weight basis). Regardless of the form used to prepare a solution, the optical rotation will change (mutarotation) until an equilibrium specific rotation of $+55\cdot7°$ is established at 20°C.[8] A working hypothesis for the occurrence of mutarotation is described elsewhere.[28] The distribution of α and β anomers in solutions of lactose is about 37·3% α and 62·7% β at 20°C.[8] The ratio α/β increases with temperature but is independent of pH.[8] Mutarotation is a first order reaction, characterized by the reaction constants, k_1 and k_2, in the equation:

$$\alpha\text{-D-Lactose} \;\underset{k_2}{\overset{k_1}{\rightleftharpoons}}\; \beta\text{-D-Lactose}$$

The mutarotation rate is defined as:

$$K = k_1 + k_2 = 10^{-4}\,\text{s}^{-1} \text{ at } 20°C$$

The rate of mutarotation has a temperature coefficient, Q_{10}, of 2·8,[8] and an apparent activation energy of 75 kJ mol^{-1}.[8] The rate is extremely rapid at pH > 2 and < 9,[5] and a combination of salts equal to that found in milk serum increases K by a factor of almost 2 compared to water.[29] The rate of mutarotation is decreased by high concentrations of sucrose as in sweetened condensed milk,[30] and it is highly dependent on moisture content.[28]

6. SOLUBILITY

The α and β forms show distinctly different solubilities, and mutarotation also manifests itself in the solubility behaviour of lactose. When α-lactose monohydrate is added in excess to water, a definite amount dissolves initially, after which an additional amount dissolves slowly until final solubility is attained. The initial solubility (~ 7 g of α-monohydrate in 100 g of water at 15°C) is the true solubility of the α form.[8] Because of mutarotation, α is converted into β, the concentration of α-lactose decreases and more α dissolves. This process continues until equilibrium is reached and final solubility is established (~ 17 g in 100 g of water) at 15°C.[8] Under similar conditions, ~ 50 g of β-lactose will dissolve in 100 g of water. A saturated solution of β-lactose yields, on mutarotation, more α-lactose than can be maintained in solution, and part of the latter crystallizes to re-establish equilibrium.[8] However, the final solubility of lactose, which is temperature dependent,[8] is identical whether α- or β-lactose is dissolved. Solubility data for lactose have been determined (cf. Ref. 8) and may be calculated for temperatures below 93·5°C using the equation:[28]

$$[\text{Solubility of lactose}] = (R+1) \times [\text{Solubility of } \alpha\text{-lactose}]$$

where $R = 1·64 - 0·0027\,T$; $T =$ temperature (°C). However, solubility values calculated by this equation are not in agreement with experimental data.

Although the solubility of lactose (among sugars) is low, lactose solutions can become highly supersaturated before spontaneous crystallization occurs. The supersolubility values for lactose are readily reproducible and, in general, the supersolubility at any temperature is equal to

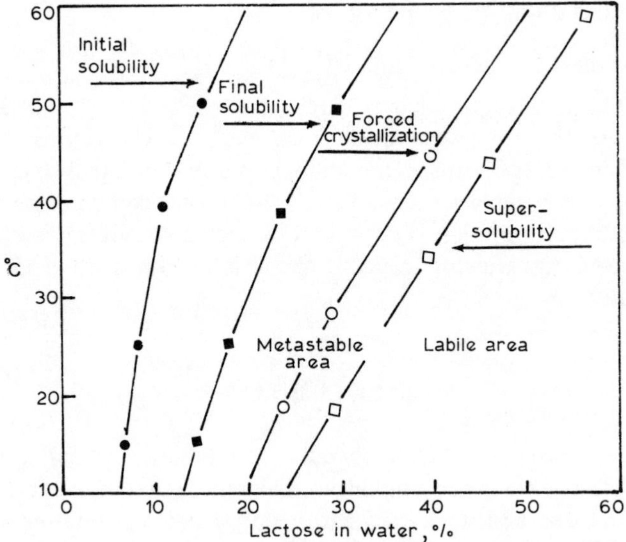

FIG. 1. Lactose solubility curves (after ref. 31).

the saturation value at a temperature 30°C higher. The relative insolubility of lactose coupled with its capacity to form supersaturated solutions is of practical importance in a range of dairy products. The phenomenon of lactose crystallization has been the subject of considerable study and discussion.[8,28,30] Solubility curves for lactose have been produced [28,31] and are based on the concept of supersaturation, metastability and lability (Fig. 1). The metastable zone occurs in the first stages of supersaturation produced by cooling a saturated solution. Crystallization does not occur readily in this zone and requires seeding with lactose crystals to induce crystallization. Induced crystallization usually occurs at concentrations in the region of 1·6 times the final solubility of lactose.[28] The labile zone is found at concentrations over 2·1 times the final solubility of lactose,[28] where spontaneous crystallization occurs.

7. CRYSTALLINE FORMS OF LACTOSE

α-Lactose monohydrate is the usual crystalline form obtained from cheese whey or aqueous lactose solutions on concentration to supersaturation at $<93.5°C$. A number of crystalline shapes may be obtained, depending on the conditions of crystallization, but the most familiar forms are the prism and tomahawk shapes.[8,28] The crystals are hard and not very soluble; they feel gritty in the mouth and produce a 'sandy' texture when the size exceeds $16\,\mu m$. Crystal growth and shape are inhibited by several substances, including β-lactose, riboflavin, gelatin and lactose monophosphate which is present in milk at a concentration of $\sim 15\,mg\,litre^{-1}$.[28]

Above $93.5°C$, β-lactose is less soluble than α-lactose; anhydrous β-lactose is the stable form crystallized from lactose solutions. Traditional processes for the manufacture of β-lactose are based on this property.[32] Modifications of these methods include heating an aqueous lactose solution under $5\text{--}40\,kg\,cm^{-2}$ pressure at $110\text{--}170°C$,[33] and heating a supersaturated solution under reflux.[34] Formation of β-lactose from α-lactose monohydrate in methanolic solutions of NaOH has been reported.[35]

β-Lactose crystallized from water has been well characterized.[8] The common form of crystal is an uneven-sided diamond. The lattice constants for the crystalline structure, as determined by X-ray diffraction, are:

$$a = 10.8\,Å;\ b = 13.34\,Å;\ c = 4.84\,Å;\ \beta = 91°15'.^{36}$$

When a solution of lactose, e.g. milk, is dried rapidly, an amorphous lactose is formed. This form of lactose, sometimes referred to as a lactose 'glass' or concentrated syrup, is present in milk powder (spray, roller, or freeze dried) and exists in an α/β ratio of 1.25.[28] This concentrated solution is stable at moisture contents of $<3.0\%$, but when water activity (A_w) is increased to ~ 0.5 ($\sim 8\%$ water) the lactose is diluted sufficiently to become mobile and form a crystal lattice.[37]

Other crystalline forms of lactose may also occur. Unstable anhydrous α-lactose is produced by heating α-monohydrate at $>100°C$ *in vacuo*. This form is very hygroscopic and rapidly takes up water to form α-monohydrate at $<93.5°C$. A stable anhydrous form, called S-lactose, can be prepared by heating α-lactose monohydrate at $\sim 150°C$ at a water vapour pressure between 6 and $80\,cm$ mercury.[38] Organic solvents, such as methanol, have been used to prepare stable anhydrous α-lactose from α-monohydrate.[39] The interrelationships between the different forms of lactose are excellently illustrated in a recent text (Fig. 2).[28]

Shaking α-lactose monohydrate with 10 times its weight of methanol containing $1\text{--}5\%$ of anhydrous hydrogen chloride causes partial mutaro-

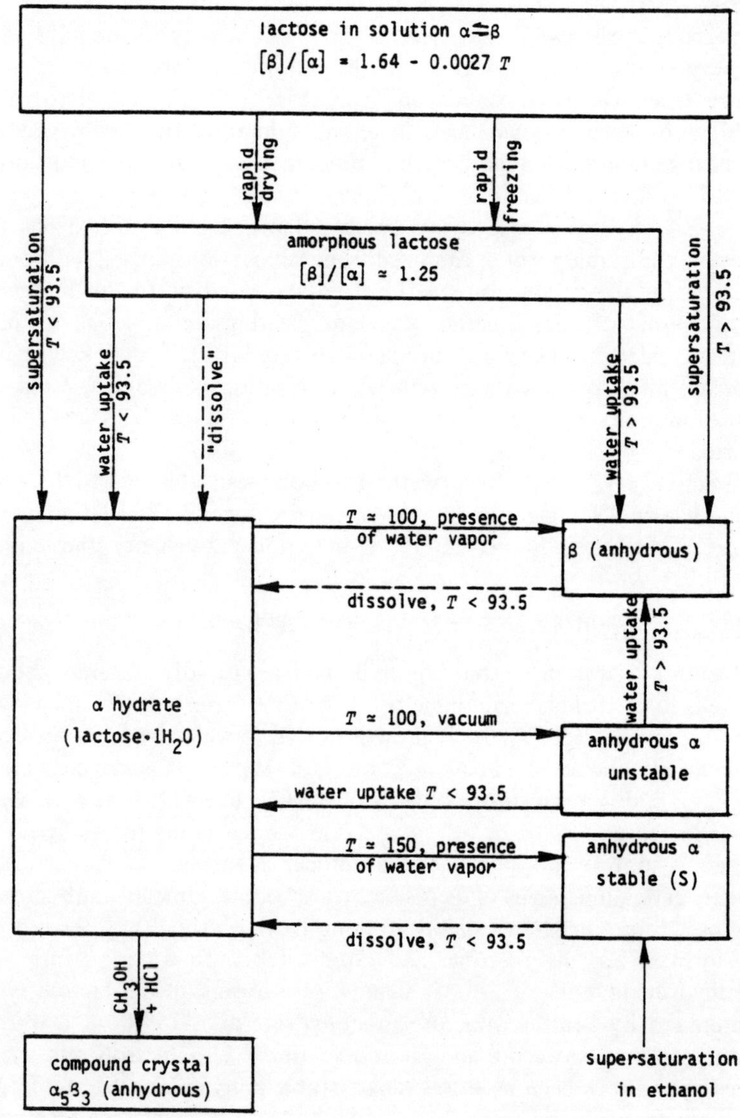

FIG. 2. Modifications of lactose $(T = {}^\circ\text{C})$.[28]

tation of α-lactose to the β form and molecular complexes precipitate with an α/β ratio of 5/3.[40] This form is not considered to be a mechanical mixture of the α and β forms of lactose, but consists of a molecular compound containing $\alpha_5\beta_3$ as determined by infrared spectroscopy and X-ray diffraction.[41] The presence of ammonium hydroxide in methanol yields a product with an α/β ratio of 3/2.[42] Some physical properties of the different crystalline forms of lactose are presented in Table 1.

TABLE 1

PHYSICAL PROPERTIES OF DIFFERENT CRYSTALLINE FORMS OF LACTOSE[a]

	Melting point (°C)	Density (g cm^{-3})	Optical rotation, $[\alpha]_D^{20}$
α-Hydrate	221 (43)		+91·1 (28)
	201·6 (44)	1·545 (46)	+89·4 (8)
	217 (45)	1·537 (45)	+85·0 (47)
α (unstable)	222·8 (8)	1·540 (46)	
α (stable)	216 (45)	1·540 (45)	
β	235 (45)	1·587 (47)	+33·5 (28)
	252·2 (8)		+35 (8)
	229·5 (34)	1·590 (46)	+35 (34)
Complex α/β (4/1)	204 (45)	1·573 (45)	
Complex α/β (5/3)	203 (45)	1·584 (45)	+67·9 (40)

[a]Numbers in brackets are the references.

8. LACTOSE IN DAIRY PRODUCTS

Lactose and lactose crystallization is of major practical importance in a range of milk products such as sweetened condensed milk, frozen products, and dehydrated milk and whey.

8.1. Sweetened Condensed Milk (SCM)

The typical composition of SCM is 9% fat, 43% sucrose, 27% water and 21% milk solids-not-fat ($\sim$10–12% lactose). In SCM, lactose is highly supersaturated and lactose crystals are formed. The solubility of lactose in SCM is only $\sim$10 g in 100 g of water compared to $\sim$20 g in 100 g of water at room temperature.[28] This is due to the high level of added sucrose and concentration of the soluble salts of milk.[28] Lactose crystallization must be controlled to ensure a smooth-bodied product free from 'sandy' or gritty texture due to large ($>20\,\mu m$) sharp crystals; this involves the formation of a very large number of small crystals ($<10\,\mu m$). The number of crystals required to avoid sandiness and sedimentation ranges from $4\times10^5\,ml^{-1}$[8] to $10^9\,ml^{-1}$.[28] The formation of the greatest possible number of small crystals is achieved by controlled crystallization, i.e. adding seed lactose before crystallization commences, because the concentration of lactose in the original solution is in the metastable concentration zone.[31] Wet or dry seeding methods are normally adopted.[48] For wet seeding, lactose from the previous batch, which is mixed with either SCM or a saturated lactose solution, is used. However, this procedure increases the risk of microbial contamination and is used infrequently.

Dry seeding is performed by addition of very finely ground lactose powder (300 mesh, with the bulk of it under $10\,\mu m$ and a significant proportion under $1\,\mu m$).[49] The seed lactose is added at a rate of about $0.5\,kg$ per $1000\,kg$ of SCM or recombined SCM. This should induce the formation of more than 10^6 crystals ml^{-1} of product, with more than 95% under $10\,\mu m$ and all, ideally, under $15\,\mu m$.[49] The addition of the seed lactose must be accompanied by thorough agitation of SCM, as good dispersion is more crucial to successful crystallization than the actual amount of seed lactose used.[50] Since the viscosity of the SCM is high, the boiling of the SCM in the vacuum cooler, where the seeding takes place, cannot be relied upon for agitation.[50]

8.2. Frozen Milk Products

Freezing of milk and cream is usually attended by progressive destabilization of casein. This change appears to be associated with crystallization of lactose.[51-54] Destabilization depends on a number of factors, including time and temperature of storage,[51] degree of milk concentration,[51] prefreezing heat treatments[51,55] and compositional changes induced by various additives.[52,54] Protein destabilization can be retarded by treatments that temporarily obviate lactose crystallization by destruction of the lactose nuclei present, or by treatments that reduce the rate of lactose crystallization, e.g. by greatly increasing the viscosity of the product. For example, when concentrated milk is heated at 38–63°C for 30 min prior to freezing, a supersaturated solution is formed on subsequent freezing, and crystallization of α-lactose at −9·4°C can be avoided for 50 days.[52] When concentrated milk is held at −28·9°C, destabilization of casein can be prevented since lactose will exist indefinitely in a metastable glass state at this temperature.[52] The degree of α-lactose crystallization at which casein instability appears is reported to range from 40%[52] to about 90%.[54]

The cause of casein destabilization is not lactose crystallization *per se*, but the high degree of concentration of serum salts in the still-unfrozen portion of the milk which ensues when crystallization of lactose occurs and is followed by freezing of the water in which lactose had previously been dissolved. Analysis of the unfrozen liquid from freshly frozen concentrated milk ultrafiltrate at $-7.5°C$ showed that ~ 11-fold concentration of milk salts occurred (Table 2) prior to freezing,[53] and up to

TABLE 2

ANALYSIS OF ULTRAFILTRATES OF SKIM MILK AND THE LIQUID PORTION OF FROZEN ULTRAFILTRATES[56]

	Concentration (mM) *in ultrafiltrate of skim milk*	*Concentration* (mM) *in ultrafiltrate of liquid portion of frozen concentrated milk*
pH	6·7	5·8
Chloride	34·9	459
Citrate	8·0	89
Phosphate	10·5	84
Sodium	19·7	218
Potassium	38·5	393
Calcium	9·1	59

30-fold concentration develops as lactose crystallizes.[54] Since normal milk serum is saturated with calcium and phosphate,[57] considerable precipitation of calcium phosphate thus takes place on freezing, as follows:

$$3\,Ca^{2+} + 2\,H_2PO_4^- \rightleftharpoons Ca_3(PO_4)_2 + H^+$$

which accounts for the fall in pH from 6·7 to 5·8 (Table 2). Thus, the structural integrity of the casein micelles can be expected to be challenged by the pH and salt variations. Fast freezing causes no immediate pH change, but a low pH develops on frozen storage, presumably because equilibrium is established slowly between liquid and solid phases.[58] However, slow freezing results in greater protein stability than rapid freezing, and this pattern prevails regardless of the solids content of the milk.[51,59] The destabilizing effect of rapid freezing could be related to the effect of the process on the rate of lactose crystallization: rapid freezing apparently increases the rate of lactose nucleation compared to slow freezing, and in turn increases the rate of lactose crystallization during frozen storage.[59] Agitation during cooling and freezing also enhances lactose crystallization and protein destabilization. Reduction in soluble constituents of milk by dialysis leads to protein stabilization.[60] The effect of ultrafiltration on the stability of frozen milk is somewhat similar to that reported for dialysis.[59] Frozen skim milk samples with permeate/retentate ratios of 10/90, 20/80 and 30/70 had at least three times the storage stability of an unmodified control sample.[39] Part of the increased stability could be due to reduction in the amount of lactose available for crystallization. However, partial removal of calcium and redistribution of milk salts between the colloidal and soluble phases may also be important. When the extent of ultrafiltration was increased to 40/60, protein stability in the frozen retentate declined somwhat compared to that of less concentrated retentates.[59] The decreased stability is reported to be associated with removal of soluble phosphate in excess of 30% rather than with lactose crystallization;[59] removal of 30% of soluble phosphate from skim milk by dialysis was previously shown to cause a significant decrease in stability.[60] This results in considerable solubilization of colloidal calcium phosphate in the retentate following ultrafiltration, thus producing a less favourable environment for protein stability.[59] Conditions governing lactose crystallization and accompanying destabilization of casein are very similar in ice cream but, in this product, defects have almost disappeared with the introduction of stabilizers which act by impeding formation of lactose nuclei.[61] Partial hydrolysis of lactose by β-galactosidase results in protein stabilization and loss of sandiness owing to the increase in the soluble sugar content of the milk. (See Chapter 3, Section 9.2.3 on the applications of lactase.)

8.3. Milk and Whey Powders

8.3.1. Milk Powders

Lactose is the major constituent of dried milk and whey products; dried whole milk contains $\sim 37\%$, dried skim milk $\sim 50\%$, buttermilk powder $\sim 49\%$, and whey powder $\sim 70\%$. Thus it must be expected that at least some of the properties of lactose will be reflected in the behaviour of the dried milk particles. The use of milk powders for recombination and manufacture of various food products has increased over the past decade or so, placing greater importance on definition and specification of functional properties.[62] The properties of spray-dried milk powders and how they relate to processing conditions have been reviewed,[63] while factors affecting the reconstitution properties of whole milk powder have also been studied.[64] A major factor in the low level of acceptability of milk powder as a consumer product is the difficulty of reconstituting the powder in cold water. The physical properties (i.e. colour, particle size and distribution, bulk density, flow properties, moisture content, reconstitution properties, sinkability, dispersibility, solubility and heat stability) are critical factors for classifying milk powders.[63,64] The reconstitutibility of milk powders is governed by a number of fundamental properties, namely wettability, penetrability, sinkability, dispersibility and solubility.[63,65] In commercial-scale reconstitution, impaired wetting and dispersibility of powders cause most problems. Dispersibility of powders is enhanced by increasing the mean size of the powder particles.[64,66] This can be achieved by manipulating the viscosity of the feed stream prior to drying, varying pressure at the drier nozzle, agglomeration and fluidized bed drying. Instantization, which gives rapid dispersibility and solubility, is best achieved by rewetting and redrying of food powders.[67] The process involves the formation of clusters of dried milk particles by agglomeration following rewetting of normal powder with steam. In this way, primary single particles produced by spray-drying are combined into bigger, porous aggregates ($\sim 200\ \mu m$).[66] Agglomeration is the most widely used method for obtaining 'instant' properties, and it involves finding the proper relationship between wettability, penetrability, dispersibility and the rate of hydration.[65] Agglomeration or sticking together of primary particles involves some crystallization of the lactose 'glass'.

The structural elements of a dried milk or whey particle are lactose, either amorphous or partially crystalline, casein micelles and whey protein particles, fat in globular or non-globular (free fat) forms,[68] and air as spherical cells. The fat, protein and air are presumably dispersed in a continuous phase of amorphous lactose.[69] The distribution of lactose between the amorphous glass and crystalline states depends on the processing conditions. Most of the lactose in hygroscopic milk and whey powders is in the amorphous state,[8] with β/α ratios of $1\cdot25$[28] to $1\cdot6$[69] in roller- and spray-dried powders. Owing to the high hygroscopicity of lactose glass, dried milk particles readily adsorb moisture from the surrounding atmosphere. When this happens, the concentration of lactose is diluted so that its molecules acquire sufficient mobility and space to orientate themselves into a crystal lattice,[37,68,69] which leads to clumping and 'caking' or compaction of powder particles to a hard mass.[69] Lactose crystallization leads to protein insolubility, de-emulsification of fat,[68] and accelerated flavour deterioration.[70]

The instantizing process for producing readily soluble powders is somewhat similar to that reported above. Optimum rewetting and hydration of dried powder to achieve agglomeration is obtained at 12% moisture.[67] The primary particle surface becomes tacky, which facilitates adhesion of adjoining particles, and partial crystallization of lactose occurs before the particles are redried. This process produces loose, spongy aggregates of low density which are free-flowing and readily dispersible.

Either of two processes is usually used to achieve agglomeration: the 'straight through' method and the 're-wet' method.[65] In the former, the powder is taken out of the drying chamber with a somewhat higher moisture than normal, and the drying process is completed in a vibro-fluidizer at suitable temperatures to achieve agglomeration and the required final moisture content. The re-wet instantizing process, as the name implies, involves controlled, slight rehydration of dried particles to facilitate adhesion; the product is redried to the desired final moisture content using vibro-fluidized systems.

8.3.2. Whey Powders

Lactose plays a dominant role in determining the properties of whey powders. In hygroscopic whey powders, most of the lactose is in the amorphous or glass-like state. Owing to its high hygroscopicity, the powders readily adsorb moisture from the atmosphere. When this happens, the dry powder becomes moist and sticky initially, and later changes into a sintered, syrupy material in which the original particles have lost their identity; finally, the material changes into a hard mass. On the other hand, a whey powder, in which part of the amorphous lactose has been allowed to crystallize from a supersaturated solution, is essentially non-hygroscopic and free-flowing. The properties of lactose in whey powders and the effects of crystallization of the properties thereof are discussed in a number of publications.[8,71,72] The drying of non-hygroscopic whey involves the following operations:

1. Preheating at about 95°C for up to 30 min.
2. Concentration to 50–60% total solids.
3. Pre-drying crystallization.

4. Spray-drying.
5. Post-drying crystallization.
6. Fluid bed drying.
7. Fluid bed/pneumatic cooling.

Concentration of whey to 50–60% total solids is essential to achieve supersaturation of lactose in the solution. If the concentrate is dried at this stage, amorphous lactose will be present in the powder, rendering it highly hygroscopic. In the pre-crystallization step, the concentrate is flash-cooled from 60°C to 30°C and pumped to crystallization tanks fitted with cooling jackets and agitators. Seeding may be carried out at this stage using either finely ground α-lactose or well-crystallized whey powder; the addition is made gradually, with agitation, during the filling of the tanks.[71] The concentrate is cooled below 25°C in the case of sweet whey, or, in the case of acid wheys, to below 10°C and held for ~16 h. Under these conditions, crystallization of ~70% of the lactose occurs and, ideally, crystal size should be less than 50 μm to prevent damage to the dryer.

After the pre-crystallization step the concentrate is spray-dried in a drier operating at a low outlet temperature, ~55°C,[71] which gives a powder with a moisture content of 8–14%. The powder is then conveyed on a crystallization belt to allow secondary crystallization to occur.[71] The excess moisture is then removed in the 'after-drying' equipment consisting of several vibro-fluidizers. Alternatively, the powder may be fed directly to the vibro-fluidizers, omitting the crystallization belt. The final powder consists of large agglomerates and has a low bulk density, with 85–100% of the lactose in the crystalline α-monohydrate form.[71,72]

The sticking of the whey solids to the hot metal surfaces of the drying chamber is a major problem and limits the drying conditions that can be applied to whey. This behaviour is known as thermoplasticity, and the temperature at which sticking occurs depends on the contents of lactic acid, amorphous lactose and moisture.[73] An increase in lactic acid concentration up to 16% (dry weight basis) shifts the sticking point linearly down from 98°C to 57°C. An acid whey with 80% of the lactose crystallized has a sticking point of 78°C while the same product with only 45% of pre-crystallized lactose has a sticking temperature of 60°C. Thus, pre-crystallization of lactose and a two-stage drying system are considered mandatory for the production of good quality whey powders.[72]

9. WATER SORPTION PROPERTIES OF LACTOSE SYSTEMS

Water sorption phenomena in foods have been extensively reviewed,[74–76] and it is well recognized that moisture sorption affects the stability and quality attributes of all dehydrated foods. The control of moisture sorption and water activity is a basic food processing technique and influences chemical deterioration, Maillard browning, lactose crystallization, gradual loss of solubility and rate of rehydration. Desorption of moisture is important in relation to the drying process and the determination of total solids by oven-drying methods,[77] while adsorption is of practical significance in selecting drying procedures, optimum packaging materials and storage conditions.

The sorption behaviour of various milk powders has been reported by a number of workers.[78–80] For most foods, the shapes of the adsorption and desorption curves are similar and sigmoidal but the desorption equilibrium moisture content is usually higher than the adsorption equilibrium value.[75] A sharp discontinuity in the water sorption isotherm in the neighbourhood of water activity (A_w) of 0·50 has been observed.[69,78–80] The shape and position of the curve vary with the composition of the sample in question; that is, the content and relative ratios of protein, lactose and salts, and the crystalline state of lactose, affect rates and extent of water vapour sorption.[81]

The moisture sorption isotherms of pure crystalline systems and non-crystalline materials have been examined and reviewed by several workers.[78,82–85] Many low molecular weight sugars and salts may be present in one of several states: crystalline solids, amorphous solids and aqueous solutions. The tightly packed molecules in a pure crystalline system (e.g. lactose or sucrose) adsorb very little water, and only on the crystal surfaces, until the water activity is high $(>0·8)$, when the sugar begins to dissolve (Fig. 3). When the amorphous glass form of the sugar is present, exposure to increasing humidities results in water uptake, reaching sorption levels far higher than that of the crystalline form. This difference is due to the larger internal area available for water sorption in the amorphous, less regular material. Some hydrogen bonds are disrupted, and at some water activity the sugar molecules are somewhat diluted and acquire sufficient mobility to undergo transformation from the metastable amorphous state to the more stable crystalline form.[65] In this process, the sugar loses water and the crystals become tightly packed. Thus, if moisture gain is measured as a function of time, a discontinuity in the isotherm occurs at constant A_w, as observed for sucrose[82] and

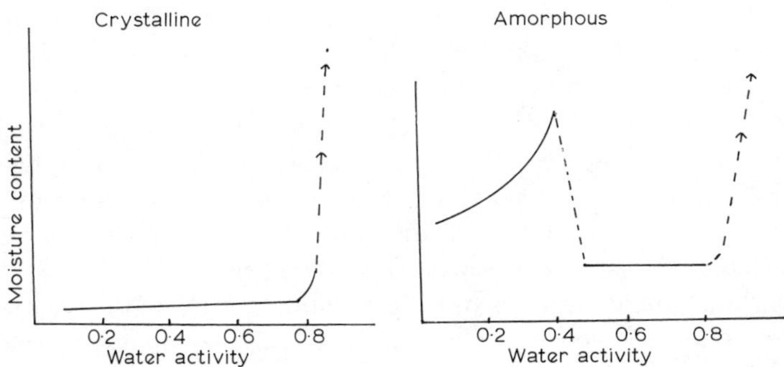

FIG. 3. Moisture sorption isotherms of pure crystalline and amorphous systems.

lactose.[79,84] The rapid drop in moisture content is illustrated in Fig. 3. The break in the curve for lactose glass occurs between A_w 0·43 and 0·52 (Fig. 4).[79,84] Beyond this point, the lactose becomes non-hygroscopic and very little water sorption occurs. The transition from the amorphous to the crystalline state has been termed 'a state of collapse'[86] because, on visual examination, the amorphous solid appears to undergo actual viscous flow immediately prior to cystallization. The water that is released on crystallization is available to interact with other components and effect chemical and physical deterioration of other constituents that may be associated with the sugar.

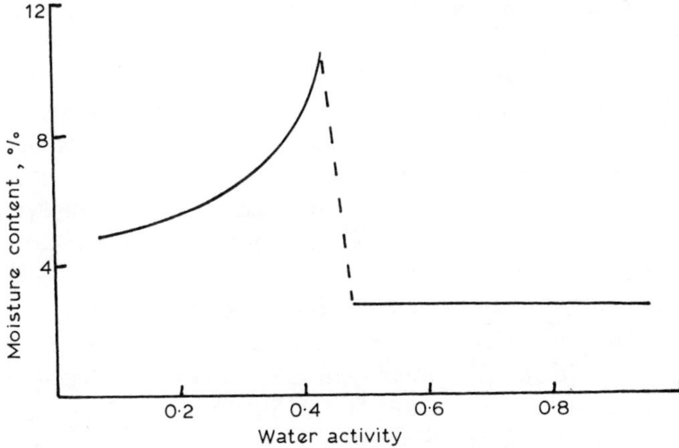

FIG. 4. Water sorption isotherm for lactose initially in the amorphous form at 25°C.[84]

The crystallization time is dependent on both A_w[82,85] and temperature.[87] Increases in A_w accelerate the rate of lactose crystallization in whey powders.[85] The accelerating effect of increased A_w on crystallization is similar to that observed for sucrose;[82] crystallization of amorphous sucrose is observed after ~ 2 days at $A_w = 0.30$, after ~ 100 days at $A_w = 0.22$, and after ~ 1000 days at $A_w = 0.16$. As the temperature is increased from 14°C to 34°C, the discontinuity in the isotherm of freeze-dried lactose shifts towards lower water activities.[87]

The shift from amorphous to the crystalline state has also been noted for lactose in milk powders. Upon initial water sorption, discontinuities have been reported for various dried milk products in the 0.50–0.60 A_w range.[69,78,79,84,85,87–89] In all cases, the isotherm discontinuities were attributed to the crystallization of amorphous lactose. The break in the curve occurs at slightly higher values of A_w for dried milk than for pure amorphous lactose.[84] However, other workers[90] obtained isotherms for non-fat dry milk which did not exhibit a discontinuity. They observed that a combination of sorption by lactose in the glass state and sorption on polar sites in proteins occurred at A_w less than 0.4. At higher A_w, moisture sorption is by the protein only, since lactose is in the crystalline state. The smooth and sigmoid isotherms obtained[90] were probably due to the limited number of A_w intervals analysed and the desorption technique employed. The amount of water adsorbed at the point of discontinuity and the magnitude of the drop in the isotherm are inversely related to the amount of lactose present in the milk powders.[78]

The discontinuity and characteristics of the sorption isotherm depend on the state of lactose in the powder.[85,88] The sorption isotherm for non-hygroscopic whey powder is continuous or smooth over the entire A_w range; however, the sorption isotherm for hygroscopic whey samples shows a break in the 0.33–0.44 A_w range.[86] At lower A_w values, i.e. below 0.44, the water content of non-hygroscopic whey powder is considerably lower than that for hygroscopic whey powder.[85] These results would be expected from the isotherm behaviour for crystalline and non-crystalline systems shown in Fig. 3.

The value of A_w at which the break in the isotherm is observed varies with the composition of the sample.[78] Generally, the break is observed at lower A_w values in whey powders than in whey protein concentrates or milk powders; it is in the A_w range 0.35–0.5 for whey powders and whey concentrates whereas for milk powders it occurs at A_w 0.6.[78] The initial sorption in the lower portion of the isotherm is much greater for milk powder than for whey powder, which reflects the preferential binding of water by protein (casein) at low A_w (<0.35).[78] Whey protein concentrate powders containing 78%, 83% and 87% of whey proteins show identical isotherms up to A_w 0.4 but above this the whey powders containing higher levels of lactose and ash (i.e. lower protein) adsorb more moisture.[80] Sorption isotherms for Cottage cheese whey powders containing $\sim56\%$ of lactose show a number of discontinuities in the A_w range 0.26–0.49.[91] The series of discontinuities is considered to be associated with various states of lactose crystallization equilibria; equilibration is slow at each point on the curve, requiring ~5 days for the mass of sorbed water to become constant.

Water sorption isotherms for different components of whey reveal that the contributions of the components are additive. Removal of lactose from milk powder results in a typical continuous isotherm, e.g. the acid and ultracentrifugal casein systems in Fig. 5.[79] Comparison of the water

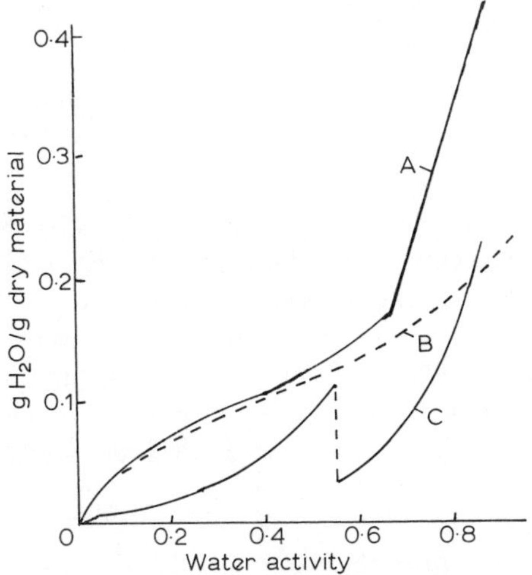

FIG. 5. Water sorption isotherms for dairy powders: A, acid casein; B, ultracentrifugal casein, i.e. casein micelles; C, acid whey.[79]

sorption isotherms for milk protein, lactose and salts (Jenness–Koops synthetic milk salts system) demonstrated that there is a progressive order of moisture sorption by components of milk powder preparations at increasing A_w.[79] Initially, casein binds moisture and swells; next, whey proteins and then lactose bind moisture and crystallize (A_w 0.2–0.6),

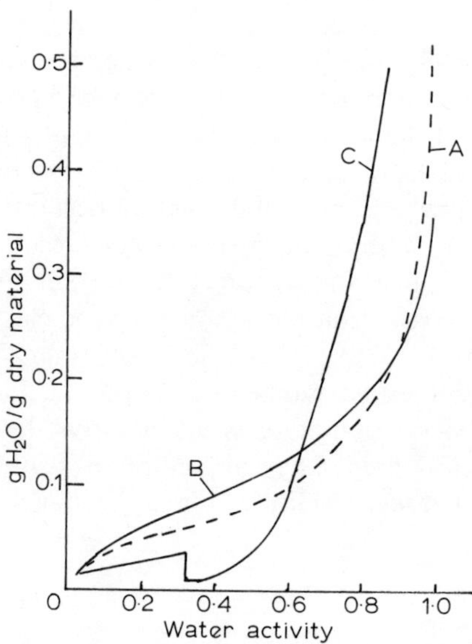

FIG. 6. Water sorption isotherm of dehydrated whey protein concentrate and dialysed fractions thereof: A, protein concentrate; B, non-dialysable components; C, dialysate.[80]

and finally, as A_w rises above 0·6, the salts rapidly adsorb moisture. Thus the low molecular components (dialysable constituents), including lactose, exert a dominant role at relatively high water activities (Fig. 6). Therefore careful control of moisture content, to prevent lactose crystallization and loss of quality, is vital in high-lactose products. It is recommended that dried milk be kept at <6% moisture content, $A_w = 0·4$, to prevent lactose crystallization,[84] and baby foods and other dry milk products of similar composition should be stored at $A_w \sim 0·25$ and below 35°C.[89]

Temperature also affects the location on the isotherm of the breakpoint caused by lactose crystallization. On elevating the temperature, the lactose crystallization-induced break-point occurs at progressively lower water activities.[84,87,89]

The desorption isotherm follows a smooth sigmoid curve and lies above the initial adsorption isotherm for the dry product;[78,89] it therefore exhibits normal hysteresis.[74] Subsequent adsorption data, obtained after the first complete adsorption–desorption cycle, yield smooth reversible sigmoid curves that are almost exactly superimposable upon the initial desorption data.[78] The most interesting phenomenon observed is

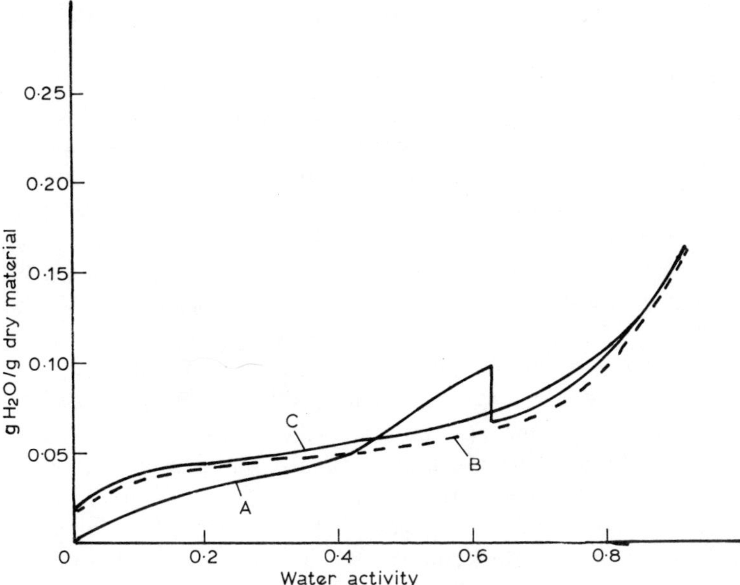

FIG. 7. Water sorption isotherm for foam-spray-dried whole milk at 24·5°C: A, adsorption; B, desorption, C, re-adsorption.[78]

that some water remains irreversibly bound and cannot be removed even under high vacuum at 25°C (Fig. 7).[78] Thus, where lactose crystallization has occurred, as in non-hygroscopic powder, complete dehydration is not possible. Lactose crystallization in milk powder aggregates affects the accuracy of moisture determination, since the water of hydration of α-lactose is only partly removed by the standard oven-drying methods.[92] The mean values obtained for dried milk by the Karl Fischer and toluene distillation methods are almost identical,[93] but other workers[94] have found that oven-drying at 102°C underestimates the moisture content of skim and whole milk powders by, on average, 0·33% and 0·49% compared with either the Karl Fischer or toluene distillation methods. A somewhat wider range (0·3–1·1% moisture) of differences between toluene and oven methods has been reported;[95] in agreement with previous findings,[96] these authors[95] suggest that the difference between the toluene and oven methods is due to the variable amounts of crystalline lactose in the powders. This conclusion appears valid in view of the data shown in Fig. 7.[78]

10. LACTOSE REACTIONS IN HEATED MILKS

Lactose in milk and milk products may be involved in a number of heat-induced changes, depending upon the heating temperature. Direct caramelization, which has a high activation energy, is not of major consequence in most milk products, although some of the carbon dioxide formed on sterilization of milk can be traced to caramelization of lactose.[97] When milk is heated at high temperatures (120–140°C), the pH decreases to 5·5–6·0 at coagulation,[98] and lactose is responsible for ~ 50% of the total developed acidity[98,99] through production of organic acids (mainly formic)[100] in the presence of oxygen.[99]

Heat treatment of milk and milk products primarily involves interaction between lactose and proteins, resulting in Maillard browning, loss of available lysine and nutritive value, and off-flavour development.[101] Classically, the Maillard reaction occurs between an aldo or a keto sugar and the amino group of an amino acid. The overall reaction has been divided into three stages.[101] The early reaction involves the formation of the Amadori compound, l-amino-1-deoxy-2-ketose, which is uncoloured but biologically unavailable,[102] and loss of nutritive value occurs without any observable colour or flavour changes.[103] The second stage involves: dehydration of the Amadori compound by loss of three molecules of water to form furfurals and reductones; fission, mainly by dealdolization and Strecker degradation, i.e. the interaction of amino acids and dicarbonyls. The final stage consists of the conversion of furfurals, fission products, reductones and Strecker aldehydes into melanoidin pigments, with further involvement of amines. Maillard browning has been treated in a comprehensive manner at a recent conference[104] which included specific sections on the Maillard reaction in milk and milk products.[101,102,105,106] However, a comprehensive review on browning in milk and its products has not been written since 1955,[107] when the mechanism of the reaction was not clearly defined. In unconcentrated milk, provided that heating conditions are not too drastic, the Maillard reaction generally ceases at the Amadori stage.[101,103,106] However, the nutritive value of the milk may be reduced because lysine, which primarily reacts with lactose to form lactulosyl-lysine, is not biologically available.[101,108] Lysine losses reported for milk and milk products depend on the type of heat treatment, the composition of the products, the manufacturing practices and the method used to determine available lysine.[101,102,106] The range of lysine losses generally observed in recent studies is summarized in Table 3. Lactase-treated milks and

TABLE 3
LYSINE DAMAGE IN MILK AND MILK PRODUCTS

Product	Range of lysine loss (%)
Pasteurized milk	0–3
UHT milk	0–3
Spray-dried milk powder	0–3
Sweetened condensed milk	0–3
In-can sterilized milk	10–15
Evaporated milk	15–20
Roller-dried milk powder	20–75

milks in which lactose is replaced by glucose are much more sensitive to heat because of the increased number of available reducing aldehyde groups; lysine losses of 55% have been reported for spray-dried lactase-treated milk compared with 0–3% for the control.[102]

Maillard reactions also occur during storage. A 30% loss of available lysine (an index of lactose reactivity) occurred in UHT milk stored at 30°C for 14 months.[106] The available lysine in a sample of dried milk was reduced by 10% following storage at 10°C for 12 months,[109] while 13% blockage of lysine was observed after 1 week and 34% after 8 weeks when spray-dried milk was stored at 60°C.[102]

Low moisture levels and moderate storage temperatures are the critical factors affecting the interaction between lactose and protein (lysine in particular). The Maillard reaction occurs with maximum velocity at $A_w = 0.3–0.7$.[110,111] In milk powder, browning reactions reach a maximum at $A_w = 0.6–0.7$,[111,112] and are generally not important from a practical viewpoint at $A_w < 0.2$. The maximum rate of browning in whey powders occurs at a somewhat lower water activity ($A_w = 0.44$) than in dried milk;[113] the lower A_w for whey powder is related to the transition of lactose from the amorphous to the crystalline form.[85]

Heating also induces changes in lactose which may be used as an index of quality. Lactulose (4-O-β-D-galactopyranosyl-D-fructose), an epimer of lactose, is formed in heated milks.[114–116] The mechanism by which lactulose is formed is not known with certainty, but it is thought to be formed by alkali epimerization (Lobry de Bruyn transformation) and catalysed by amino compounds. It has been proposed[117] that the lactulose content of a milk sample may be used to assess the thermal treatment used during processing; for example, the lactulose content is capable of differentiating between pasteurized, UHT and in-container sterilized milks.[116]

11. MANUFACTURE OF LACTOSE

The principal raw material for lactose production is cheese whey,[8] although casein whey may also be used. With the application of new membrane processing techniques in dairy technology, ultrafiltration permeate from the production of cheese and whey protein is now becoming a major dairy waste in the USA, Canada, Europe and New Zealand.[118] In membrane processing operations the problem of waste is reduced almost entirely to milk salts and lactose. Table 4 shows the composition of 'whey type' dairy waste compared to that of milk. Expansion in the use of ultrafiltration will depend largely on the profitable use of permeate which may be used for pig feeding, as a crystalline cattle lick or as a source of non-protein nitrogen in the form of lactosyl-urea.[118] A general solution to the problem of utilizing permeate rests with the human food industry.[119]

The manufacture of lactose[8,120–122] is similar to that of sucrose. It is basically a five-step process: concentration, coagulation and removal of

TABLE 4

AVERAGE COMPOSITION OF DAIRY WASTES COMPARED TO THAT OF MILK

Component	Protein (%)	Fat (%)	Lactose (%)	Minerals (%)
Milk	3·3	3·8	5·0	0·9
Cheese whey	0·9	0·05	4·8	0·6
Permeate	<0·1	0·01	4·8	0·6

TABLE 5

SOME TYPICAL PHYSICAL AND CHEMICAL DATA FOR VARIOUS GRADES OF LACTOSE[8]

Component	Fermentation	Crude	Edible	USP[a]
Lactose (%)	98·0	98·4	99·0	99·85
Moisture, non-hydrate (%)	0·35	0·3	0·5	0·1
Protein (%)	1·0	0·8	0·1	0·01
Ash (%)	0·45	0·40	0·2	0·03
Lipid (%)	0·2	0·1	0·1	0·00
Acidity, as lactic acid (%)	—[b]	—[b]	<2	<1
Specific rotation, $[\alpha]_D^{25}$	—[b]	—[b]	52·4°	52·5°

[a] USP = US Pharmacopoeia grade.
[b] Not normally determined.

whey proteins, further concentration, crystallization and recovery of crystals (usually by centrifugation). The crystals are usually washed and may be dissolved and recrystallized to yield a product of high purity. The normal commercial lactose is α-monohydrate, but there is a limited market for β-lactose which is produced by crystallizing above 93·5°C. Commercial lactose is available in a number of grades (Table 5).[8]

12. USES OF LACTOSE

Lactose is used extensively in the pharmaceutical and fermentation industries as well as in infant foods, dietetic foods, fruit, vegetable and dairy products. Lactose may be used as an additive in the foods listed in Table 6.

Human milk contains $\sim 7\cdot0\%$ of lactose (bovine $\sim 5\cdot0$) and $1\cdot0\%$ of protein (bovine $\sim 3\cdot5\%$). Most modern baby foods are essentially 'humanized' cow's milk, i.e. the lactose content is increased and the protein content decreased. Most of the lactose used for baby food preparations is added in the form of demineralized whey powder.[123]

Lactose may be used as an additive in powdered foods as a 'free flowing' agent. The hygroscopicity of lactose glass may also be exploited to advantage as a means of adsorbing free moisture from low-moisture foods and fixing it as water of crystallization of lactose.[123] For example, food particles may be coated in a solution of flavouring and/or colouring materials, and the coated particles then tumbled in a powdered lactose glass which adsorbs surface moisture and forms a capsule around the

TABLE 6
SOME HUMAN FOODS IN WHICH LACTOSE IS USED

Dairy products	Sweet drinks
Sweets	Fruit juice
Cocoa and chocolate	Liqueurs
Ice cream	Instant drinks
Bakery goods	Meats
Bread	Canned sausages
Flour	Spice and flavourings
Noodles	Mustard
Snacks	Dry soups and sauces
Canned fruits	Mayonnaise and dips
Jam, marmalade	Frozen desserts

food particles. Lactose glass is also used in 'instantizing' or increasing the dispersibility of certain foods. Products are prepared containing 5–50% of lactose, spray-dried and then 'instantized' by moistening and redrying.[8] Alternatively, powdered lactose glass is added to the powdered food, which is then wetted slightly, causing the lactose to crystallize and form agglomerates which entrap the other food components. Such products are free-flowing and capable of dispersing rapidly, similar to instant milk powders.

Lactose is used extensively in the formulation of various types of candy, fudges and caramels.[124] It has certain advantages in these products, e.g. less sweetness, better colour binding and better mouth-feel properties, but it also has the disadvantage of tending to crystallize and cause grittiness.[132] Lactose, along with sucrose, may be used to sugar-coat chocolate buttons and hazelnuts;[125] the lactose suppresses sucrose crystallization, thereby allowing the coating to be effected at relatively low temperatures, and reduces the sweetness of the coating.

Lactose is used in the bakery and confectionary industries where it reacts with protein amino groups to give Maillard browning which improves colour and flavour.[126,127] Lactose enhances the emulsifying and creaming properties of shortenings, which improves product quality, facilitates baking operations,[128] gives increased loaf volume and external appearance score, and extends the shelf-life freshness by 50–100% as compared to the shelf-life of standard formulations.[129] Lactose may also be used in sugar coating and as a humectant in bakery products.[130] If proper precautions are taken, lactose is a very suitable icing sugar for many purposes; it gives reduced sweetness and better body with less chipping and cracking.

13. LACTOSE HYDROLYSIS

One of the most interesting technological developments has been the hydrolysis of lactose to sweetening syrups.[119,131–135] Lactose hydrolysis has become industrially applicable, and it appears that it will become the preferred industrial way of solving the problem of utilizing permeate and high-lactose streams. There are three possible approaches to lactose hydrolysis: mineral acid; cation exchange resins, H^+ form; enzymes, both free and immobilized.

A high-temperature acid hydrolysis system including a specially constructed heat exchanger has been used for the hydrolysis of deproteinized liquids, such as ultrafiltration permeate of cheese whey. The pH of the permeate is brought to ~1·5 by added acid or in a strong-acid cation exchange column.[132,133] Hydrolysis is achieved by heating at 90–150°C. In this system, technology and/or engineering problems seem to be formidable, and no existing industrial uses are known.[136]

Probably the most significant advance in lactose utilization has been the development of β-galactosidase technology which is reviewed in Chapter 3. β-Galactosidase has numerous applications in dairy technology:[137] production of lactose-reduced milk and dairy products for patients with lactase deficiency; pre-hydrolysis of lactose to accelerate acid production and ripening of cheese[138] and yoghurt; prevention of lactose crystallization in ice cream and concentrated milks, and reduction of hygroscopicity in dehydrated dairy products; modification of the functional properties of lactose (solubility, sweetness, reducing power, fermentability) to increase its range of uses in non-dairy foods.

The use of free β-galactosidase is economically feasible only for small industrial plants for some specialized applications.[132] However, the process became economically feasible with the development of the immobilized enzyme Corning process.[139] The Corning Glass Works Inc. and the British Milk Marketing Board developed a joint research programme on the feasibility of using immobilized β-galactosidase to hydrolyse lactose. The success of the pilot-scale trials led to the installation of a small industrial-scale unit at a cheese plant in Cheshire, England.[134] Purified β-galactosidase from *Aspergillus niger* is immobilized within silica beads by means of a silane/glutaraldehyde linkage to give total immobilization of the enzyme without impairing its activity. The liquid to be treated (whey, demineralized whey, lactose-rich permeate from ultrafiltered whey) flows down through a column of enzyme-coated glass beads and emerges at the bottom with up to 95% of the lactose hydrolysed.

Hydrolysis of lactose to galactose/glucose markedly changes the two properties of greatest commercial importance in sugars, i.e. sweetness and solubility (Table 7).[134]

The hydrolysed lactose syrup is expected to find applications in non-alcoholic beverages, ice cream, yoghurt, salad dressings, jams, pectin jellies, toffees, fudge and boiled sweets. Novel applications of hydrolysed lactose have been discussed in a recent article.[123] The technology exists for the commercial conversion of glucose (from starch hydrolysis) into fructose via immobilized glucose isomerase.[140] Thus, using two linked immobilized enzyme reactors, it is possible to produce a sweet glucose–

                         P. A. MORRISSEY

TABLE 7
COMPARISON OF SOME PROPERTIES OF GALACTOSE/GLUCOSE SYRUPS WITH THOSE
OF OTHER SUGARS

	Sucrose	*Lactose*	*Galactose/ Glucose*	*Fructose*
Solubility (25°C) (%)	68	22	60	—
Sweetness	100	15	70	123
Viscosity (60% solution) (cp at 20°C)	481	—	380	—

fructose–galactose syrup from lactose:

$$\text{Lactose} \xrightarrow{\text{β-Galactosidase}} \text{Glucose} + \text{Galactose}$$

$$\xrightarrow{\text{Glucose isomerase}} \text{Glucose} + \text{Fructose} + \text{Galactose}$$

Lactose may be converted into lactobionic acid by lactose dehydrogenase, and this may have application in the acidification of dairy products and other foods.[141] The glucose from hydrolysed lactose may be converted via the action of immobilized glucose oxidase into glucono-δ-lactone and hence into gluconic acid. Gluconic acid is an important food acidulant and glucono-δ-lactone is an important food acidogen which may be used for direct acidification of dairy products.[142]

Lactose has served as an extender for spices and volatile aromas.[12] It is used to trap the volatiles emitted during the manufacture of instant coffee; the adsorbate, when added to the powdered coffee, improves its flavour. The adsorptive properties of α-lactose anhydride are better than those of β-lactose.[143,144] The compounds adsorbed include hydrocarbons, alcohols, esters and ketones, and the interactions between lactose and the adsorbates can be summarized as shown in Table 8.[145] Lactose is a very effective flavour enhancer in products such as barbecue sauces, salad dressings, fruit drinks and pie fillings.[128] Its flavour enhancement properties are exploited in the use of lactose as an extender in sodium monoglutamate and protein hydrolysates.[146]

Some derivatives of lactose may have industrial or nutritional potential. Lactulose (prepared by rearrangement of lactose in alkali solution at ∼35°C) is reported to promote the growth of *Lactobacillus bifidus* in the intestine of infants with beneficial effects.[123] Lactitol, the lactose equivalent of sorbitol, is a polyhydric alcohol produced by reduction of lactose

TABLE 8
INTERACTIONS BETWEEN LACTOSE AND ADSORBATES[145]

Adsorbate	Type of interaction
Hydrocarbons	Non-specific interactions, van der Waals type
Esters, aldehydes, and ketones	Van der Waals attractions, as with hydrocarbons, plus alignment of dipoles and possibly some specific interactions; plus hydrogen bonding between oxygen of adsorbate and lactose hydroxyl group proton
Alcohols	Van der Waals attractions, plus alignment of dipoles, as above; plus hydrogen bonding, as above; plus second hydrogen bonding, involving alcoholic proton and oxygen of lactose; minus possible decrease in stability due to strained ring from simultaneous formation of two hydrogen bonds

using a nickel catalyst at about 100°C under pressures exceeding 40 atm.[147] It is a non-nutritive sweetener, much sweeter than the parent lactose and roughly one-third as sweet as sucrose; it is more stable than lactose and does not participate in the Maillard reaction. Lactitol palmitate is produced by direct esterification with fatty acids at 100–160°C and has good emulsifying and detergent properties.[147,148] In general, lactose-derived surfactants are considered to be as effective as analogous sucrose derivatives.[147] Numerous chemical modifications of lactose are described in Chapter 2 of this book; it is very likely that some of these derivatives will be found to have technologically interesting properties, and research in this area should be encouraged.

A novel use for lactose is as a binder for a variety of powders or pellets such as iron ore pellets and iron–steel pellets produced from iron fines captured in pollution control equipment.[149] The powders cannot be fed directly to blast-furnaces and have to be compacted into pellets. Lactose, with its low solubility, high melting point and high heat of combustion, acts as a binder and effects savings in the amount of energy expended. Perhaps the most significant development in recent years is the production of potable ethanol from whey, permeate or lactose, commercialized by the Express Dairy Company. An industrial scale plant has been in operation at Cork, Ireland, for about 6 years. This process may also be used for vinegar manufacture. However, the production of whey-derived ethanol for motor fuels (gasohol) is considered to be uneconomic.[118]

14. DIETARY SIGNIFICANCE OF LACTOSE

The adverse reactions that may develop after ingestion by a lactase-deficient individual of foods containing lactose are discussed in Chapter 4. However, lactose is also recognized as having a positive effect on calcium absorption in rats[150] and humans.[151] The effect of lactose is not specific and is shared with other carbohydrates if they are present in the ileum together with calcium.[150] However, owing to its low digestibility, lactose is the only carbohydrate that normally reaches the ileum intact, where it interacts with the brush border membrane, thereby increasing its permeability to calcium.[152] Lactose fed to vitamin D deficient rats increases the plasma calcium concentration and improves body weight gain and bone mineralization,[153] while a high incidence of lactase deficiency has been observed among patients with osteoporosis.[154] Thus it appears that lactose intake may reduce the need for vitamin D, and it has been suggested[155] that the high content of lactose may contribute to the antirachitic activity of human milk which is higher than one would expect on the basis of its vitamin D content.

REFERENCES

1. MESSER, M. and KERRY, K. R., *Science*, 1973, **180**, 201.
2. MALPRESS, F. H. and HYTTEN, F. E., *Biochem. J.*, 1958, **68**, 708.
3. JENNESS, R., REGEHR, E. A. and SLOAN, R. E., *Comp. Biochem. Physiol.*, 1964, **13**, 339.
4. TYNDALE-BISCOE, H., *Life of Marsupials*, 1973, Edward Arnold, London, p. 82.
5. JENNESS, R. and SLOAN, R. E., *Dairy Sci. Abstr.*, 1970, **32**, 599.
6. ASHWORTH, U. S., RAMAIAH, G. D. and KEYES, M. C., *J. Dairy Sci.*, 1966, **49**, 1206.
7. PILSON, M. E. Q. and KELLEY, A. L., *Science*, 1962, **135**, 104.
8. NICKERSON, T. A., In: *Fundamentals of Dairy Science*, B. H. Webb, A. H. Johnson and J. A. Alford (eds), 1974, Avi Publishing Co., Westport.
9. WATKINS, W. M. and HASSID, W. Z., *J. Biol. Chem.*, 1962, **237**, 1432.
10. BRODBECK, U. and EBNER, K. E., *J. Biol. Chem.*, 1966, **241**, 762.
11. HILL, R. L. and BREW, K., *Adv. Enzymol.*, 1975, **103**, 420.
12. BRODBECK, U., DENTON, W. L., TENAHASHI, N. and EBNER, K. E., *J. Biol. Chem.*, 1967, **242**, 1391.
13. BREW, K., *Nature*, 1969, **222**, 671.
14. LEY, K. M. and JENNESS, R., *Arch. Biochem. Biophys.*, 1970, **138**, 464.
15. SHUKLA, T. P., *CRC Crit. Rev. Food Technol.*, 1973, **3**, 241.
16. KUHN, N. J., CARRICH, D. T. and WILDE, C. J., *J. Dairy Sci.*, 1980, **63**, 328.

17. WHEELOCK, J. V., ROOK, J. A. F. and DODD, F. H., *J. Dairy Res.*, 1965, **32**, 79.
18. SHIPE, W. F., *J. Dairy Sci.*, 1959, **42**, 1745.
19. LINZELL, J. L. and PEAKER, M., *Physiol. Rev.*, 1971, **51**, 564.
20. PEAKER, M., International Dairy Federation, Doc. 125, 1980, p. 159.
21. PINKERTON, F. and PETERS, I. I., *J. Dairy Sci.*, 1956, **39**, 916.
22. PINKERTON, F. and PETERS, I. I., *J. Dairy Sci.*, 1958, **41**, 392..
23. PETERS, I. I., MULAY, C. A. and SHRODE, R. R., *J. Milk Food Technol.*, 1959, **22**, 71.
24. BARRY, J. M. and ROWLAND, J. J., *Biochem. J.*, 1953, **54**, 575.
25. WHITTIER, E. O., *J. Dairy Sci.*, 1944, **27**, 505.
26. PARRISH, F. W., TALLEY, F. B. and PHILLIPS, J. G., *J. Food Sci.*, 1981, **46**, 933.
27. SHARP, P. F. and DOOB, H. Jr, *J. Dairy Sci.*, 1941, **24**, 589.
28. WALSTRA, P. and JENNESS, R., *Dairy Chemistry and Physics*, 1984, John Wiley & Sons, New York.
29. HASSE, G. and NICKERSON, T. A., *J. Dairy Sci.*, 1966, **49**, 127.
30. PATEL, K. N. and NICKERSON, T. A., *J. Dairy Sci.*, 1970, **53**, 1654.
31. HUNZIKER, D. F., *Condensed Milk and Milk Products*, 1926, Hunziker, La Grange, Ill.
32. HUDSON, C. S. and BROWN, F. C., *J. Am. Chem. Soc.*, 1908, **30**, 960.
33. ASANO, Y., AOKI, Y. and YAMAZAKI, N., *Chem. Abstr.*, 1980, **93**, 1505906.
34. ITOH, T., KATOH, M. and ADACHI, S., *J. Dairy Res.*, 1978, **45**, 363.
35. OLANO, A. and RIUS, J. J., *J. Dairy Sci.*, 1978, **61**, 300.
36. BUMA, T. J. and WIEGERS, G. A., *Neth. Milk Dairy J.*, 1967, **21**, 208.
37. BUMA, T. J., *Neth. Milk Dairy J.*, 1966, **20**, 81.
38. SHARP, P. F., US Patent, 2319562, 1943.
39. LIM, S. G. and NICKERSON, T. A., *J. Dairy Sci.*, 1973, **56**, 843.
40. HOCKETT, R. C. and HUDSON, C. S., *J. Am. Chem. Soc.*, 1931, **53**, 4455.
41. BUSHILL, H., WRIGHT, W. B., FULLER, C. H. F. and BELL, A. V., *First Congr. Food Sci. Technol.*, 1962, 237.
42. OLANO, A., BERNHARD, R. A. and NICKERSON, T. A., *J. Food Sci.*, 1977, **42**, 1066.
43. ITOH, T., SATOH, M. and ADACHI, S., *J. Dairy Sci.*, 1977, **60**, 1280.
44. GILLIS, J., *Rec. Trav. Chim.*, 1920, **39**, 88.
45. SIMPSON, T. D., PARRISH, F. W. and NELSON, M. L., *J. Food Sci.*, 1982, **47**, 1948.
46. BUMA, T. J., *Neth. Milk Dairy J.*, 1965, **19**, 249.
47. SMITS, A. and GILLIS, J., *Proc. Acad. Sci. Amsterdam*, 1918, **20**, 520.
48. NEWSTEAD, D. F., International Dairy Federation, Doc 142, 1980, p. 59.
49. CHWAT, T., International Dairy Federation, Doc 116, 1979, p. 23.
50. MULLER, L. L. and KIESEKER, F. G., *Proc. 17th Intern. Dairy Congress*, 1966, **E/F**, 141.
51. BELL, R. W. and MUCHA, T. J., *J. Dairy Sci.*, 1952, **35**, 1.
52. TUMERMAN, L., FRAM, H. and CORNELY, K. W., *J. Dairy Sci.*, 1954, **37**, 830.
53. TESSIER, H., ROSE, D. and LUCENA, C. V., *Can J. Tech.*, 1956, **34**, 131.
54. DESAI, I. D., NICKERSON, T. A. and JENNINGS, W. G., *J. Dairy Sci.*, 1961, **44**, 215.

55. ROSE, D. and TESSIER, H., *Can. J. Tech.*, 1956, **34**, 139.
56. ROSE, D. and TESSIER, H., *J. Dairy Sci.*, 1959, **42**, 898.
57. PYNE, G. T., *J. Dairy Res.*, 1962, **29**, 101.
58. VAN DEN BERG, L., *J. Dairy Sci.*, 1961, **44**, 26.
59. KOSCHAK, M. S., FENNEMA, O., AMUNDSON, C. H. and LEE, J. Y., *J. Food Sci.*, 1981, **46**, 1211.
60. EL-NEGOUMY, A. M. and BOYD, J. C., *J. Dairy Sci.*, 1965, **48**, 23.
61. NICKERSON, T. A., *J. Dairy Sci.*, 1960, **43**, 847.
62. International Dairy Federation, Brussels, Doc 142, 1982.
63. WOODHAMS, D. J. and MURRAY, M. J., *N. Z. J. Dairy Sci. Technol.*, 1974, **9**, 172.
64. BALWIN, A. J. and SANDERSON, W. B., *N. Z. J. Dairy Sci. Technol.*, 1973, **8**, 92.
65. WULFF, J., In: *Milk and Whey Powders*, 1980, Society of Dairy Technology, London, p. 33.
66. NEFF, E. and MORRIS, H. A., *J. Dairy Sci.*, 1968, **51**, 330.
67. SANDERSON, W. B., *N. Z. J. Dairy Sci. Technol.*, 1978, **13**, 137.
68. KING, N., *Dairy Sci. Abstr.*, 1965, **27**, 91.
69. TROY, H. C. and SHARP, P. F., *J. Dairy Sci.*, 1930, **13**, 140.
70. COULTER, S. T., JENNESS, R. and CROWE, L. K., *J. Dairy Sci.*, 1949, **32**, 986.
71. HYND, J., *J. Soc. Dairy Technol.*, 1980, **33**, 52.
72. PARKINSON, J., In: *Milk and Whey Powders*, 1980, Society of Dairy Technology, London, p. 49.
73. PALLANSCH, M. J., Proceedings of Whey Products Conference, Agricultural Research Service, US Dept. of Agriculture, Eastern Regional Laboratory, 1972, Publication No. 3779.
74. LABUZA, T. P., *Food Technol.*, 1968, **22**, 263.
75. LABUZA, T. P., In: *Water Relations of Foods*, R. B. Duckworth (ed.), 1975, Academic Press, London, p. 155.
76. ROCKLAND, L. B. and STEWART, G. F. (eds), In: *Water Activity: Influences on Food Quality*, 1981, Academic Press, London.
77. KARMAS, E., *Food Technol.*, 1980, **34**(4), 52.
78. BERLIN, E., ANDERSON, B. A. and PALLANSCH, M. J., *J. Dairy Sci.*, 1968, **51**, 1339.
79. BERLIN, E., ANDERSON, B. A. and PALLANSCH, M. J., *J. Dairy Sci.*, 1968, **51**, 1912.
80. BERLIN, E., KLIMAN, P. J., ANDERSON, B. A. and PALLANSCH, M. J., *J. Dairy Sci.*, 1973, **56**, 984.
81. BERLIN, E., In: *Water Activity: Influences on Food Quality*, L. B. Rockland and G. F. Stewart (eds), 1981, Academic Press, London, p. 467.
82. MAKOWER, B. and DYE, W. B., *J. Agric. Food Chem.*, 1956, **4**, 72.
83. KAREL, M., In: *Water Relations of Foods*, R. B. Duckworth (ed.), 1975, Academic Press, London, p. 639.
84. WARBURTON, S. and PIXTON, S. W., *J. Stored Prod. Res.*, 1978, **14**, 143.
85. SALTMARCH, M. and LABUZA, T. P., *J. Food Sci.*, 1980, **45**, 1231.
86. TO, E. C. and FLINK, J. M., *J. Food Technol.*, 1978, **13**, 567.
87. BERLIN, E., ANDERSON, B. A. and PALLANSCH, M. J., *J. Dairy Sci.*, 1970, **53**, 146.

88. SHARP, P. F. and DOOB, H. Jr, *J. Dairy Sci.*, 1941, **24**, 679.

89. VARSHNEY, N. N. and OJHA, T. P., *J. Dairy Res.*, 1977, **44**, 93.

90. HELDMAN, D. R., HALL, C. W. and HEDRICK, T. I., *J. Dairy Sci.*, 1965, **48**, 845.

91. BERLIN, E. and ANDERSON, B. A., *J. Dairy Sci.*, 1975, **58**, 25.

92. THOMASON, V. J., MROWETZ, G. and DELFS, E. M., *Milchwissenschaft*, 1972, **27**, 76.

93. DELLA-MONICA, E. S. and HOLDEN, T. E., *J. Dairy Sci.*, 1968, **51**, 40.

94. HENDRICK, H., MOOR, H. DE and CORDIER, J. DE, *Dairy Sci. Abstr.*, 1969, **31**, 462.

95. WALLEY, B., O'CONNOR, C. B. and NAGLE, M., *Irish Agric. Creamery Rev.*, 1977, **30**(4), 11.

96. BOON, P. M. and WOODHAMS, D. J., *N. Z. J. Dairy Sci. Technol.*, 1974, **9**, 151.

97. DUTRA, R. C., TARASSUK, N. P. and KLEIBER, M., *J. Dairy Sci.*, 1958, **41**, 1017.

98. PYNE, G. T. and MCHENRY, K. A., *J. Dairy Res.*, 1955, **22**, 60.

99. SWEETSUR, A. W. M. and WHITE, J. C. D., *J. Dairy Res.*, 1975, **42**, 73.

100. GOULD, I. A., *J. Dairy Sci.*, 1945, **28**, 379.

101. MAURON, J., In: *Progress in Food and Nutrition Science: Maillard Reaction in Foods*, C. Eriksson (ed.), Vol. 5, 1981, Pergamon Press, Oxford, p. 5.

102. FINOT, P. A., DEUTSCH, R. and BUGARD, E., In: *Progress in Food and Nutrition Science: Maillard Reaction in Foods*, C. Eriksson (ed.), Vol. 5, 1981, Pergamon Press, Oxford, p. 345.

103. HURRELL, R. F. and CARPENTER, K. J., *Brit. J. Nutr.*, 1974, **32**, 589.

104. ERIKSSON, C. (ed.), *Progress in Food and Nutrition Science: Maillard Reaction in Foods*, Vol. 5, 1981, Pergamon Press, Oxford.

105. SALMARCH, M., VAGNINI-FERRARI, M. and LABUZA, T. P., In: *Progress in Food and Nutrition Science: Maillard Reaction in Foods*, C. Ericksson (ed.), Vol. 5, 1981, Pergamon Press, Oxford, p. 331.

106. MOLLER, A. B., In: *Progress in Food and Nutrition Science: Maillard Reaction in Foods*, C. Eriksson (ed.), Vol. 5, 1981, Pergamon Press, Oxford, p. 357.

107. PATTON, S., *J. Dairy Sci.*, 1955, **38**, 457.

108. HURRELL, R. F. and CARPENTER, K. J., In: *Progress in Food and Nutrition Science: Maillard Reaction in Foods*, C. Eriksson (ed.), Vol. 5, 1981, Pergamon Press, Oxford, p. 159.

109. LUOUET, F. M., MOUILLET, L., BOUDIER, T. F. and VINCENT, J. P., International Dairy Federation, Doc. 142, 1982, p. 154.

110. LABUZA, T. P., TANNENBAUM, S. R. and KAREL, M., *Food Technol.*, 1970, **24**, 543.

111. LONCIN, M., BIMBENET, J. J. and LENGES, J., *J. Food Technol.*, 1968, **3**, 131.

112. BEN-GARA, I. and ZIMMERMAN, G., *J. Food Sci. Technol.*, 1972, **9**, 113.

113. LABUZA, T. P. and SALTMARCH, M., *J. Food Sci.*, 1981, **47**, 92.

114. MARTINEZ-CASTRO, I. and OLANO, A., *Milchwissenschaft*, 1980, **35**, 5.

115. GEIER, H. and KLOSTERMEYER, H., *Milchwissenschaft*, 1983, **38**, 475.

116. ANDREWS, G. R., *J. Soc. Dairy Technol.*, 1984, **37**, 92.

117. BURTON, H., International Dairy Federation, Doc. 157, 1983, p. 3.
118. COTON, S. G., *J. Soc. Dairy Technol.*, 1980, **33**, 89.
119. ALLUM, D., *J. Soc. Dairy Technol.*, 1980, **33**, 59.
120. HULL, M. E., *J. Dairy Sci.*, 1958, **41**, 330.
121. KAVANAGH, J. A., *N. Z. J. Dairy Sci. Technol.*, 1975, **10**, 132.
122. BRINKMAN, G. E., *J. Soc. Dairy Technol.*, 1976, **29**, 101.
123. FOX, P. F., International Dairy Federation, Doc. 125, 1980, p. 22.
124. WEAL, B. C. and SOUTHWARD, C. R., *N. Z. J. Dairy Sci. Technol.*, 1974, **9**, 2.
125. BOESIG, W., German Federal Republic Patent Application 2936040, 1981 (*Dairy Sci. Abstr.*, 1982, **44**, 400).
126. HOFSTRAND, J. T., ZAEHRINGER, M. V. and HIBBS, R. A., *Cereal Sci. Today*, 1965, **10**, 212.
127. GUY, E. J., *Bakers Dig.*, 1971, **45** (2), 34.
128. REGER, J. V., *Cereal Sci. Today*, 1958, **3**, 270.
129. ZENNER, S. F. and STANBERRY, D. C., US Patent 4233321, 1980 (*Dairy Sci. Abstr.*, 1981, **43**, 606).
130. ASH, D. J., *Food Prod. Dev.*, 1976, **10** (6), 85.
131. SHUKLA, T. P., *CRC Crit. Rev. Food Technol.*, 1975, **5**, 325.
132. MACBEAN, R. D., *N. Z. J. Dairy Sci. Technol.*, 1979, **14**, 113.
133. BOER, R. DE and ROBERTSEN, T., *Neth. Milk Dairy J.*, 1981, **35**, 95.
134. DICKER, R., *Dairy Ind. Intern.*, 1982, **47** (4), 19, 21.
135. MANN, E. J., *Dairy Ind. Intern.*, 1983, **48** (1), 11.
136. JELEN, P., *Food Technol.*, 1983, **37** (2), 81.
137. FOX, P. F. and MORRISSEY, P. A., In: *Industrial and Clinical Enzymology*, L. Vitale and V. Simeon (eds), 1980, Pergamon Press, Oxford, p. 39.
138. RIDHA, S. H., CRAWFORD, R. J. M. and TAMINE, A. Y., *Dairy Ind. Intern.*, 1983, **48** (12), 19.
139. LEVINE, G. A., In: *Proceedings, 1980 Whey Products Conference*, 1981, Philadelphia, US Department of Agriculture, p. 111.
140. BARKER, S. A., *Process Biochem.*, 1975, **10** (10), 39.
141. WRIGHT, D. G. and RAND, A. G. Jr, *J. Food Sci.*, 1973, **38**, 1132.
142. FOX, P. F., *Dairy Sci. Abstr.*, 1978, **40**, 727.
143. NICKERSON, T. A. and DOLBY, R. M., *J. Dairy Sci.*, 1971, **54**, 1212.
144. LEE, I., NICKERSON, T. A. and BERNARD, R. A., *J. Dairy Sci.*, 1975, **58**, 319.
145. NICKERSON, T. A., *J. Agric. Food Chem.*, 1979, **27**, 672.
146. WELCH, D., *Food Tech. Aust.*, 1965, **17**, 318.
147. VAN VELTHUIZSEN, J. A., *J. Agric. Food Chem.*, 1979, **27**, 68.
148. SCHOLNIK, F., SUCHARSHI, M. K. and LINFIELD, M. V., *J. Am. Oil Chem. Soc.*, 1974, **51**, 8.
149. FERRETTI, A. and CHAMBERS, J. V., *J. Agric. Food Chem.*, 1979, **27**, 687.
150. VAUGHAN, D. W. and FILER, L. J., *J. Nutr.*, 1960, **71**, 10.
151. ZIEGLER, E. E. and FOMAN, S. J., *Pediat. Res.*, 1980, **14**, 513.
152. AMBRECHT, H. J. and WASSERMAN, R. H., *J. Nutr.*, 1976, **106**, 1265.
153. HALLORAN, B. P. and DE LUCA, H. F., *Arch. Biochem. Biophys.*, 1981, **209**, 7.
154. BIRGE, S. J. JR, KEUTMAN, H. J., CAUTRECASAS, P. and WHEDON, G. D., *N. Engl. J. Med.*, 1967, **276**, 445.
155. SCHAAFSMA, G. and WAARD, H., *Voeding*, 1982, **43**, 398.

Chapter 2

DEVELOPMENTS IN THE CHEMISTRY AND CHEMICAL MODIFICATION OF LACTOSE

L. A. W. THELWALL

*Tate and Lyle Group Research and Development,
Philip Lyle Memorial Research Laboratory, Reading, UK*

1. INTRODUCTION

Developments in the chemistry of lactose have progressed rapidly over the last two decades. With the current interest in the syntheses of oligosaccharides of milk and blood group substances, together with more sophisticated synthetic and analytical techniques emerging, this trend is likely to continue. Despite this, lactose as a raw material has yet to reach its full potential. Some progress has been achieved in the use of lactose and its hydrolysates in foodstuffs,[1,2] but as a chemical feedstock this disaccharide is still in its infancy.

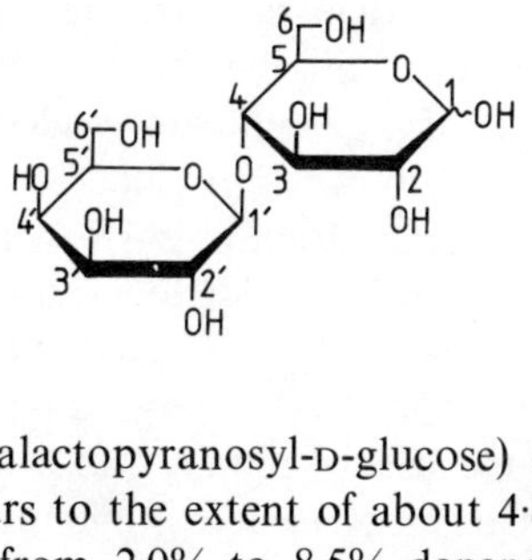

Lactose (4-*O*-β-D-galactopyranosyl-D-glucose) (**1**) is the predominant sugar of milk. It occurs to the extent of about 4·8% (w/v) in cow's milk and has a range of from 2·0% to 8·5% depending on the species of mammal.[3,4] The usual commercial source of lactose is whey, a by-product of the cheese industry. Comprising over 70% of whey solids, these are often used directly as a source of lactose in many foods, although some formulations require a purer form of the sugar.

A number of review articles have appeared on various aspects of lactose.[3-19] It is the aim in this chapter to supplement these by highlighting the recent advances in its chemistry, and it is therefore not intended to be exhaustive in its coverage.

2. STRUCTURE

Early work[20,21] on the constitution of lactose showed, by hydrolysis studies, that it was composed of D-glucose and D-galactose. The non-reducing part of the molecule was attributed to D-galactose from the results of the sequential oxidation of lactose to lactobionic acid and hydrolysis to give D-galactose and gluconic acid.[22] In addition, the carboxylic acid formed by application of the Kiliani reaction[23] to lactose afforded glucoheptonic acid and D-galactose after hydrolysis. Similarly, hydrolysis of lactosone yielded D-galactose and glucosone.[24] Hydrolysis of octa-*O*-methyllactose[25] and methyl octa-*O*-methyllactobionate[26] afforded 2,3,6-tri-*O*-methyl-D-glucose and 2,3,5,6-tetra-*O*-methyl-D-gluconic acid, respectively, as part of the hydrolysates. These observations are consistant with a 1→4 linkage in lactose. Further evidence for the location of the linkage was provided by Zemplen[27] who subjected lactose to two successive 'descents of the series' which gave an *O*-β-D-galactopyranosyl-D-erythrose which failed to give a phenylosazone, indicating that C-2 of the erythrose moiety (C-4 of the original disaccharide) was blocked. Hence it was suggested[27] that a 1→4 linkage is present in lactose. The principal evidence for a β linkage was provided from the hydrolysis of lactose by the galactosidases of almond emulsin, specific enzymes known to catalyse the hydrolysis of β-galactopyranosides without affecting the α-anomers. Lactose is therefore 4-*O*-β-D-galactopyranosyl-D-glucose.

Modern methods of structural analysis are heavily dependent upon physical techniques such as X-ray crystallography and nuclear magnetic resonance spectroscopy. These methods have the advantage of being non-destructive to the sample, and a large amount of detailed information about a compound can be obtained in a relatively short period. The X-ray crystal structures of α-lactose monohydrate and β-lactose have been determined by Rao *et al.*[28] and Hirotsu and Shimada[29] respectively. Both molecules exhibit intramolecular hydrogen bonding between 3-OH and 5'-O, with a notable shortening of the hydrogen bond in β-lactose (2·707 Å) compared to that in α-lactose (2·811 Å) and β-cellobiose (isomorphous with β-lactose). All the oxygen atoms in β-lactose, with the exception of the bridge oxygen atom, are involved in a hydrogen-bonding network which comprises two terminating chains.

3. SYNTHESIS

The first definitive synthesis of lactose was carried out in 1942 by Hudson and co-workers[30] who condensed 1,6-anhydro-2,3-O-isopropylidene-D-mannopyranose with tetra-O-acetyl-α-D-galactopyranosyl bromide. Subsequent hydrolysis of the O-isopropylidene and O-acetyl groups, and acetolysis of the 1,6-anhydro ring afforded 'α-epi-lactose octaacetate'. This compound was converted by means of the glycal synthesis into lactose in an overall yield of 8% (based on the acetobromolactose). This was improved to 35% by Curtis and Jones[31] who replaced the anhydro mannose derivative by 2,3:5,6-di-O-isopropylidene-D-glucose diethyl acetal. Shapiro *et al.*[32] used 2,3-di-O-acetyl-1,6-anhydro-β-D-glucopyranose in the synthesis of lactose to overcome the low reactivity of the 4-OH of D-glucopyranose derivatives in the $^4C_1$ conformation. This was condensed with tetra-O-acetyl-α-D-galactopyranosyl bromide, to give 1,6-anhydro-β-lactose hexaacetate in 41% yield. Lactose has also been prepared[33,34] from cellobiose (C-4' epimer of lactose) via the 4',6'-di-O-mesyl derivative. A very recent method for the synthesis of lactose is that used in the synthesis of oligosaccharides.[35] This involves the condensation of methyl 2,3,6-tri-O-benzyl-D-glucopyranosides (α and β) with tetra-O-acetyl-α-D-galactopyranosyl bromide in a 1:1 mixture of nitromethane and benzene containing mercuric cyanide. Conventional removal of the protecting groups afforded α-lactose monohydrate.

4. BIOSYNTHESIS

The biosynthesis of lactose was first achieved using particulate preparations from lactating guinea-pig or bovine mammary glands.[36] The process involved the condensation of UDP-galactose and D-glucose by D-galactosyl transferase to yield lactose and UDP. A soluble preparation was later isolated from bovine milk capable of synthesizing lactose from

UDP-galactose and D-glucose by the same reaction.[37,38]

In a recent investigation of the biosynthesis of lactose and its deoxy derivatives, UDP-galactose was used as the donor and D-glucose or its monodeoxy derivatives as acceptors.[39] The reaction was catalysed by lactosynthetase (UDP-D-galactose:D-glucose-4-β-D-galactosyl trans-ferase, EC 2.4.1.22) in an enzyme system isolated from goat milk and in the presence of α-lactalbumin. With the exception of 4-deoxy-D-glucose, monodeoxyglucose entered the transglycosylation reaction of biosynthesis of lactose. The affinity of the enzyme and the rate of biosynthesis of lactose and its deoxy derivatives decreased in the order: lactose, 2-deoxylactose, 6-deoxylactose, 3-deoxylactose.

5. ESTERIFICATION

Selective acylation of hydroxyl groups in carbohydrates, in addition to being of theoretical interest, can have significant practical use. Benzoylation of lactose was originally performed using benzoyl chloride in a 20% sodium hydroxide solution.[40] This invariably led to the formation of mixtures of compounds, which were difficult to separate by crystallization. However, Vasquez et al.[40a] repeated the benzoylation procedure and were able to isolate a crystalline hepta-*O*-benzoyllactose in 24% yield. This compound was shown by methylation experiments to be 1,2,6,2′,3′,4′,6′-hepta-*O*-benzoyl-β-lactose (**2**). The low reactivity of the 3-OH towards esterification has also been observed by others. Detailed studies on the selective esterification of lactose have been carried out to determine the relative reactivities of the hydroxyl groups towards the reagents used. In 1974, Hough and co-workers[41,42] determined the order of reactivity of the hydroxyl groups in methyl β-lactoside (**3**) towards benzoyl chloride in pyridine to be: $6′ > 3′ > 6 > 2 > 2′,4′ > 3$. These results were surprising because, in general, primary hydroxyl groups are more reactive than secondary hydroxyl groups.[43] The enhanced reactivity of the 3′-OH is thought to be due to an activating effect of the *cis*-orientated 4′-OH. The 'inner hydroxyl groups' 3-OH, 2-OH and 6-OH, however, display unexpectedly low reactivity, and it was proposed[42] that a more hindered location exists for these hydroxyls in close proximity to the glycosidic linkage. This relative inactivity of the 3-OH is also found in other $1 \rightarrow 4$ linked disaccharides such as β-maltose,[44] methyl β-maltoside,[45] and β-cellobiose.[46] Exploitation of the low reactivity of the 3-OH in lactose towards benzoylation provided a convenient route for

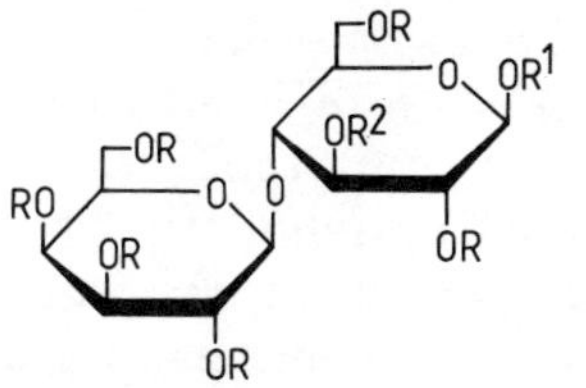

2 R = R¹ = Bz; R² = H
3 R = R² = H; R¹ = Me
4 R = Bz; R¹ = Me; R² = H

the synthesis of its 3-epimer.[47] Regioselective benzoylation of methyl β-lactoside gave the 2,6,2′,3′,4′,6′-hexabenzoate (**4**) in 33% yield. Mesylation of this hexabenzoate using mesyl chloride in pyridine followed by nucleophilic displacement of the sulphonate ester by the benzoate anion furnished the compound **5**. Conventional deprotection of **5** afforded the 3-epimer of lactose, 4-O-β-D-galactopyranosyl-D-allose (**6**). The

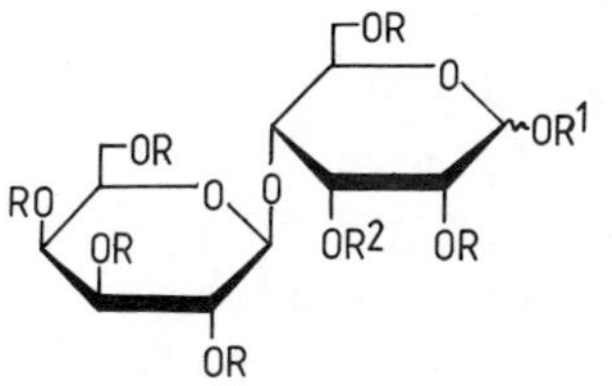

5 R = R² = Bz; R¹ = Me(β)
6 R = R¹ = R² = H

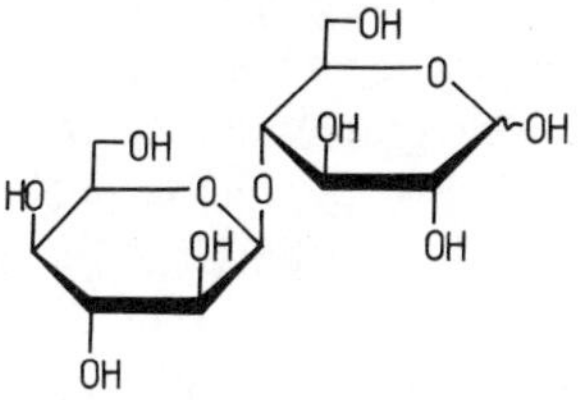

7

isomerization of secondary hydroxyl groups in lactose is well known, with examples such as 'epi-lactose'[48,49] (4-O-β-D-galactopyranosyl-D-mannose) and 'neo-lactose'[50,51] (4-O-β-D-galactopyranosyl-D-altrose). Until quite recently very little work had been carried out on the isomerization of hydroxyl groups in the galactopyranosyl residue. In 1976, Chiba and Tejima[52] reported on the synthesis of 4-O-β-D-idopyranosyl-D-glucose (**7**) from lactose. The reducing disaccharide was prepared from 1,6-anhydro-4′,6′-O-benzylidene-β-lactose (**8**) via selective benzoylation, epoxide formation at C-2′/C-3′, alkaline cleavage of the 'talo-epoxide', and then removal of the protecting groups.

A study of the relative reactivities of the secondary hydroxyl groups in 1,6-*O*-anhydro-4′,6′-*O*-benzylidene-β-lactose (**8**) by selective benzoylation was also carried out by Tejima and co-workers.[53] By using various molar equivalents of benzoyl chloride in pyridine at −20°C, they found the order of reactivity of the secondary hydroxyl groups in **8** to be 3′>2>3>2′. Compound **8** is of particular interest as it contains two pairs of hydroxyl groups, of which those in the glucose moiety are *trans*-diaxial whereas the pair in the galactopyranosyl residue are *trans*-diequatorial. A later investigation[54] of the relative reactivities of the secondary hydroxyl groups in **8** towards toluene-*p*-sulphonylation showed an order of 2>3′>3>2′.

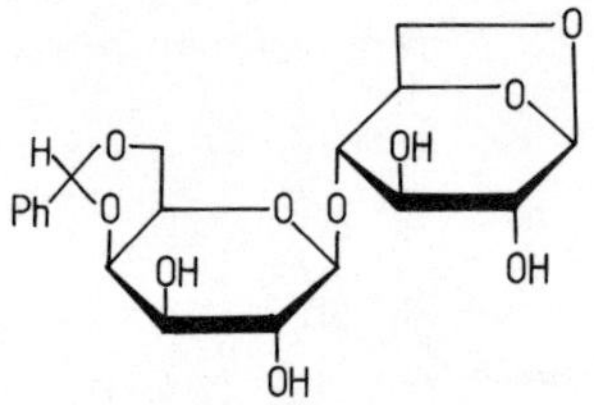

8

The regioselective mono-esterification of α-lactose has been described.[55] By using reactive acylating reagents such as *N*-acyl-thiazolidinethiones or esters of *p*-nitrophenol, mercaptobenzothiazole and 8-hydroxyquinoline, the anomeric hydroxyl group of lactose was selectively esterified in high yield. Other methods of regioselective esterification of lactose have employed the coordination control method of stannylation followed by acylation.[56] The procedure involves the partial stannylation of the hydroxyl groups in lactose with tri-*t*-butyltin(II) oxide. These trialkyltin complexes are believed to be stabilized by intramolecular coordination with the oxygen of the neighbouring hydroxyl groups or the ring oxygens. This regioselectivity enhances the nucleophilicity of the bound hydroxyls which are then selectively esterified. The 2,6,3′,6′-tetrabenzoate **9** was prepared by this method in 72% yield. Acetylation of a primary hydroxyl in the presence of a secondary hydroxyl group was demonstrated in 2,3:5,6:3′,4′-tri-*O*-isopropylidenelactose dimethyl acetal. Matta and co-workers[57] selectively acetylated the 6-OH using ethyl acetate in the presence of Woelm neutral alumina. This procedure was extended to various sugar diols to give moderate yields of the primary acetates.

Ammonolysis of hepta-*O*-benzoyl-β-lactose (**10**) (3-OH free) leads to the formation of both lactose (75·1%) and 6-*O*-benzoyllactose (21·5%).[58]

9 $R^1 = R^2 = R^3 = R^4 = H; R = Bz$
10 $R^1 = Bz(\beta); R = R^2 = R^4 = Bz; R^3 = H$

When the reaction was performed on octa-O-benzoyllactose, 1,1-bis-(benzamido)-1-deoxy-4-O-β-D-galactopyranosyl-D-glucitol (6·7%) and lactose (82%) were the major products, although N-benzoyllactosylamine (0·7%) was later identified as a component of the reaction mixture.[59] The absence of nitrogenous products in the ammonolysis of **10** is a consequence of no benzoyl group being present at 0-3, a requirement necessary for the migration reaction[60] that leads to the formation of compounds nitrogenated at C-1. Comparative studies of the ammonolysis of maltose[44] and lactose[58] heptaacetates (3-OH free in both) showed that 6-O-benzoylmaltose and 6-O-benzoyllactose were formed in 40% and 21·5% yields, respectively, together with the parent disaccharide. The difference in the yields of monobenzoate produced for each disaccharide has been rationalized on the basis of stabilizing interactions of the carbonyl function of the benzoyl group at C-6 with the glycosidic and ring oxygens of the respective sugars. The $(1\rightarrow4)$-β-D-glycosidic linkage in lactose allows fewer interactions than with maltose which has $(1\rightarrow4)$-α-D-, and consequently the benzoyl group is less stable to ammonolysis. The cooperative stabilizing interaction in 6-O-benzoyllactose is indicated by the longer period (4 days) required for removal of the benzoyl group (at C-6) in lactose compared to that of 6-O-benzoyllactitol (4 hours).

An additional example of selective deacylation in lactose derivatives is provided by the reaction of 2,3,6,2′,3′,4′,6′-hepta-O-acetyllactosyl bromide with pyridine;[61] the products obtained were 3,3′,4′-tri-O-acetyl- and 3,6,3′,4′,6′-penta-O-acetyllactosylpyridinium bromides.

The sulphonate ester group in carbohydrate chemistry[62,63] is most often used in structural analysis and in synthetic intermediates for the preparation of a wide variety of other functional groups. The ability of the sulphonyloxy group to stabilize a negative charge makes it an ideal leaving group in nucleophilic displacement reactions.[62] This feature has been used[34] successfully in the synthesis of amino derivatives of lactose, whereby benzyl 4′,6′-di-O-mesyl-β-cellobioside pentaacetate (**11**) was converted (via its azide intermediates) into the 4′,6′-diamino-, 4′-amino- and 6-amino-deoxylactoses. This latter product required inversion of

11 R = R^2 = OAc; R^1 = OBn; R^3 = R^5 = OMs; R^4 = H
12 R = R^4 = OH; R^1 = OMe; R^2 = R^5 = OTs; R^3 = H

configuration at C-4′ by an S_N2 displacement of the 4′-mesyl group by the benzoate anion, to give a lactose analogue. It is interesting to note that the first claim of the use of a displacement reaction of the sulphonyloxy group by an acyloxy anion in the disaccharide field was demonstrated in the conversion of cellobiose into lactose.[33] Treatment of **11** (derived from the 4′,6′-cyclic phenylboronate **13**) with sodium benzoate in boiling *N,N*-dimethylformamide gave, after removal of the protecting groups, crystalline α-lactose hydrate in an overall yield of 2% (from cellobiose octaacetate).

13 R = OAc; R^1 = OBn

The preparation and use of methyl 6,6′-di-*O*-tosyl-β-lactoside (**12**) has recently been described.[64] The ditosylate was prepared in high yield from methyl β-lactoside by reaction of **3** with *p*-toluenesulphonyl chloride in pyridine, and was further used to prepare 6,6′-diacetamido-6,6′-dideoxylactose.

6. HALOGENATED DERIVATIVES

The halogenation of carbohydrates can be carried out by a variety of methods. In general the halogenation at primary positions occurs readily, whilst at secondary positions reaction may proceed if the stereoelectronic factors that govern these reactions are favourable.[65] As the introduction of a halogen into a carbohydrate usually operates by way of a displacement type of mechanism, inversion of configuration occurs at secondary carbons.

The chlorination of methyl β-lactoside (**3**) using mesyl chloride–*N*,*N*-dimethylformamide, and also sulphuryl chloride, has been extensively studied.[66] Treatment of **3** with 30 equivalents of mesyl chloride–*N*,*N*-dimethylformamide at 94°C for 9 days gave five components after acetylation and fractionation of the mixture on silica gel. These were characterized as the 3,6,3′,4′,6′-pentachloro (**14**) (11%), 3,6,3′,6′-tetrachloro (**15**) (10%), and 3,6,4′,6′-tetrachloro (**16**) (18%) derivatives. A mixture of two isomeric trichlorides was also isolated from the mixture which underwent further separation procedures to give the 3,6,6′-trichloro (**17**) (2·5%) and 6,3′,6′-trichloro (**18**) (0·8%) compounds. When the reaction was repeated using a lower proportion of mesyl chloride (10 equivalents) and at 60°C, two major components were detected in the less complex mixture. These were subsequently characterized as the 6,6′-dichloride **19** (25%) and a product of cleavage of the interglycosidic bond, methyl 6-chloro-6-deoxy-β-D-glucopyranoside (isolated as their peracetates).

14 $R^1 = R^3 = R^5 = H; R^2 = R^4 = R^6 = Cl$
15 $R^1 = R^3 = R^6 = H; R^2 = R^4 = Cl; R^5 = OAc$
16 $R^1 = R^4 = R^5 = H; R^2 = R^6 = Cl; R^3 = OAc$
17 $R^1 = R^4 = R^6 = H; R^2 = Cl; R^3 = R^5 = OAc$
18 $R^1 = R^5 = OAc; R^2 = R^3 = R^6 = H; R^4 = Cl$
19 $R^1 = R^3 = R^5 = OAc; R^2 = R^4 = R^6 = H$

Chlorination of methyl β-lactoside using 5 molar equivalents of sulphuryl chloride gave, on the whole, poor yields of chlorinated compounds (2% for 6,6′-dichloro). However, when 10 molar equivalents of sulphuryl chloride was used, the 3,6,3′,4′,6′-pentachloro derivative **14** was obtained in 16% yield.

As cellobiose is a C-4′ epimer of lactose, an alternative method of obtaining derivatives of lactose is by reaction of cellobiose whereby inversion of configuration at C-4′ occurs. Chlorination of benzyl-β-cellobioside using mesyl chloride–*N*,*N*-dimethylformamide gave, amongst other products, the 6,4′,6′-trichloro-lactoside **20**. A good yield of the

20 $R = R^2 = OAc; R^1 = R^4 = Cl; R^3 = H$
21 $R = R^1 = R^2 = OBz; R^3 = H; R^4 = Cl$
22 $R = R^1 = R^4 = OBz; R^2 = H; R^3 = Cl$

4′,6′-dichloro-lactoside **21** (50%) has also been obtained using a suitably protected cellobiose derivative.[67] Treatment of benzyl 2,3,6,2′,3′-penta-*O*-benzoyl-β-cellobioside with sulphuryl chloride in pyridine afforded a mixture of two dichlorides. The major component was the expected benzyl 4′,6′-dichloro-β-lactoside pentabenzoate, isolated in 50% yield. The minor component (4·3%) was identified as benzyl 2,3,6-tri-*O*-benzoyl-4-*O*-(2,4-di-*O*-benzoyl-3,6-dichloro-3,6-dideoxy-β-D-gulopyrano-syl)-β-D-glucopyranoside (**22**), a product of neighbouring group partici-pation by the benzoyloxy group at C-3′ during the loss of the 4′-chlorosulphate group. An unequivocal synthesis of the 4′,6′-dichloride **21** was subsequently achieved in high yield by the action of lithium chloride on the corresponding 4′,6′-dimesylate, and was identical with the 4′,6′-dichloride obtained from the sulphuryl chloride reaction.

Very few examples of brominated and iodinated derivatives of lactose have been reported. Hepta-*O*-acetyl-6-deoxy-6-iodo-β-lactose, a precur-sor to unsaturated and deoxy lactose analogues, was obtained by reaction of hepta-*O*-acetyl-6-*O*-tosyl-β-lactose with sodium iodide in boiling acetonitrile.[68] Other 6-iodo compounds have been prepared by treatment of the corresponding 6-*O*-mesylates with sodium iodide in acetonitrile.[69] Bromination has occurred at the C-6′ position in lactose. Using the procedure of Hanessian,[70] Tejima and co-workers[71] were able to convert tetra-*O*-acetyl-1,6-anhydro-4′,6′-*O*-benzylidene-β-lactose (**23**) into tetra-*O*-acetyl-1,6-anhydro-4′-*O*-benzoyl-6′-bromo-6′-deoxy-β-lac-tose (**24**), by heating **23** with *N*-bromosuccinimide and barium carbonate in tetrachloromethane and dichloromethane for 2·5 h. A 52% yield of the 6′-bromide **24** was obtained after chromatography.

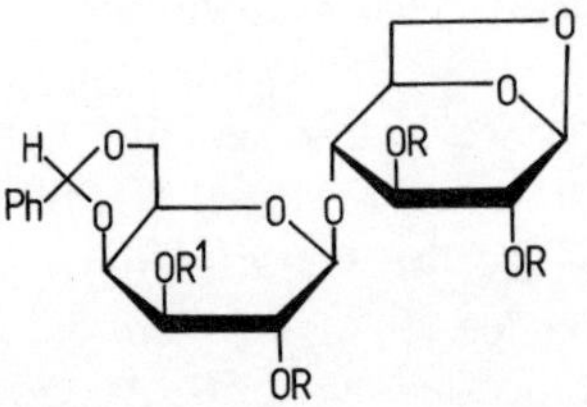

23 R = R¹ = Ac
34 R = H; R¹ = Ts

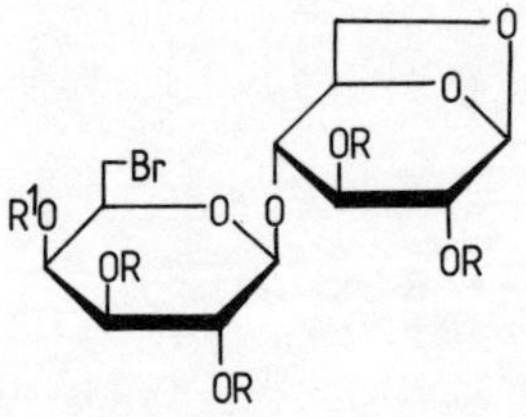

24 R = Ac; R¹ = Bz

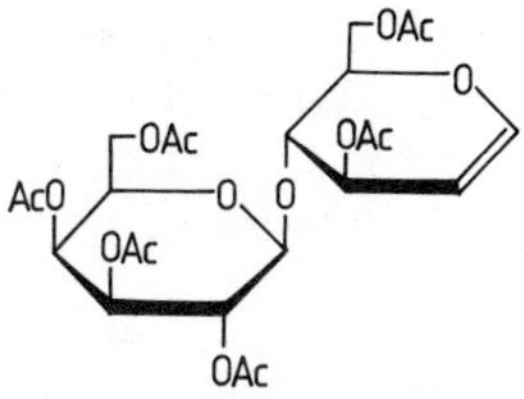

25

In common with the bromide and iodide derivatives of lactose, very little is known about the lactose fluorides. In 1977, Kent and Dimetrijevich[72] described the use of fluoroxytrifluoromethane as a source of 'electrophilic fluorine'. They applied this reagent to hexa-*O*-acetyl-D-lactal (**25**) which produced a number of fluorinated disaccharides (at C-1 and/or C-2 positions).

7. LACTOSYL HALIDES

The hepta-*O*-acetyllactosyl halides (**26–29**) are most important as precursors of lactosides. They may be prepared by treatment of lactose octaacetate with the appropriate hydrogen halide. The thermodynamically more stable α-anomer is usually formed; in common with other sugars, the stability of lactosyl halides decreases as one ascends the series from fluoro to iodo.

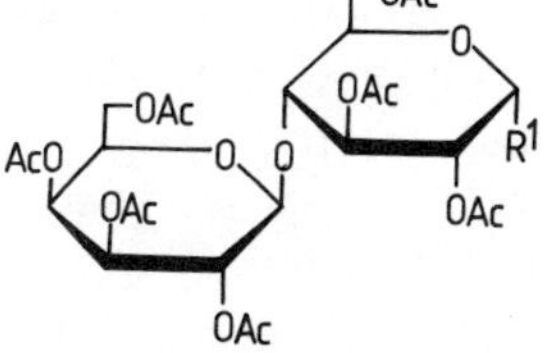

26 $R^1 = F$
27 $R^1 = Cl$
28 $R^1 = Br$
29 $R^1 = I$

Hepta-*O*-acetyl-α-lactosyl fluoride (**26**),[73] chloride (**27**),[74] bromide (**28**)[75,76] and iodide (**29**)[75] have all been obtained by reaction of lactose octaacetate with the respective hydrogen halide in acetic acid or acetic anhydride. Later methods for the synthesis of hepta-*O*-acetyl-α-lactosyl chloride relied on the use of phosphorus pentachloride–aluminium chloride in chloroform,[75] titanium tetrachloride[77] or liquid hydrogen chloride[78] as chlorinating agents. Some of these methods tend to produce mixtures of the α- and β-lactosyl chlorides. An efficient procedure for the synthesis of 1,2-*trans*-*O*-acetyl glycosyl chlorides of lactose, cellobiose and glucose has been described.[79] Hepta-*O*-acetyl-β-lactosyl chloride was obtained in 76% yield from the reaction of hepta-*O*-acetyl-α-lactosyl bromide with lithium chloride in hexamethylphosphoramide.

8. LACTOSIDES

The preparation of methyl β-lactoside (**3**) was reinvestigated in 1952, by Smith and Van Cleve,[80] who claimed that earlier reports of its synthesis by Ditmar[81] were probably inaccurate. In their account, crystalline methyl β-lactoside was prepared by the reaction of α-acetobromolactose with methanol, in the presence of excess silver carbonate, followed by deacetylation with sodium methoxide. Recrystallization of the crude material from 96% ethanol afforded methyl β-lactoside as its monohydrate. The structure was confirmed by methylation and periodate oxidation studies. In a later paper by the same authors,[82] a direct preparation of methyl β-lactoside was described using α-acetobromolactose and magnesium ethoxide in methanol. A variety of lactosides have been prepared by use of the Koenigs–Knorr reaction. These include 2-chloroethyl,[83,84] 2-bromoethyl,[84] 2-iodoethyl,[84] 3-chloropropyl,[83] benzyl, 1-menthyl,[85] myristyl,[86] cholesteryl,[87] and deoxycorticosterone β-lactosides.[88]

Various aryl glycosides of lactose have been prepared in good yields. Dea[89] used a modification of the procedure of Helferich and Griebel,[90] who had earlier reported the synthesis of phenyl β-lactoside by treating α-acetobromolactose with potassium phenoxide in aqueous acetone. This method was adapted by Dea[89] to prepare the *o*-, *m*-, and *p*-bromophenyl and -chlorophenyl β-lactosides and also the *o*- and *p*-iodophenyl analogues. In the search for a facile method for the synthesis of lactosan, Tejima and Chiba[69] investigated the alkaline hydrolysis of aryl lactosides. The *o*-chlorophenyl glycoside, deemed to be the most suitable precursor for the formation of lactosan, was obtained by fusing lactose octaacetate with *o*-chlorophenol in the presence of *p*-toluenesulphonic acid and acetic anhydride. This method gave a 23% yield of *o*-chlorophenyl β-lactoside heptaacetate. In an analogous fashion, phenyl β-lactoside heptaacetate was synthesized from lactose and phenol in 31% yield. When octa-*O*-acetyl-β-lactose was fused with phenol in the presence of anhydrous zinc chloride, the corresponding α-glycoside was obtained in 55% yield. The methodology for stereoselective glycosidation has developed considerably in recent years.[90a] According to Tejima and co-workers,[91] α-glycosides of reducing disaccharides can be prepared in preference to the β-anomers, by alcoholysis of the appropriate β-glycosyl dimethyldithiocarbamates. Thus methyl α-lactoside was formed on methanolysis of β-lactosyl N,N-dimethyldithiocarbamate, and isolated (as its peracetate) from the anomeric glycoside acetates by fractional crystallization in 54% yield. Gas chromatographic analysis of the corresponding benzyl alcoholysis reaction revealed an α:β ratio of 2:1.

9. ETHERS

Classical methods for determining constitution and structural features of carbohydrates relied greatly on a study of the degradation of their permethylated derivatives. The structure of lactose was investigated[25] using this method. Hydrolysis of a crystalline methyl hepta-O-methyl-lactoside (**30**) gave 2,3,4,6-tetra-O-methyl-D-galactose and 2,3,6-tri-O-methyl-D-glucose. On the basis of this result, the linkage was judged to be either 1→4 or 1→5 (galactose–glucose) in lactose.

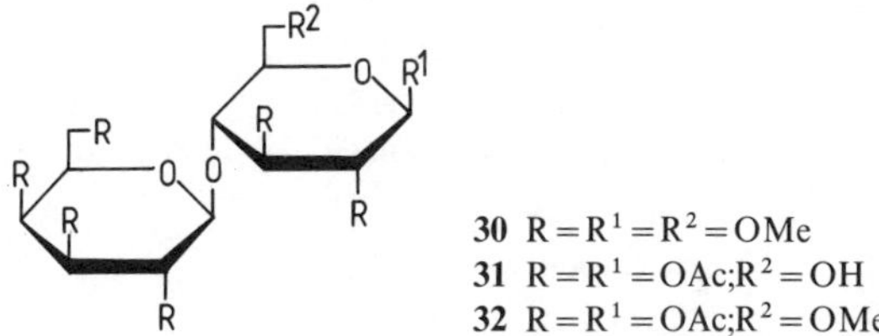

30 R = R¹ = R² = OMe
31 R = R¹ = OAc; R² = OH
32 R = R¹ = OAc; R² = OMe

Several methods exist for preparing methyl ethers. Common methylating reagents used are dimethyl sulphate–sodium hydroxide,[92,93] methyl iodide–silver oxide–acetone,[94] sodium hydride–methyl iodide–N,N-dimethylformamide[93] or diazomethane–boron trifluoride etherate.[95] Methylation of 1,2,3,2′,3′,4′,6′-hepta-O-acetyl-β-lactose (**31**) using diazomethane–boron trifluoride etherate reagent gave crystalline hepta-O-acetyl-6-O-methyl-β-lactose (**32**) in 72% yield.[69] Other positions that have been methylated in lactose derivatives are C-3,[53,54] C-2′,[53,54] C-4′,[96] C-6′,[97] C-3,2′,[53,54] C-4′,6′,[96,97] C-2,3,2′,[53,54] C-3,2′,3′,[54] and C-2,3,2′,3′,4′,6′.[97]

The benzyl ether protecting group has found widespread use in the synthesis of higher oligosaccharides. They are conveniently prepared

33 R = Bn;R[1] = H

using the reagent systems benzyl chloride–potassium hydroxide,[98] benzyl bromide (or chloride)–sodium hydride–N,N-dimethylformamide,[99] or benzyl bromide–sodium hydroxide–dimethyl sulphoxide.[100]

Takamura *et al.*[101,102] used the partially benzylated compound 1,6-anhydro-2,3,2′,4′-tetra-O-benzyl-β-lactose (**33**) in the synthesis of 3′,6′-di-β-N-acetylglucosaminyl-β-lactose[101] and lacto-N-neohexaose.[102] Compound **33** was prepared from 1,6-anhydro-4′,6′-O-benzylidene-3′-O-tosyl-β-lactose (**34**) (see p. 44) in an overall yield of 48%. Their method of synthesis involved the benzylation of **34** (at C-2,3,2′) followed by debenzylidenation. The resultant 4′,6′-diol was selectively tosylated at C-6′ and then benzylated at C-4′. Detosylation at C-3′ and C-6′, using a sodium/mercury amalgam afforded the required diol (**33**).

An alternative procedure for obtaining partially benzylated derivatives of lactose uses the method for stereoselective benzylidene ring cleavage.[103] Thus, treatment of benzyl penta-O-benzyl-4′,6′-O-benzylidene-β-lactoside with an equimolar amount of LiAlH₄–AlCl₃ reagent gave an isomeric mixture of two hepta-O-benzyl derivatives. These products were identified as benzyl 2,3,6,2′,3′,4′-hexa-O-benzyl-β-lactoside (**35**) and benzyl 2,3,6,2′,3′,6′-hexa-O-benzyl-β-lactoside (**36**), and were isolated in 71% and 14·4% yields respectively.[104]

35 R = R[1] = Bn;R[2] = H
36 R = R[2] = Bn;R[1] = H

Unlike ester blocking groups, benzyl ethers are relatively stable in both acid and basic conditions, and do not normally migrate to, or participate with, neighbouring positions. These features, in addition to its ease of removal by catalytic hydrogenation, make it an attractive protecting group in oligosaccharide syntheses.[35,101,102,105–108]

10. ANHYDRO DERIVATIVES

An important linkage in this class of compounds is the 1,6-anhydro derivative (**37**) of lactose, commonly known as lactosan. A number of groups have reported[109,110] on the preparation of 1,6-anhydro-lactose. In a recent study of its formation via based-catalysed elimination of aryl glycosides, the *o*-chlorophenyl glycoside of lactose was found to be the most favourable precursor for 1,6-anhydro-β-lactose.[69]

37

The synthetic value of three-membered anhydro derivatives (i.e. epoxides) in carbohydrate chemistry is well known. They are usually formed by alkaline hydrolysis of suitably orientated substituents such as sulphonic acid esters. Once formed, their value is apparent because of the ability to introduce new substituents stereospecifically into the carbohydrate ring, by nucleophilic attack of the epoxide. Configurational changes can be brought about by this latter reaction owing to the nature of cleavage of the anhydro ring. This has been demonstrated[52,111] by the formation of a new disaccharide, 4-*O*-β-D-idopyranosyl-D-glucose (**7**) from the epoxide intermediate **38**. Compound **38** was synthesized by reaction of the 2'-mesyl or the 2'-tosyl derivative of 1,6-anhydro-4',6'-*O*-benzylidene-2,3,3'-tri-*O*-benzoyl-β-lactose with 1·1M methanolic sodium methoxide solution at reflux temperature. The 2',3'-epoxide was then stereospecifically ring-opened using aqueous potassium hydroxide to give the *trans*-diaxial product. Conventional removal of the protecting groups afforded 4-*O*-β-D-idopyranosyl-D-glucose (**7**).

38

11. CYCLIC ACETAL DERIVATIVES

A number of cyclic acetal derivatives of lactose have been prepared in recent years. An early report of the use of the isopropylidene acetal as a protecting group in lactose was by Flowers and co-workers,[112] who reacted benzyl β-lactoside with acetone in the presence of an acid catalyst. The type of catalyst was crucial to the yield of reaction. A moderate yield of the 3′,4′-acetal **39** was achieved with sulphuric acid as

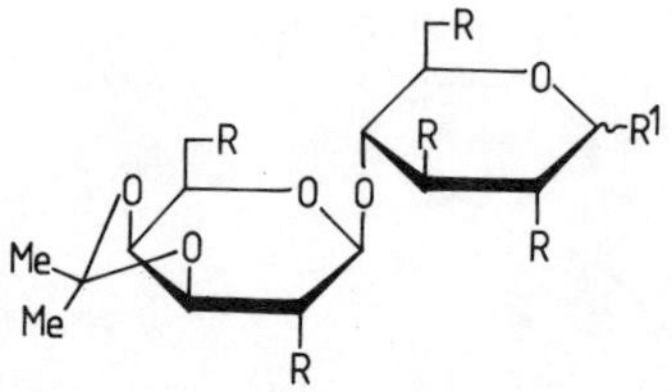

39 $R^1 = OBn(\beta); R = OH$
41 $R = R^1 = OBz$

catalyst, whereas the use of anhydrous copper sulphate failed to cause reaction. The preferred catalyst, however, was *p*-toluenesulphonic acid, which gave a 76% yield of benzyl 3′,4′-*O*-isopropylidene-β-lactoside. This compound was further used as an intermediate in a trisaccharide synthesis concerning a study of glycosphingolipids. The corresponding 3′,4′-acetal of methyl β-lactoside has also been prepared in good yield by the application of 2,2-dimethoxypropane and an acid catalyst as the acetalating reagent.[113] Acetal exchange has found widespread use in the protection of glycol systems in mono- and di-saccharides. The most commonly used reagent is 2,2-dimethoxypropane in *N*,*N*-dimethylformamide with a catalytic amount of *p*-toluenesulphonic acid. Application of this reagent system to lactose has given moderate yields of the 4′,6′-acetal **40**[114,115] and the 3′,4′-acetal **41**.[115] Lower temperatures (ambient) and longer reaction time (3 h) appear to favour the formation of the kinetic product, i.e. the 4′,6′-acetal, whereas higher temperatures (80–85°C) and short reaction times (45 min) produce the thermodynamic product (3′,4′-

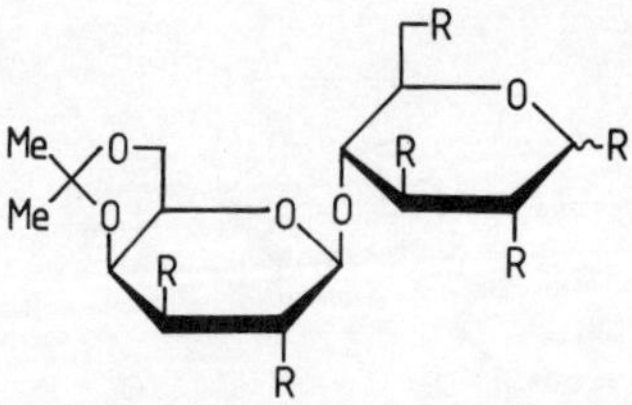

40 $R = OBz$

acetal). In contrast, a reaction which operates exclusively under kinetic control is the reaction of lactose with 2-methoxypropene in N,N-dimethylformamide with p-toluenesulphonic acid.[116] A 30–35% yield of the novel diacetal 2′,6:4′,6′-di-O-isopropylidenelactose (**42**) was obtained

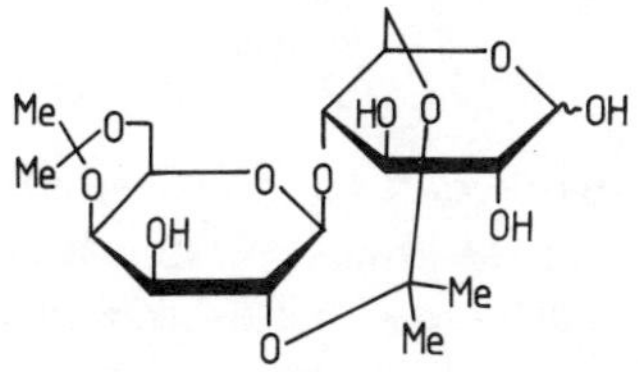

42

by this method. This derivative allows a convenient route for the further modification of lactose at the C-1, C-2, C-3 and C-3′ positions. Even higher derivatization by acetal formation has been achieved[117] in lactose by its reaction with 2,2-dimethoxypropane containing p-toluene-sulphonic acid at reflux temperature. Isopropylidene acetals were formed between O-3′ and O-4′ in the galactopyranosyl residue, and between O-5, O-6 and O-2,O-3 in which the glucose moiety of lactose had reacted in the acyclic form and with concomitant acetalation at C-1. The resultant tetraacetal, 2,3:5,6:3′,4′-tri-O-isopropylidenelactose dimethyl acetal (**43**),

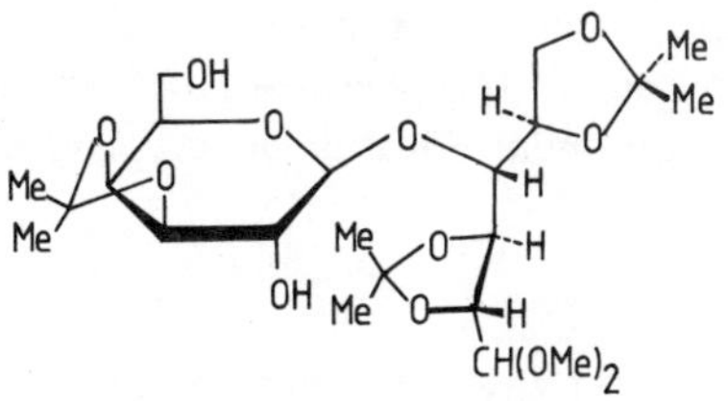

43

is a unique precursor to derivatives of lactose modified at the C-2′ and/or C-6′ positions.[118] The preparation of the analogous tri-O-benzylidenelactose dimethyl acetal (**44**) was recently reported.[119] Treatment of lactose in anhydrous methanol with benzaldehyde dimethyl acetal and p-toluenesulphonic acid, followed by removal of methanol by azeotropic distillation, caused concomitant acetalation and glycosidation of lactose. An inherent disadvantage of this procedure, however, is the introduction of new chiral centres (due to the acetal carbons), and consequently it yielded the tetraacetal as a mixture of diastereoisomers.

Monobenzylidene acetals of lactose have been prepared as intermediates in the synthesis of amino-sugars[89] and 6-deoxy-hexosyl-

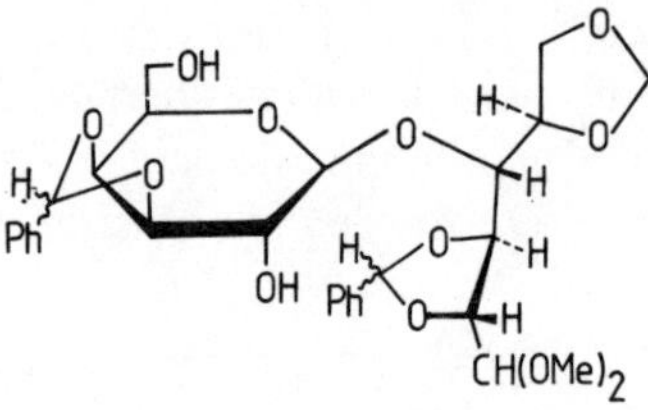

CH(OMe)$_2$ **44**

hexoses.[104] Hough and co-workers[119a] utilized methyl 4′,6′-*O*-benzylidene-β-lactoside to gain access to the 4′ and 6′ positions. This was prepared by reaction of lactose in benzaldehyde with crushed anhydrous zinc chloride. By a series of reactions performed on the 4′,6′-acetal, involving per-benzoylation, deacetalation, mesylation of the resulting 4′,6′-diol, azide displacement of the mesylate groups, deacylation, catalytic reduction of the amino functions, peracetylation and finally acetolysis, 4′,6′-dia-cetamido-4′,6′-dideoxycellobiose hexaacetate was formed.

An elegant use of the 4′,6′-*O*-benzylidene acetal of lactose was dem-onstrated by Liptak *et al.*[104] who required protected disaccharides having only their 6′-OH free. Benzylidenation of benzyl β-lactoside using benzaldehyde–zinc chloride followed by perbenzylation gave benzyl penta-*O*-benzyl-4′,6′-*O*-benzylidene-β-lactoside. Hydrogenolysis of this compound using LiAlH$_4$–AlCl$_3$ caused stereoselective cleavage of the benzylidene ring, to give a benzyl ether and free hydroxyl group. Two isomeric hepta-*O*-benzyl derivatives were isolated and identified as the 4′- and 6′-*O*-benzyl compounds (**35** and **36**) in 71% and 14·4% yields respectively.

12. UNSATURATED DERIVATIVES

The earliest report of an unsaturated derivative of lactose was by Fischer and Thierfelder.[120] By treating hepta-*O*-acetyl-α-lactosyl bromide (**28**) with zinc and acetic acid, they obtained hexa-*O*-acetyl lactal (**25**). A more recent method for its preparation involves the use of bicyclic amines.[121] Reaction of hepta-*O*-acetyl-α-lactosyl bromide with either 1,8-diaza-bicyclo[5,4,0]undec-7-ene (**45**) or 1,5-diazabicyclo[4,3,0]non-5-ene (**46**) in *N*,*N*-dimethylformamide gave good yields of hexa-*O*-acetyl lactal.

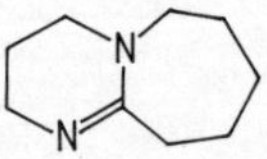

45 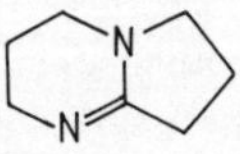   **46**

In common with alkenes, lactal can undergo addition reactions, and it has been used in the preparation of fluorinated derivatives of lactose.[72] Unsaturation at other positions in lactose has occurred between C-4' and C-5',[119a] and at both the primary positions to produce exocyclic double bonds at C-5',C-6'[71] and C-5,C-6.[122] These latter unsaturated derivatives were obtained by dehydrohalogenation of the appropriate halogeno (iodide or bromide) compound, using silver fluoride in dry pyridine. The use of sodium azide in hexamethylphosphoramide in the reaction with 2,3,2',3',4'-penta-O-acetyl-1,6-anhydro-6'-bromo-6'-deoxy-β-lactose can also produce the C-5',C-6' ene, in addition to the expected product of nucleophilic displacement, the 6'-azide.[71]

13. DEOXY DERIVATIVES

Deoxy-sugars[123] comprise an important class of compounds and are found widely in Nature as constituents of polysaccharides and to a smaller extent as the free sugars. Terminal-deoxy-hexoses and the dideoxy-hexoses are often found as components of cardiac glycosides and as antigenic determinants in bacterial polysaccharides. Their preparation can be accomplished by a number of methods, most commonly by hydride cleavage of epoxides,[124] reductive dehydrohalogenation[71] and, in the case of terminal deoxy derivatives, by catalytic reduction of the exocyclic vinyl ether.[125]

6-Deoxylactose (**47**) was prepared from 6-iodo-β-lactose heptaacetate (**48**) by catalytic reduction, using Raney nickel as the catalyst, and in the presence of triethylamine.[125] Deacetylation of the resultant 6-deoxy-heptaacetate afforded **47** as hygroscopic crystals. This deoxy-sugar was also obtained as a minor product by an alternative procedure. When the heptaacetate **48** was stirred at ambient temperature with silver fluoride in anhydrous pyridine in the absence of light, for 22 h, the 5,6-ene was produced in 43% yield. Catalytic hydrogenation of this product using a palladium catalyst afforded, in addition to the expected 6-deoxylactose

47 $R = R^1 = R^4 = R^5 = OH; R^2 = R^3 = H$

48 $R = R^4 = R^5 = OAc; R^1 = OAc(β); R^2 = I; R^3 = H$

49 $R = R^1 = R^2 = R^3 = OH; R^4 = R^5 = H$

heptaacetate, a major compound which was subsequently characterized as 4-*O*-β-D-galactopyranosyl-6-deoxy-α-L-idopyranose heptaacetate. This product was found to be the major component of the hydrogenation reaction, irrespective of the choice of catalyst used. In contrast, hydrogenation of the corresponding 5,6-glucosene has been reported to give predominantly the 6-deoxy-D-gluco isomer.[126]

Synthesis of '3'-deoxy-*ido*lactose' (**49**)[124] [4-*O*-(3-deoxy-β-D-*lyxo*-hexopyranosyl)-D-glucose] was achieved by treatment of 1,6-*O*-anhydro-4-*O*-(2,3-anhydro-4,6-*O*-benzylidene-β-D-talopyranosyl)-β-D-glucose with LiAlH$_4$ in boiling tetrahydrofuran. *Trans*-diaxial cleavage of the epoxide proceeded predominantly according to the Furst–Plattner rule. Deprotection gave **49** as a hygroscopic amorphous powder, which was reported to have slight sweetness properties, although this was not quantified. In continuance of their chemical studies of reducing disaccharides, Tejima and co-workers[125] prepared 6'-deoxylactosan by reductive dehydrobromination of tetra-*O*-acetyl-1,6-*O*-anhydro-4'-*O*-benzoyl-6'-bromo-6'-deoxy-β-lactose. The free deoxy-sugar was, however, not reported in this communication.

14. NITROGEN-CONTAINING COMPOUNDS

In the past decade, a number of aminodeoxy derivatives of lactose have been synthesized. Amino-sugars often occur as components of many antibiotics[127,128] and bacterial polysaccharides,[129] and are therefore of considerable interest. Rolland *et al.*[130] have recently described the total synthesis of α-disaccharidyl-2,5,6-trideoxy-streptamines related to aminoglycoside antibiotics derived from lactose and maltose. The glycosides, however, were found to be devoid of significant antibacterial activity.

N-Glycosylasparagine derivatives of lactose, maltose and cellobiose, required as model compounds in a study of glycoproteins, were synthesized from their respective octaacetates via the 1-bromo, 1-azido and 1-amino derivatives.[131] Condensation of the 1-amines with 1-benzyl *N*-benzyloxycarbonyl-L-aspartate, using dicyclohexylcarbodiimide or 2-ethoxy-*N*-ethoxycarbonyl-1,2-dihydroquinoline as coupling reagents, afforded, after removal of the protecting groups, the required *N*-glycosylasparagines. The *N*-lactosyl- and *N*-maltosyl-asparagines were found to be very hygroscopic and decomposed after 4–6 weeks. Spinola and Jeanloz[132] have also reported the formation of some lactose/lactosamine derivatives linked to asparagine.

Synthetic glycoproteins can be produced[133] by reductive amination of protein and reducing disaccharide in the presence of sodium cyanoborohydride.[134] The reductive amination of lactose with selected alkylamines and sodium cyanoborohydride in boiling methanol and in the presence of a weak organic acid afforded N-alkyl-(1-deoxylactitol-1-yl)amines.[135] Sodium cyanoborohydride[134] was used to minimize the formation of Amadori rearrangement products, by selectively reducing the imine initially produced by the condensation of an alkylamine with lactose.

The introduction of the amino function into the lactose molecule has been the subject of work for a number of groups. In 1973, Tejima and Chiba[69] reported the synthesis of 6-acetamido-6-deoxylactose (**50**). Their synthetic route to this compound involved the cleavage of the 1,6-anhydro ring in lactosan hexaacetate followed by mesylation to give the hepta-O-acetyl-6-O-mesyllactose **51**. Benzyl or methyl glycosidation of **51** (via the 1-bromide) gave moderate yields of the corresponding β-glycosides. These products were readily converted into their respective 6-azides by reaction with sodium azide in hexamethylphosphoramide at 100°C. Raney nickel hydrogenation of benzyl hexa-O-acetyl-6-azido-6-deoxy-β-lactoside, followed by acetylation, afforded benzyl hexa-O-acetyl-6-acetamido-6-deoxy-β-lactoside together with hepta-O-acetyl-6-acetamido-6-deoxy-β-lactose (**52**). De-O-acetylation of **52** using methanolic sodium methoxide yielded 6-acetamido-6-deoxylactose (**50**) as a hygroscopic powder which failed to show any antibacterial activity against twenty microorganisms. In later papers, Tejima and co-workers described the synthesis of 4′-acetamido-4′-deoxy-α-lactose (**53**)[136] and 6′-acetamido-6′-deoxy-α-lactose (**54**).[71] Compound **53** was prepared in seven steps from 1,6-anhydro-β-cellobiose, whereas a longer sequence of reactions was required for the synthesis of **54** from 1,6-anhydro-β-lactose.

50 R = R^1 = R^3 = OH;R^2 = NHAc
51 R = R^1 = R^3 = OAc;R^2 = OMs
52 R = R^3 = OAc;R^1 = OAc(β);R^2 = NHAc
53 R = R^2 = OH;R^1 = OH(α);R^3 = NHAc

Edwards *et al.*[34] have also used cellobiose derivatives as precursors to amino-lactoses. The key intermediate, benzyl 2,3,6,2′,3′-penta-O-benzoyl-β-cellobioside, was obtained in 27% overall yield from cellobiose, via the 4′,6′-O-benzylidene acetal intermediate. This pentabenzoate was converted into its 4′,6′-dimesylate, which was further used for conventional syntheses of the 6′-amino-, 4′-amino- and 4′,6′-diamino-lactoses.

In 1977, Jezo[64] reported the preparation of methyl 6-acetamido-6-deoxy-, 6′-acetamido-6′-deoxy- and 6,6′-diacetamido-6,6′-dideoxy-lactosides from substituted methyl β-lactoside and methyl α-D-gluco-pyranoside. The 6,6′-diacetamido derivative **55** was obtained by standard procedures involving the nucleophilic displacement of the tosyl groups in methyl penta-*O*-acetyl-6,6′-di-*O*-tosyl-β-lactoside by azide anion. Sequential hydrogenation, acetylation and de-*O*-acetylation of the resultant diazide afforded the 6,6′-diacetamide **55** as a hygroscopic solid. Synthesis of the 6′-acetamide **56** was accomplished via the 4′,6′-*O*-benzylidene acetal derivative, in which the derived 4′,6′-diol was selectively tosylated at C-6′. In a similar manner to that used in the preparation of **55**, the 6′-tosylate **57** was converted into methyl 6′-acetamido-6′-deoxy-β-lactoside (**56**). An alternative approach was used for the synthesis of the 6-acetamide **58**. Condensation of methyl 2,3-di-*O*-benzoyl-6-*O*-mesyl-α-D-glucopyranoside with tetra-*O*-acetyl-α-D-galacto-pyranosyl bromide yielded the 6-*O*-mesyl-α-lactoside, which underwent conversion into **58** by conventional chemical methods.

54 R = R^2 = OH;R^1 = OH(α);R^3 = NHAc
55 R = R^1 = OH;R^2 = R^3 = NHAc
56 R = R^2 = OH;R^1 = OMe(β);R^3 = NHAc
57 R = R^2 = OAc;R^1 = OMe(β);R^3 = OTs
58 R = R^3 = OH;R^1 = OMe(α);R^2 = NHAc

The widespread occurrence of *N*-acetyllactosamine as a component of the oligosaccharide groups of many *O*- and *N*-linked glycoproteins[137] and the oligosaccharides of milk[138–140] is well known. It is also a building unit for certain Ii human blood group activities,[141–143] as well as of the ABH Type 2 determinants.[144] Consequently there exists a number of methods for its preparation,[145–147] with starting compounds as diverse as 1,6-anhydro-4′,6′-*O*-benzylidene-β-lactose[148] and 1-*N*-benzyl-3-*O*-D-galactopyranosyl-D-arabinosylamine.[149,150]

15. LACTOSE-CONTAINING OLIGOSACCHARIDES

Oligosaccharides are a class of carbohydrates which comprise from two to ten monosaccharide units. They are found in hormones, antibodies and growth factors, and are also a component of cell membranes. It has been suggested[151] that they may act as biological 'name tags' enabling cells and molecules to recognize one another.

During an early study of complex glycosphingolipids incorporating the sequence 4-*O*-β-D-galactopyranosyl-D-glucose (lactose), the glycosidation of lactose at C-3' or C-4' was investigated by Beith-Halahmi *et al.*[112] Access to these positions was readily achieved by acetonation of benzyl β-lactoside with acetone in the presence of *p*-toluenesulphonic acid. The 3',4'-acetal **39** obtained gave a pentaacetate upon peracetylation, and on further treatment with hot aqueous acetic acid it afforded the 3',4'-diol. Reaction of this diol with α-acetobromogalactose in nitromethane–benzene solution in the presence of mercuric cyanide gave, after deprotection, the expected β-(1″→3')-galactopyranosyllactose in an overall yield of 19% (from benzyl β-lactoside). This trisaccharide has since been discovered to be a major carbohydrate component of the milk of the tammar wallaby and the grey kangaroo, existing in concentrations of up to 16 g litre^{-1}.[152]

Synthesis of α-(1″→4')-galactopyranosyllactose (as its β-methyl glycoside) was recently carried out as part of a programme concerned with a ceramide trisaccharide related to Fabry's disease.[153]. In order to study certain aspects of the disease, a model compound was required, containing a terminal α-D-galactopyranosyl moiety that could be suitably labelled with ^{14}C. A partially protected lactose derivative with only the 4'-OH free was needed. This product could not be easily prepared by a direct selective method, so a stepwise synthesis was adopted. Benzylidenation of methyl β-lactoside gave, after perbenzoylation, methyl 4',6'-*O*-benzylidene-β-lactoside pentabenzoate. The acetal function was removed using trifluoroacetic acid, and the resultant 4',6'-diol was selectively benzoylated with benzoyl chloride in pyridine to give a 62% yield of methyl 2,3,6,2',3',6'-hexa-*O*-benzoyl-β-lactoside. Halide-catalysed condensation of this hexabenzoate with 2,3,4,6-tetra-*O*-benzyl-α-D-galactopyranosyl bromide gave (after removal of the protecting groups) a 3% yield of methyl 4-*O*-(4-*O*-α-D-galactopyranosyl-D-galactopyranosyl)β-D-glucopyranoside.

6'-Galactosyllactose is the main oligosaccharide synthesized enzymatically from lactose by the transgalactosylase of *Penicillium chrysogenum* Thom.[154] The chemical synthesis of this trisaccharide was recently reported by Tejima and co-workers.[97] In their approach to its synthesis, 2,3,2',3'-tetra-*O*-acetyl-1,6-anhydro-4',6'-*O*-benzylidene-β-lactose   was used as starting material. Cleavage of the benzylidene acetal ring by hydrogenolysis on palladium black afforded the 4',6'-diol in 86% yield. Glycosidation of this diol with α-acetobromogalactose using a modified

Koenigs–Knorr reaction followed by peracetylation gave the fully protected trisaccharide as a crystalline solid in 62% yield. The free trisaccharide, $\beta(1''\rightarrow6')$-galactosyllactose, obtained as a white powder, was
shown to have the suggested structure by methylation studies (in con-
junction with chromatographic techniques). The new galactosidic linkage
was determined from optical and molecular rotation values which were
consistent with a β linkage.

There is increasing interest in lactose-containing oligosaccharides that
contain the fucose moiety. 2-α-L-Fucopyranosyllactose,[155] a component
of human milk, was found to have a modest ability to inhibit haemagglutination of human O(H) blood cells by eel serum.[156] It has also been
shown to be an effective inhibitor of the precipitation of human H
substance by certain lectins.[157] Synthesis of isomeric fucosyllactoses was
therefore initiated in 1980, by Baer and Abbas,[114] in the hope that these
products may be useful in future studies into the specificity of combining-
sites on antibodies and lectins.

The first of the fucosyllactoses to be prepared was 3-O-β-L-
fucopyranosyllactose.[114] The method involved the kinetically controlled
isopropylidenation of lactose using 2,2-dimethoxypropane, which gave a
4',6'-O-isopropylidene derivative. Acetylation of this product followed by
deacetalation provided 1,2,3,6,2',3'-hexa-O-acetyl-β-lactose. Condensation of this hexaacetate with tri-O-acetyl-α-L-fucopyranosyl bromide
under Koenigs–Knorr conditions caused concomitant $3'\rightarrow6'$ acetyl
migration, with the attachment of the L-fucopyranosyl group at O-3'
and in the β disposition. An improved route for the synthesis of this
$\beta(1''\rightarrow3')$ trisaccharide was later demonstrated by the same authors[115]
who replaced the 4',6'-O-isopropylidene derivative by the isomeric 3',4'-
O-isopropylidene compound. The $\alpha(1''\rightarrow3')$-fucosyllactose was also prepared from this acetal derivative by using the glycosidation method of
Lemieux,[158,159] by which α-fucosylation is achieved without complica-
tion. Thus, the 3',4'-diol (derived from the 3',4'-acetal by sequential
acetylation and deacetalation) was treated with 2,3,4-tri-O-benzyl-α-L-
fucopyranosyl bromide under bromide ion catalysis. A 43% yield of the
protected 3'-O-α-L-fucopyranosyllactose was achieved after isolation by
column chromatography. Zemplén deacetylation of this product with
subsequent hydrogenolysis of the benzyl groups gave the $\alpha(1''\rightarrow3')$
trisaccharide as an amorphous powder. The method of α-fucosylation
was also applied to the 4',6'-diol derived from 4',6'-O-isopropylidene-
lactose hexaacetate.[160] A crystalline 6'-O-(tri-O-benzyl-α-L-fuco-
pyranosyl)-β-lactose hexaacetate was obtained in over 80% yield.

Conventional deprotection yielded the free trisaccharide as a white amorphous powder. The position of the fucosyl group was determined by sequential borohydride reduction, permethylation of the resulting alditol, and acid hydrolysis. Identification of the three components of the hydrolysate by thin-layer chromatography as 2,3,4-tri-*O*-methyl-L-fucose, 1,2,3,5,6-penta-*O*-methyl-D-glucitol and 2,3,4-tri-*O*-methyl-D-galactose indicated a $1'' \rightarrow 6'$ linkage. The α configuration was attributed on the basis of its optical rotation value. In a similar fashion to that used in the preparation of 3'-*O*-α- and 6'-*O*-α-fucosyllactoses, 2'-*O*-α-L-fucosyllactose was synthesized[161] from the suitably protected lactose derivative 2,3:5,6:3',4'-tri-*O*-isopropylidenelactose dimethyl acetal.

In an alternative approach for the synthesis of 3'-*O*-α- and 3'-*O*-β-L-fucopyranosyllactoses, Takamura *et al.*[108] have described a method using 1,6-anhydro-4',6'-*O*-benzylidene-3'-*O*-tosyl-β-lactose. This compound was converted into 1,6-anhydro-2,3,2'-tri-*O*-benzyl-4',6'-*O*-benzylidene-β-lactose in 74% yield by benzylation and then detosylation. Glycosidation at the 3' position was readily accomplished using 2,3,4-tri-*O*-acetyl-α-L-fucopyranosyl bromide in benzene–nitromethane in the presence of mercuric cyanide and molecular sieves. A mixture of the α- and β-linked fucosyl derivatives was obtained and separated by column chromatography to give 37·1% and 43·1% yields respectively. The free trisaccharides were obtained after hydrogenolysis of the benzylidene and benzyl groups, acetylation, cleavage of the 1,6-anhydro-β ring to the β-acetate, and finally deacetylation.

The importance of oligosaccharides in biological systems is highlighted by their connection with blood group substances. In a recent account by Lemieux[162] the relationship between human blood groups and carbohydrate chemistry is discussed in terms of the syntheses and conformational properties of determinants for the ABO and Lewis systems.

16. USES AND POTENTIAL APPLICATIONS OF LACTOSE DERIVATIVES

Streamlined manufacturing methods currently make lactose competitive with other sugars for use in the food industry, and a number of outlets exist for its direct use in foodstuffs and also for the products obtained from the fermentation and hydrolysis of lactose. The applications of lactose are indeed diverse, from a flavour enhancer of raw fish products[163] to a potential power source in bio-fuel cells.[164] Despite current uses, large quantities of lactose (as a component of whey) are disposed of annually. The chemical modification of lactose may help to alleviate this problem by producing derivatives of potential use. A variety of possible uses exist for the chemical derivatives of lactose; the following speculations are based upon parallels found in other areas of carbohydrate chemistry.

Esterification of disaccharides[165] with one or more fatty acid groups has produced compounds that exhibit surface activity and have applications which include industrial detergents,[166] antimicrobial agents,[167] food emulsifiers[168] and surface coatings.[169] In general, reducing sugars are not used as they decompose under the reaction conditions used to produce the ester (mono- and di-palmitates of lactose have been synthesized[118] from partially protected derivatives, and were reported to exhibit a degree of surface activity). This precludes the use of lactose, but not its reduced form, lactitol. Lactitol fatty acid monoesters[170] exhibit physical properties not unlike those of sucrose analogues. Properties that are likely to change according to the nature of the substituent ester group are detergency, foam performance, surface and interfacial tension, emulsification and wetting properties. The ratio of polar hydroxyl groups to non-polar fatty acid chains within the sugar is also an important factor in determining the overall property of the surfactant and the hydrophilic–lipophilic balance (HLB) value is determined from this ratio. Surface active agents with HLB values less than 8, in a range from 1 (non-polar) to 20 (polar), would be expected to emulsify water-in-oil, whereas those with values of 13 and above emulsify oil-in-water. Synthesis of mono- and di-myristoyl esters of methyl β-lactoside has recently been reported,[171] although the properties of these esters were not described. Myristic acid esters of disaccharide polyols are likely to have high HLB values, and consequently would be better foamers and detergents than the longer-chain esters. The longer-chain stearates and palmitates have lower HLB values and would probably have better emulsification properties at the interface of two liquids than the C_8–C_{12} homologues.[172] Sugar esters are excellent emulsifiers, and a substantial range of products such as frozen desserts, solid flavour concentrates, fats, margarine, ointments, lotions and creams include the use of emulsifying agents. Lactose esters also have potential uses as humectants, plasticizers, additives for lacquers and hot-melt adhesives, and bittering agents. Sucrose octaacetate, for example, is widely used as a denaturant for ethyl alcohol and a modifier for adhesives. A drawback of using esters in certain formulations is their lack of stability at high pH. This could be overcome by use of long-chain alkyl ethers which are more stable under basic conditions than esters. Lactose ethers such as allyl and other unsaturated ethers may find applications in surface coatings, as cross-linking agents in polyester resins, modifying agents for alkyds or co-polymers with acrylics.

Lactose can be converted by alkaline treatment into the disaccharide lactulose (4-*O*-β-D-galactopyranosyl-D-fructose). This sugar has been used to treat constipation, especially in the elderly. It appears to act as an extremely mild purgative and, because of its stimulation of the colonic lactobacilli, helps to minimize the formation of ammonia-producing organisms. Lactulose is therefore valuable in the treatment of the condition known as chronic hepatoportal encephalopathy, in which nitrogenous substances formed in the colon bypass the liver and affect the brain.

Another important class of derivatives of lactose includes the halodeoxy compounds. These products may have properties of biological importance. Chlorinated sugars, for instance, can display a range of properties depending on the number and position of the chlorine atoms within the molecule. For example, certain monochlorinated disaccharides are potential antifertility drugs,[173] whereas higher substituted derivatives of sucrose can give compounds that are intensely sweet.[174] This diversity in properties makes it difficult to predict uses for the chemical derivatives.

Whilst the study of the chemistry of lactose will continue to produce a variety of chemical substances, it may be some time before the value of these products is fully realized. Ideally, a system of evaluating the physical and biochemical properties of the compounds is needed but, of course, its efficacy would ultimately be governed by economic constraints.

REFERENCES

1. GUY, E. J., *Bakers Dig.*, 1971, **45**, 34.
2. MOORE, K., *Food Prod. Dev.*, 1978, **12**, 72.
3. WHITTIER, E. O., *Chem. Rev.*, 1926, **2**, 85.
4. WHITTIER, E. O., *J. Dairy Sci.*, 1944, **27**, 503.
5. NICKERSON, T. A., In: *Developments in Food Carbohydrate-1*, Birch, G. G. and Shallenberger, R. S. (eds), 1977, Applied Science Publishers, London, p. 77.
6. SIMAN, J., *Prumysl Potravin*, 1952, **3**, 283.

7. KANTSCHER, H., *Pharm. Ind.*, 1954, **16**, 281; 1956, **18**, 136.

8. KUHN, R., BAER, H. H. and GAUHE, A., *Chem. Ber.*, 1955, **88**, 1135.

9. CLAMP, J. R., HOUGH, L., HICKSON, J. L. and WHISTLER, R. L., *Adv. Carbohydr. Chem.*, 1961, **16**, 159.

10. THELWALL, L. A. W., In: *Developments in Food Carbohydrate—2*, LEE, C. K. (ed.), 1980, Applied Science Publishers, London, p. 275.

11. THELWALL, L. A. W., *J. Dairy Res.*, 1982, **49**, 713.

12. NICKERSON, T. A., In: *Fundamentals of Dairy Chemistry*, Webb, B. H., Johnson, A. H. and Alford, J. A. (eds), 1974, AVI Publishing Company Inc., p. 273.

13. JENNESS, R. and PATTON, S., In: *Principles of Dairy Chemistry*, 1959, J. Wiley & Sons, New York, p. 73.

14. HARJU, M. and KREULA, M., In: *Carbohydrate Sweeteners in Foods and Nutrition*, Koivistoinen, P. and Hyvonen, L. (eds), 1980, Academic Press, London, p. 233.

15. SCHUKLA, T. P., *Crit. Rev. Food Technol.*, 1975, **5**, 325.

16. BAYLESS, T. M., PAIGE, D. M. and FERRY, G. D., *Gastroenterology*, 1971, **60**, 605.

17. PAIGE, D. M., BAYLESS, T. M., HUANG, S-S. and WEXLER, R., In: *Physiological Effects of Food Carbohydrates*, Jeanes, A. and Hodge, J. (eds), 1975, Amer. Chem. Soc. Symp. Series 15, Washington, p. 191.

18. MULLER, L. L., *Aust. J. Dairy Tech.*, 1976, **31**, 92.

19. ROETMAN, K., *Neth. Milk Dairy J.*, 1982, **36**, 1.

20. VOGEL, H. A., *Ann. Physik*, 1812, **42**, 129.

21. ERDMANN, E. O., *Jahresber. Chem.*, 1855, 671.

22. FISCHER, E. and MEYER, J., *Ber.*, 1889, **22**, 361.

23. KILIANI, H., *Ann.*, 1893, **272**, 198.

24. FISCHER, E., *Ber.*, 1888, **21**, 2633.

25. HAWORTH, W. N. and LEITCH, G. C., *J. Chem. Soc.*, 1918, **113**, 188.

26. HAWORTH, W. N., and LONG, C. W., *J. Chem. Soc.*, 1927, 544.

27. ZEMPLEN, G., *Ber.*, 1926, **59**, 2402; 1927, **60**, 1309.

28. RAO, S. T., FRIES, D. C. and SUNDARALINGAM, M., *Acta Crystallogr.*, 1971, **B27**, 994.

29. HIROTSU, K. and SHIMADA, A., *Bull. Chem. Soc. Japan*, 1974, **47**, 1872.

30. HASKINS, W. T., HANN, R. M. and HUDSON, C. S., *J. Am. Chem. Soc.*, 1942, **64**, 1490, 1852.

31. CURTIS, E. J. C. and JONES, J. K. N., *Can. J. Chem.*, 1959, **37**, 358.

32. SHAPIRO, D., RABINSON, Y. and DIVER-HABER, A., *Biochem. Biophys. Res. Commun.*, 1969, **37**, 28.

33. KUZUHARA, H. and EMOTO, S., *Agric. Biol. Chem.*, 1966, **30**, 122.

34. EDWARDS, R. G., HOUGH, L. and RICHARDSON, A. C., *Carbohydr. Res.*, 1977, **55**, 129.

35. TAKEO, K., OKUSHIO, K., FUKYAMA, K. and KUGE, T., *Carbohydr. Res.*, 1983, **121**, 163.

36. WATKINS, W. M. and HASSID, W. Z., *J. Biol. Chem.*, 1962, **237**, 1432.

37. BABAD, H. and HASSID, W. Z., *J. Biol. Chem.*, 1964, **239**, PC946.

38. BABAD, H. and HASSID, W. Z., *J. Biol. Chem.*, 1966, **241**, 2672.

39. ZEMEK, J., KUCAR, S., ZAMOCKY, J. and AUGUSTIN, J., *Coll. Czech. Chem. Commun.*, 1979, **44**, 1992.

40. SKRAUP, H., *Monatsh. Chem.*, 1889, **10**, 389; L. Kueney, *Z. Physiol. Chem.*, 1890, **14**, 330.

40a VASQUEZ, I. M., THIEL, I. M. E. and DEFERRARI, J. O., *Carbohydr. Res.*, 1973, **26**, 351.

41. BHATT, R. S., HOUGH, L. and RICHARDSON, A. C., *Carbohydr. Res.*, 1974, **32**, C4-6.

42. BHATT, R. S., HOUGH, L. and RICHARDSON, A. C., *J. Chem. Soc., Perkin 1*, 1977, 2001.

43. HAINES, A. H., *Adv. Carbohydr. Chem. Biochem.*, 1976, **33**, 11.

44. THIEL, I. M. E., DEFERRARI, J. O. and CADENAS, R. A., *Ann.*, 1969, **723**, 192.

45. TAKEO, K. and OKANO, S., *Carbohydr. Res.*, 1977, **59**, 379.

46. VASQUEZ, I. M., THIEL, I. M. E. and DEFERRARI, J. O., *Carbohydr. Res.* 1976, **47**, 241.

47. BHATT, R. S., HOUGH, L. and RICHARDSON, A. C., *Carbohydr. Res.*, 1976, **51**, 272.

48. BERMANN, M., *Ann.*, 1923, **434**, 79.

49. WATTERS, A. J. and HUDSON, C. S., *J. Am. Chem. Soc.*, 1930, **52**, 3472.

50. KUNZ, A. and HUDSON, C. S., *J. Am. Chem. Soc.*, 1926, **48**, 1978, 2345.

51. RICHTMEYER, N. K. and HUDSON, C. S., *J. Am. Chem. Soc.*, 1935, **57**, 1716.

52. CHIBA, T. and TEJIMA, S., *Chem. Pharm. Bull.* (Tokyo), 1976, **24**, 1684; 1977, **25**, 1049.

53. CHIBA, T., HAGA, M. and TEJIMA, S., *Carbohydr. Res.*, 1975, **45**, 11.

54. TAKAMURA, T. and TEJIMA, S., *Chem. Pharm. Bull.* (Tokyo), 1978, **26**, 1117.

55. ROULLEAU, F., PLUSQUELLEC, D. and BROWN, E., *Tetrahedron Lett.*, 1983, **24**, 719.

56. OGAWA, T. and MATSUI, M., *Tetrahedron*, 1981, **37**, 2363.

57. RANA, S., BARLOW, J. and MATTA, K., *Tetrahedron Lett.*, 1981, **22**, 5007.

58. VASQUEZ, I. M., THIEL, I. M. E. and DEFERRARI, J. O., *Carbohydr. Res.*, 1973, **26**, 351.

59. DEFERRARI, J. O., THIEL, I. M. E. and CADENAS, R. A., *Carbohydr. Res.*, 1973, **29**, 141.

60. GROS, E. G. and DEULOFEU, V., *J. Org. Chem.*, 1964, **29**, 3647.

61. PISKORSHA-CHLEBOSKA, A., *Rocz. Chem.*, 1973, **47**, 49 (*Chem. Abstr.*, 1973, **79**, 32217y).

62. TIPSON, R. S., *Adv. Carbohydr. Chem.*, 1953, **8**, 107.

63. BALL, D. H. and PARRISH, F. W., *Adv. Carbohydr. Chem.*, 1968, **23**, 233; 1969, **24**, 139.

64. JEZO, I., *Chem. Zvesti*, 1978, **32**, 493.

65. RICHARDSON, A. C., *Carbohydr. Res.*, 1969, **10**, 395.

66. BHATT, R. S., HOUGH, L. and RICHARDSON, A. C., *Carbohydr. Res.*, 1976, **49**, 103.

67. EDWARDS, R. G., HOUGH, L., RICHARDSON, A. C. and TARELLI, E., *Carbohydr. Res.*, 1974, **35**, 111.

68. TEJIMA, S., *Carbohydr. Res.*, 1971, **20**, 123.

69. TEJIMA, S. and CHIBA, T., *Chem. Pharm. Bull.* (Tokyo), 1973, **21**, 546.

70. HANESSIAN, S., *Carbohydr. Res.*, 1966, **2**, 86.
71. CHIBA, T., HAGA, M. and TEJIMA, S., *Chem. Pharm. Bull.* (Tokyo), 1975, **23**, 1283.
72. KENT, P. W. and DIMETRIJEVICH, S. D., *J. Fluor. Chem.*, 1977, **10**, 455.
73. HELFERICH, B. and GOOTZ, R., *Ber.*, 1929, **62**, 2505.
74. SKRAUP, Z. H. and KREMANN, R., *Monatsh.*, 1901, **22**, 375.
75. HUDSON, C. S. and KUNZ, A., *J. Am. Chem. Soc.*, 1925, **47**, 2052.
76. FISCHER, E. and FISCHER, H., *Ber.*, 1910, **43**, 2521.
77. PACSU, E., *Ber.*, 1928, **61**, 1508.
78. FISCHER, E. and ARMSTRONG, E. F., *Ber.*, 1902, **35**, 833.
79. DICK, W. E. and WEISLEDER, D., *Carbohydr. Res.*, 1976, **46**, 173.
80. SMITH, F. and VAN CLEVE, J. W., *J. Am. Chem. Soc.*, 1952, **74**, 1912.
81. DITMAR, R., *Monatsh.*, 1902, **23**, 870; *Ber.*, 1902, **35**, 1951.
82. SMITH, F. and VAN CLEVE, J. W., *J. Am. Chem. Soc.*, 1955, **77**, 3159.
83. COLES, H. W., DODDS, M. L. and BERGEIM, F. H., *J. Am. Chem. Soc.*, 1938, **60**, 1020.
84. KORBETT, W. M. and KIDD, J., *J. Chem. Soc.*, 1959, 1594.
85. FROSCHL, N., ZELLNER, J. and ZAK, H., *Monatsh.*, 1930, **55**, 25.
86. VEINBERG, A. Y., POTAPOV, V. M., VAKALOVA, L. A., DEM'YANOVIC, V. M. and SOMOKHVALOV, G. I., *Zh. Obshch. Khim.*, 1966, **36**, 31.
87. LETTRE, H. and HAGEDORN, A., *Z. Physiol. Chem.*, 1936, **242**, 210.
88. MIESCHER, K. and MEYSTRE, C., *Helv. Chim. Acta*, 1943, **26**, 224.
89. DEA, I. C. M., *Carbohydr. Res.*, 1969, **11**, 363; 1970, **12**, 297.
90. HELFERICH, B. and GRIEBEL, R., *Anm.*, 1940, **544**, 191.
90a LEMIEUX, R. U., HENDRIKS, K. B., STICK, R. V. and JAMES, K., *J. Am. Chem. Soc.*, 1975, **97**, 4056.
91. CHUNG, T. G., ISHIHARA, H. and TEJIMA, S., *Chem. Pharm. Bull.* (Tokyo), 1978, **26**, 1570.
92. PERCIVAL, E. G. V., *J. Chem. Soc.*, 1935, 648.
93. BREDERECK, H., HAGELLOCH, G. and HAMBSCH, E., *Chem. Ber.*, 1954, **87**, 35.
94. McKEOWN, G. G. and HAYWARD, L. D., *Can. J. Chem.*, 1957, **35**, 992.
95. LINDLEY, M. G., BIRCH, G. G. and KHAN, R., *Carbohydr. Res.*, 1975, **43**, 360.
96. LIPTAK, A., JODAL, I. and NANASI, P., *Carbohydr. Res.*, 1976, **52**, 17.
97. CHUNG, T. G., ISHIHARA, H. and TEJIMA, S., *Chem. Pharm. Bull.* (Tokyo), 1978, **26**, 2147.
98. FLETCHER, H. G., In: *Methods in Carbohydrate Chemistry*, Whistler, R. L. and Wolfrom, M. L. (eds), 1963, Academic Press, New York and London, p. 166.
99. BRIMACOMBE, J. S., PORTSMOUTH, D. and STACEY, M., *J. Chem. Soc.*, 1964, 5614.
100. IWASHIGE, T. and SACKI, H., *Chem. Pharm. Bull.* (Tokyo), 1967, **15**, 1803.
101. TAKAMURA, T., CHIBA, T. and TEJIMA, S., *Chem. Pharm. Bull.* (Tokyo), 1981, **29**, 2270.
102. TAKAMURA, T., CHIBA, T. and TEJIMA, S., *Chem. Pharm. Bull.* (Tokyo), 1981, **29**, 587, 1027.
103. LIPTAK, A., JODAL, I. and NANASI, P., *Carbohydr. Res.*, 1975, **44**, 1.

104. LIPTAK, A., JODAL, I. and NANASI, P., *Carbohydr. Res.*, 1976, **52**, 17.
105. MILAT, M-L. and SINAY, P., *Carbohydr. Res.*, 1981, **92**, 183.
106. MATSUDA, H., ISHIHARA, H. and TEJIMA, S., *Chem. Pharm. Bull.* (Tokyo), 1979, **27**, 2564.
107. TAKAMURA, T., CHIBA, T., ISHIHARA, H. and TEJIMA, S., *Chem. Pharm. Bull.* (Tokyo), 1979, **27**, 1497.
108. TAKAMURA, T., CHIBA, T. and TEJIMA, S., *Chem. Pharm. Bull.* (Tokyo), 1981, **29**, 1076.
109. KARRER, P. and HARLOFF, J. C., *Helv. Chim. Acta*, 1933, **16**, 962.
110. MONTGOMERY, E. M., RICHTMEYER, N. K. and HUDSON, C. S., *J. Am. Chem. Soc.*, 1943, **65**, 1848.
111. CHIBA, T. and TEJIMA, S., *Chem. Pharm. Bull.* (Tokyo), 1977, **25**, 1049.
112. BEITH-HALAHMI, D., FLOWERS, H. M. and SHAPIRO, D., *Carbohydr. Res.*, 1967, **5**, 25.
113. THELWALL, L. A. W., HOUGH, L. and RICHARDSON, A. C., unpublished data.
114. BAER, H. H. and ABBAS, S. A., *Carbohydr. Res.*, 1979, **77**, 117.
115. BAER, H. H. and ABBAS, S. A., *Carbohydr. Res.*, 1980, **84**, 53.
116. FANTON, E., GELAS, J. and HORTON, D., *Chem. Commun.*, 1980, 21.
117. HOUGH, L., RICHARDSON, A. C. and THELWALL, L. A. W., *Carbohydr. Res.*, 1979, **75**, C11-12.
118. THELWALL, L. A. W., HOUGH, L. and RICHARDSON, A. C., British Patent Application 2048853a, 1980.
119. FLORENT, J-C. and MONNERET, C., *Synthesis*, 1982, 29.
119a. BHATT, R. S., HOUGH, L. and RICHARDSON, A. C., *Carbohydr. Res.*, 1975, **43**, 57.
120. FISCHER, E. and THIERFELDER, H., *Ber.*, 1894, **27**, 2031.
121. RAO, D. R. and LERNER, L. M., *Carbohydr. Res.*, 1971, **19**, 133; 1972, **22**, 345.
122. CHIBA, T., HAGA, M. and TEJIMA, S., *Chem. Pharm. Bull.* (Tokyo), 1974, **22**, 398.
123. HANESSIAN, S. *Adv. Carbohydr. Chem.*, 1966, **21**, 143.
124. CHIBA, T. and TEJIMA, S., *Chem. Pharm. Bull.* (Tokyo), 1978, **26**, 3426.
125. CHIBA, T., HAGA, M. and TEJIMA, S., *Chem. Pharm. Bull.* (Tokyo), 1974, **22**, 398.
126. HOUGH, L., KHAN, R. and OTTER, B. A., In: *Deoxy Sugars*, 1968, Amer. Chem. Soc., Washington, p. 123.
127. DUTCHER, J. D., *Adv. Carbohydr. Chem.*, 1963, **18**, 259.
128. GERO, S., CLEOPHAX, J., MERCIER, D. and OLESKER, A., In: *Synthetic Methods for Carbohydrates*, El Khadem, H. S. (ed.), 1976, Amer. Chem. Soc. Symp. Series **39**, Washington, p. 64.
129. SHARON, N., In: *The Amino Sugars*, Balazs, E. A. and Jeanloz, R. W. (eds), Vol. IIA, 1965, Academic Press, New York, p. 1.
130. ROLLAND, N., VASS, G., CLEOPHAX, J., SEPULCHRE, A-M. and GERO, S., *Helv. Chim. Acta*, 1982, **65**, 1627.
131. DUNSTAN, D. and HOUGH, L., *Carbohydr. Res.*, 1972, **23**, 17.
132. SPINOLA, M. and JEANLOZ, R. W., *Carbohydr. Res.*, 1970, **15**, 361.

133. GRAY, G. R., *Arch. Biochem. Biophys.*, 1974, **163**, 426.
134. BORCH, R. F., BERNSTEIN, M. D. and DURST, H. D., *J. Am. Chem. Soc.*, 1971, **93**, 2897.
135. HOAGLAND, P. D., PFEFFER, P. E. and VALENTINE, K. M., *Carbohydr. Res.*, 1979, **74**, 145.
136. TEJIMA, S., OKAMORI, Y. and HAGA, M., *Chem. Pharm. Bull.* (Tokyo), 1973, **21**, 2538.
137. ARNARP, J. and LONNGREN, J., *J. Chem. Soc.*, *Perkin 1*, 1981, 2070.
138. LEMIEUX, R. U., ABBAS, S. Z. and CHUNG, B. Y., *Can. J. Chem.*, 1982, **60**, 58.
139. KUHN, R. and GAUHE, A., *Chem. Ber.*, 1962, **95**, 518.
140. YAMASHITA, K., TACHIBANA, Y. and KOBATA, A., *J. Biol. Chem.*, 1977, **252**, 5408.
141. LEMIEUX, R. U., ABBAS, S. Z., BURZYNSKA, M. H. and RATCLIFFE, R. M., *Can. J. Chem.*, 1982, **60**, 63.
142. WOOD, E. and FEIZI, T., *FEBS Lett.*, 1979, **104**, 135.
143. VEYRIERES, A., *J. Chem. Soc.*, *Perkin 1*, 1981, 1626.
144. LLOYD, K. O., KABAT, E. A. and LICERIO, E., *Biochemistry*, 1968, **7**, 2976.
145. KUHN, R. and KIRSCHENLOHR, W., *Ann.*, 1956, **600**, 135.
146. OKUYAMA, T., *Tohoku J. Exp. Med.*, 1958, **68**, 313.
147. RABINSOHN, Y., ACHER, A. J. and SHAPIRO, D., *J. Org. Chem.*, 1973, **38**, 202.
148. TAKAMURA, T., CHIBA, T. and TEJIMA, S., *Chem. Pharm. Bull.* (Tokyo), 1979, **27**, 721.
149. ALAIS, J. and VEYRIERES, A., *Carbohydr. Res.*, 1981, **93**, 164.
150. LEE, R. T. and LEE, Y. C., *Carbohydr. Res.*, 1979, **77**, 270.
151. BREEZE, P., *New Scientist*, 1984, **101**, 18.
152. MESSER, M., TRIFONOFF, E., STERN, W., COLLINS, J. G. and BRADBURY, J. H., *Carbohydr. Res.*, 1980, **83**, 327.
153. COX, D. D., METZNER, E. K. and REIST, E. J., *Carbohydr. Res.*, 1978, **63**, 139.
154. BULLIO, A. and RUSSI, S., *Tetrahedron*, 1960, **9**, 125.
155. KUHN, R., BAER, H. H. and GAUHE, A., *Chem. Ber.*, 1955, **88**, 135; 1956, **89**, 2513.
156. KUHN, R. and OSMAN, H. G., *Hoppe-Seyler's Z. Physiol. Chem.*, 1956, **303**, 1.
157. PEREIRA, M. E. A. and KABAT, E. A., *Biochemistry*, 1974, **13**, 3184.
158. LEMIEUX, R. U. and DRIGUEZ, H., *J. Am. Chem. Soc.*, 1975, **97**, 4069.
159. LEMIEUX, R. U., BUNDLE, D. R. and BAKER, D. A., *J. Am. Chem. Soc.*, 1975, **97**, 4076.
160. BAER, H. H. and ABBAS, S. A., *Carbohydr. Res.*, 1980, **83**, 146.
161. ABBAS, S. A., BARLOW, J. J. and MATTA, K. L., *Carbohydr. Res.*, 1981, **88**, 51.
162. LEMIEUX, R. U., *Chem. Soc. Rev.*, 1978, **7**, 423.
163. UENO PHARMACEUTICAL CO. LTD., Japanese Patent 5105461, 1976.
164. ROLLER, S. D., BENNETTO, H. P., DELANEY, G. M., MASON, J. R., STIRLING, J. L., THURSTON, C. F. and WHITE, D. R., *Biotech 83* (Proc. Internat. Conf. on Commercial Applications and Implications of Biotechnology, 1983, Online Publishers, Northwood, UK, p. 655.
165. HURFORD, J. R., In: *Developments in Food Carbohydrate—2*, Lee, C. K. (ed.), 1980, Applied Science Publishers, London, p. 327.

166. PARKER, K. J., JAMES, K. and HURFORD, J. R., In: *Sucrochemistry*, Hickson, J. L. (ed.), 1977, Amer. Chem. Soc. Symposium Series 41, Washington, p. 97.
167. NISHIZAWA, K., INAGAKI, Y. and MORI, Z., Japanese Patent 7454524, 1974.
168. PEROTTI, A., *Int. Flavours Food Addit.*, 1977, **8**, 149.
169. FAULKNER, R. N., In: *Sucrochemistry*, Hickson, J. L. (ed.), 1977, Amer. Chem. Soc. Symposium Series 41, Washington, p. 176.
170. TAKEMURA, M., IIJIMA, B., TAKEUCHI, K. and HATTORI, S., Japanese Patent 7435318, 1974; Scholnik, F., Ben-Et, G., Sucharski, M. K., Maurer, E. W. and Lingfield, W. M., *J. Am. Oil Chem. Soc.*, 1975, **52**, 256; Van Velthuijsen, J. A., Heeson, J. G. and Kuipers, P. K., In: *Sucrochemistry*, Hickson, J. L. (ed.), 1977, Amer. Chem. Soc. Symposium Series 41, Washington, p. 136.
171. MUNAVU, R. M., NASSERI-NOORI, B. and SZMANT, H. H., *Carbohydr. Res.*, 1984, **125**, 253.
172. KOLLONITSH, V., In: *Sucrose Chemicals*, 1970, International Sugar Research Foundation Inc.
173. WAITES, G. M. H., FORD, W. C. L., KHAN, R. and JONES, H. F., British Patent 1595941, 1981.
174. HOUGH, L. and PHADNIS, S. P., *Nature*, 1976, **263**, 800.

Chapter 3

MODIFICATION OF LACTOSE AND LACTOSE-CONTAINING DAIRY PRODUCTS WITH β-GALACTOSIDASE

R. R. MAHONEY

*Department of Food Science and Nutrition,
University of Massachusetts, Amherst, USA*

1. INTRODUCTION

β-Galactosidase (β-D-galactoside galactohydrolase; EC 3.2.1.23) typically catalyses the hydrolysis of lactose to its component monosaccharides, glucose and galactose. This conversion is of considerable interest from the standpoint of food technology and nutrition because the major products of hydrolysis are, in combination, sweeter, more soluble, more easily fermented and directly absorbed from the mammalian intestine. These changes are the bases for the production of new foodstuffs such as lactose-hydrolysed milk and whey, and products derived therefrom.

The enzyme β-galactosidase is often called by the trivial name lactase. While all lactases are β-galactosidases, the converse is not true since some cellular β-galactosidases have essentially no activity on lactose which is a very specific β-galactoside. In this chapter we will be concerned only with β-galactosidases that are also lactases.

Use of β-galactosidase to hydrolyse the lactose in milk and produce small quantities of milk suitable for children and adults with a severe lactase deficiency has been practised for many years. Commercial interest in the process developed in the 1960s with the realization that lactose intolerance was, in varying degrees, a world-wide condition and that cheese whey, which contains about 70% lactose on a dry weight basis, was both a major disposal problem and an underutilized food resource.

Hydrolysis of lactose offers potential solutions to these problems. While hydrolysis can be achieved with acid[1] or cation exchange resins,[2] enzymatic conversion with β-galactosidase is preferable because of the high specificity of its action. Use of chemical catalysts leads to off-colours and off-flavours which originate from side reactions. The major disadvantage of enzymatic conversion, apart from cost factors, is the production of oligosaccharides by transferase activity, but this can be overcome by suitable process conditions.

This chapter includes a review of the enzyme β-galactosidase (lactase), the technology associated with its use, its commercial potential and the applications of lactose-hydrolysed products. Nutritional aspects of lactose digestion and of lactose hydrolysates are reviewed elsewhere in this book (Chapter 4).

2. SOURCES OF THE ENZYME

β-Galactosidases are widely distributed in nature, in accordance with their multiple functions.[3] As a result of recent commercial interest in β-galactosidase, a vast number of microbial species have been examined as potential sources. Table 1 gives a partial listing (which needs continual updating) of these sources, together with non-microbial sources of the enzyme.

The enzyme from several of these sources has been purified to homogeneity and well characterized, especially that from *Escherichia coli* which serves as a model for our understanding of the action of β-galactosidase. The ready availability of this particular enzyme has also led to its wide use in studies of β-galactosidase immobilization, but it is not considered suitable for commercial use in foods owing to potential toxicity problems associated with the use of coliforms. Its main value is in analytical work such as the enzymatic determination of lactose.[4]

2.1. Commercial Sources

For use in processing dairy foods, the enzyme must be derived from a source that is 'recognized as safe' and acceptable to regulatory bodies in each of the countries of interest. As a result, the enzymes available commercially are derived from: the yeasts *Kluyveromyces fragilis*, *Kluyveromyces lactis* and *Candida pseudotropicalis*; the fungi *Aspergillus niger* and *Aspergillus oryzae*; and bacteria that are closely related to

TABLE 1
SOURCES OF β-GALACTOSIDASE WHICH ACT ON LACTOSE

Animal organs	Intestine	
Plants	Alfalfa seed	Coffee berries
	Almond	Peach
	Apricot	
Fungi	*Alternaria alternata*	*Fusarium moniliforme*
	Alternaria palmi	*Mucor pusillus*
	Aspergillus foetidis	*Mucor meihei*
	Aspergillus niger	*Neurospora crassa*
	Aspergillus oryzae	*Scopulariopsis* sp.
	Curvularia inaequalis	
Yeasts	*Candida pseudotropicalis*	
	Kluyveromyces fragilis	
	Kluyveromyces lactis	
	Kluyveromyces marxianus	
Bacteria	*Bacillus acidocaldarius*	
	Bacillus coagulans	
	Bacillus megaterium	
	Bacillus stearothermophilus	
	Bacillus subtilis	
	Escherichia coli	
	Lactobacillus bulgaricus	
	Lactobacillus helveticus	
	Lactobacillus thermophilus	
	Leuconostoc citroverum	
	Streptococcus cremoris	
	Streptococcus lactis	
	Streptococcus thermophilus	
Kefir grains	Lactobacilli plus yeast	

Bacillus stearothermophilus. Although an extensive technology has been developed using enzymes from these organisms, there is a continuous search for new enzymes with superior properties, e.g. better thermostability at neutral pH, for use in milk. In this respect, the β-galactosidases from some thermophilic lactic acid bacteria seem especially promising, since the source organisms are generally acceptable and regarded as safe.

3. MECHANISM OF ACTION

β-Galactosidases have been extensively characterized, and there is a wealth of information regarding the kinetic behaviour of the enzyme from a variety of sources. Knowledge of the protein structural features is less complete, although in the case of the *E. coli* enzyme both the sub-unit structure and primary sequence have been determined.[5,6] Despite all this, very little is known about the mechanism of action of the enzyme or about its active site.

A generalized mechanistic scheme has been proposed[7] which involves a minimum of three steps and allows for the transfer reaction leading to oligosaccharide formation, as well as hydrolysis, viz:

1. Enzyme + Lactose → Enzyme–Lactose complex
2. Enzyme–Lactose → Galactosyl–Enzyme + Glucose
3. Galactosyl–Enzyme + H_2O → Galactose + Enzyme

   or

   Galactosyl–Enzyme + Acceptor sugar → Oligosaccharide + Enzyme

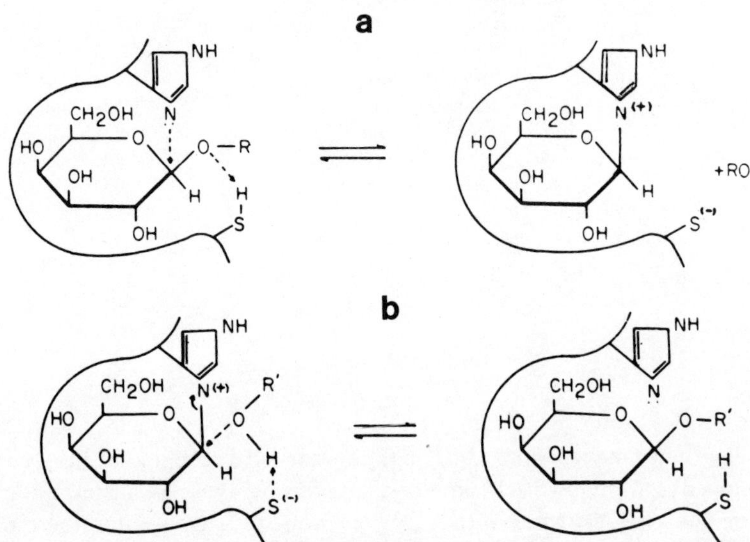

FIG. 1. A possible mechanism for neutral-pH β-galactosidase: (a) formation of the galactosyl–enzyme complex and release of glucose (ROH); (b) transfer of the galactosyl moiety to an acceptor (R'OH) to form free galactose or an oligo-saccharide. Reproduced with permission from reference (7), Academic Press, 1960.

Based on inhibition experiments with sulphydryl reagents and pH–activity studies, it has been suggested that the neutral-pH enzymes from *E. coli*[7] and *K. fragilis*[8] both contain a sulphydryl group acting as a general base and an imidazole group acting as a nucleophile, to facilitate cleavage of the glycosidic linkage. This mechanism of action is depicted in Fig. 1. However, other studies have indicated that the sulphydryl group, while essential for maintaining an active conformation, does not participate in the catalytic step.[8,9] Essentially nothing is known about the active site of the fungal β-galactosidases; in these enzymes it is possible that a carboxyl group acts as a nucleophile and that an imidazole group acts as an electrophile, but there is no direct evidence for this at present.

4. TRANSFERASE ACTIVITY

The mechanism of β-galactosidase activity outlined above indicates that the enzyme will transfer galactose to any acceptor containing an hydroxyl group. Where the acceptor is water, free galactose is formed. However, all of the sugars present in the reaction system can also act as acceptors, and this gives rise to various di-, tri- and higher saccharides, collectively termed oligosaccharides. They can be readily separated from the monoses by paper, liquid and gas–liquid chromatography (after derivatization), and by gel filtration.[10–14]

4.1. Amount and Nature of the Transferase Products

The amount and nature of the oligosaccharides formed depend primarily on the enzyme source, the substrate concentration and the reaction time. In the case of the *E. coli* enzyme, it is also affected by the pH, magnesium ions and the anomeric configuration of lactose.[10] For a given enzyme, the level of oligosaccharides increases with substrate concentration.[10–12] At high starting lactose concentrations (15–20%), maximum oligosaccharide levels can account for between 30% and 45% of the total sugars present, depending upon the enzyme.[10,11] Under these conditions, between 10 and 12 different oligosaccharides have been detected in the reaction mixture.[11,14,15] The major products that have been identified are the disaccharides β-D-galactose-(1→6)-D-glucose (allolactose) and β-D-galactose-(1→6)-D-galactose (galactobiose), and the trisaccharides β-D-galactose-(1→6)-allolactose, β-D-galactose-lactose and β-D-galactose-galactobiose. Disaccharides consisting of β-D-galactose linked 1→2 and

$1 \rightarrow 3$ with D-glucose, and tetrasaccharides consisting of β-D-galactose linked $1 \rightarrow 6$ to one of the above trisaccharides have also been identified.[10,12-15] Quantitatively, allolactose is the major transferase product.

At lower lactose levels, such as those found in milk and whey, transferase activity is reduced but *maximum* oligosaccharide levels can still reach 22–25% of total sugars with neutral-pH enzymes;[11,13] lower values (5%) have been reported where only tri- and higher saccharides were measured.[16] In contrast, oligosaccharide production in whey by the acid-pH enzyme from *A. niger* is only 1–2% of total sugars.[23] This is a decided advantage when converting lactose in this medium, though whether it is due to the pH itself or reflects some special feature of the particular enzyme is not known.

Our knowledge of the transfer mechanism is extremely limited though it is clear that most enzymes have a marked specificity for synthesis of the $1 \rightarrow 6$ linkage. In the case of the *E. coli* β-galactosidase it has been shown that allolactose can be formed by direct transfer of galactose from position 4 to position 6 of the glucose moiety without prior release of the glucose from the enzyme, as well as by transfer to position 6 of the free glucose.[10]

4.2. Kinetics of the Transferase Reaction

An example of the kinetics of oligosaccharide production during the hydrolysis of lactose in milk is shown in Fig. 2. Oligosaccharides (defined here as all sugars other than lactose, glucose and galactose) appeared as soon as the monosaccharides were liberated. The actual concentration of oligosaccharides at any time depends upon their relative rates of formation and breakdown. In Fig. 2, the peak level was reached after about 1 h when more than 90% of the lactose had been hydrolysed. At this time the oligosaccharides represented 25% of the total sugars present. Thereafter the level fell, presumably because the rate of breakdown, while slow, exceeded the rate of formation from the almost depleted supply of lactose. In this system, the major oligosaccharides detected were allolactose and galactobiose, which accounted for 60% and 30%, respectively, of the total.

The measured concentration of each species during the reaction is shown in Table 2. During most of the first hour, when 94% of the lactose was hydrolysed, the ratio of free glucose to free galactose was close to 2:1 rather than the normally assumed ratio of 1:1. It was only when the reaction had gone essentially to completion that this latter ratio was

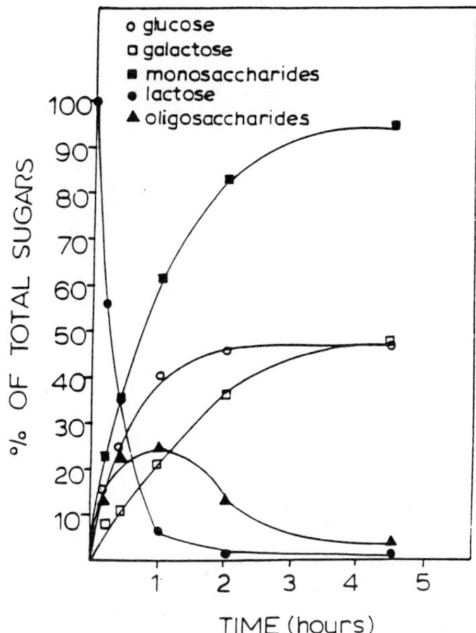

FIG. 2. Profile of oligosaccharides and other sugars during hydrolysis of lactose in milk by β-galactosidase from *S. thermophilus*. Reproduced with permission from reference (13), Food Chemistry, 1983.

achieved. The composition of the oligosaccharides also changed during the reaction, as indicated by the glucose/galactose ratio which continuously declined, suggesting that galactose-rich oligosaccharides were hydrolysed more slowly than those containing equal amounts of both glucose and galactose.

4.3. Consequences of Transferase Activity

Quite apart from the theoretical aspects, there are a number of practical consequences of oligosaccharide formation:

1. Judging from the data in Table 2, oligosaccharide levels are highest when the extent of lactose hydrolysis is 50–90% (the range of practical interest). Complete removal of oligosaccharides can only be achieved by prolonging the reaction time far beyond that required for removal of 95% of the lactose, thereby increasing processing costs.

2. Concentration of lactose solutions such as whey permeate, prior to hydrolysis, has advantages but will lead to higher levels of oligo-

TABLE 2

DISTRIBUTION OF SUGARS DURING HYDROLYSIS OF LACTOSE IN MILK BY *S. thermophilus* β-GALACTOSIDASE AT 37°C[13]

Time (h)	Lactose (%)	Glucose (%)	Galactose (%)	Glucose/galactose ratio	Oligosaccharides (%)	Glucose/galactose ratio in oligosaccharides
0·17	55·7	14·6	8·7	1·67	13·1	0·36
0·42	35·3	25·0	11·4	2·19	22·7	0·26
1·00	6·16	40·2	21·3	1·89	24·8	0·21
2·00	1·74	46·4	37·1	1·25	13·8	0·18
4·50	1·17	46·7	47·5	0·98	3·61	—

saccharides if processed with neutral-pH enzymes. The low solubility of these products may cause crystallization problems during storage of concentrated syrups. These problems can be lessened by the use of fungal enzymes at acid pH where oligosaccharide production is minimized.

3. Oligosaccharides are hydrolysed very slowly, if at all, by the digestive β-galactosidase of the human intestinal mucosa. For example, allolactose (a major transferase product in milk) is hydrolysed at less than 10% of the rate for lactose by mucosal homogenates, and even this may be due to intracellular rather than digestive brush-border enzymes.[17] As a result, these products will escape digestion in the small intestine and will pass into the colon where they will be fermented and can contribute to the symptoms of lactose intolerance.

The nutritional significance of this is not entirely clear. The oligosaccharide level in lactose-hydrolysed milk is usually less than 25% of the sugars or 1·25% on a wet-weight basis, but even this level may cause problems for those who are lactose intolerant. The oligosaccharides can best be removed by prolonging the time of enzymatic hydrolysis. This can be easily done with batch processes, but tends to negate some of the advantages of using immobilized enzymes in a continuous process.

4. Routinely, the presence of 1 mol of glucose (or galactose) is taken to mean that 1 mol of lactose has been hydrolysed. The data in Table 2 show that this is not the case. There are substantial differences in the concentration of free glucose and galactose as a result of incorporation of these sugars into oligosaccharides; this in turn leads to errors if determination of an individual sugar is used to follow the reaction. At higher residual lactose levels (>35%), the value for glucose leads to overestimation of the amount of lactose remaining, but gives a fairly good estimate of the total glycosides remaining to be hydrolysed. At lower residual lactose levels (<35%), the value of glucose tends to underestimate the remaining glycosides. Use of the value for galactose leads to similar errors.

5. ACTIVITY MEASUREMENTS AND ANALYSIS OF PRODUCTS

5.1. Analysis of Product Sugars

Analysis of the action of the enzyme on lactose, including lactose in milk and whey, is complicated by the existence of both hydrolytic and transferase activities. When this occurs, determination of glucose tends to overestimate the amount of lactose remaining, since some of the glucose will have been incorporated into oligosaccharides. Determination of galactose gives an even worse estimate since oligosaccharides *in toto* contain more galactose than glucose. As a result, the degree of hydrolysis of lactose cannot be determined by estimation of a single monosaccharide, except where the former approaches 100%.

In practice the whole purpose of hydrolysing lactose is to produce monosaccharides. It therefore follows that a good practical definition, which is widely accepted, is:

$$\text{Percentage hydrolysis} = \frac{\text{Total monosaccharides}}{\text{Original lactose concentration}} \times 100$$

Total monosaccharides in the presence of disaccharides can be determined by simple colorimetric methods[18,19] or even by specific enzymatic analysis for glucose and galactose separately.[10]

A complete analysis of the reaction mixture requires separation of the products. This can be done by paper chromatography,[11,13] gas–liquid chromatography[10,14] or high performance liquid chromatography (HPLC).[20-22] The first two are time-consuming but permit separation of the oligosaccharides which can then be quantified. HPLC allows very rapid analysis of glucose and galactose, with a minimum of sample preparation. The columns used also allow separation of di- from trisaccharides, but at present lactose is not effectively separated from other disaccharides.

5.2. Analysis via Freezing Point Depression

A completely different approach to estimating lactose hydrolysis is to follow changes in the physical properties of the solution as the reaction proceeds. Thus, freezing point depression, osmotic pressure and optical rotation all change as monosaccharides are produced. Freezing point depression can be measured very accurately in milk and is linearly related to the degree of hydrolysis.[23-25] This method has the further advantage that the sample can be analysed directly without deproteinization, as is required for sugar analysis. Given that the freezing point depression can be measured in less than 5 min, this method seems ideal for routine quality control of both lactose hydrolysis and the activity of β-galactosidase preparations.[25] An interesting side-effect of the freezing point depression is that adulteration of lactose-hydrolysed milk by up to 30% added water cannot be readily detected by the cryoscopic method.[26] However, 10% added water can be detected by the Quevenne lactometer.[26]

6. CHARACTERISTICS OF MICROBIAL β-GALACTOSIDASES

The major commercial sources of β-galactosidase are the yeasts *K. fragilis* and *K. lactis* and the fungi *A. niger* and *A. oryzae*. Recently a thermostable enzyme from a *Bacillus* species has also become available. Some of the important characteristics of these enzymes, and of others which are potential commercial sources, are shown in Table 3.

The primary characteristic that determines the application of a given enzyme is the operational pH range. On this basis, the enzymes can be divided into two groups: the acid-pH enzymes from fungi which are suitable for processing of acid whey and acid whey permeate, and the neutral-pH enzymes from yeasts and bacteria which are suitable for processing of milk and sweet whey. Within the two groups, the choice of an individual enzyme depends on factors such as price and those characteristics that dictate its operating performance, e.g. temperature stability and inhibition by reaction products.

6.1. Fungal Enzymes

These enzymes have pH optima in the range 3–5, which makes them suitable for processing of acid whey and its permeate. They have relatively high temperature optima and are typically used at temperatures of up to 50°C where they are reasonably stable. The combination of high temperature and low pH used for processing acid whey effectively discourages microbial growth, particularly in immobilized enzyme reactors.

In contrast to the neutral-pH enzymes, the fungal β-galactosidases do not require ionic activators and are not inhibited by sulphydryl reagents such as *p*-chloromercuribenzoate, indicating that sulphydryl groups do not play an essential role in catalysis or structure.

Of the two major sources, the *A. niger* enzyme has been preferentially used in commercial applications because it is more thermostable and has a lower pH optimum than the *A. oryzae* enzyme. These factors make it more suitable for immobilization. Its main disadvantages are that it is intracellular (compared with the *A. oryzae* enzyme, which is extracellular) and that it is strongly inhibited by galactose. Since it has only a weak affinity for lactose (a general feature of the fungal enzymes) the ratio K_m(lactose)/K_i(galactose) is, relatively, very high (Table 3). This ratio is a measure of the susceptibility of the enzyme to product inhibition. In the case of the *A. niger* enzyme, the build-up of product

TABLE 3

CHARACTERISTICS OF SOME MICROBIAL β-GALACTOSIDASES

Source	Molecular weight $\times 10^{-3}$	pH optimum	Temperature optimum (°C)	Stability at 55°C at pH optimum[a]	Activators[b]	Ionic inhibitors	K_m lactose (mM)	K_i galactose (mM)	K_m/K_i
A. niger	124	3·0–4·0	55–60	v. good	none needed	none	20–85	4	5–20
A. oryzae	90	5·0	50–55	good	none needed	none	50	57	0·88
K. lactis	135	6·5–7·3	35	v. poor	K^+, Mg^{2+}	Ca^{2+}, Na^+	12–17	42	0·38
K. fragilis	201	6·6	37	v. poor	K^+, Mg^{2+}, Mn^{2+}	Ca^{2+}, Na^+	14	28	0·5
E. coli	540	7·2	40	poor	Na^+, K^+, Mg^{2+}	—	2	21	0·1
B. subtilis	—	6·5	50	fair	none needed	—	700	40	17
B. stearothermophilus	215–230	5·8–6·4	65	excellent	Mg^{2+}	—	2	20	0·1
S. thermophilus	530	7·1	55	fair	K^+, Mg^{2+}, Mg^+	Ca^{2+}	7	60	0·11
L. thermophilus	540	6·2	55	—	—	—	6	—	—

[a] In buffer in the absence of substrate.
[b] Ionic material likely to be found in dairy products.

causes the reaction rate to slow markedly with time and makes complete conversion into monosaccharides difficult to achieve. In spite of this, the enzyme is regarded favourably because of its other superior characteristics.

6.2. Yeast Enzymes

The enzymes from *K. fragilis* and *K. lactis* are generally very similar. With neutral pH optima, they are well suited to hydrolysis of lactose in milk. Milk also supplies the potassium and magnesium ions needed for activity. However, both enzymes are strongly inhibited by the high levels of calcium in milk and slightly by the sodium.[27,28] They are also inhibited by galactose, but the effect on conversion rates is not as marked as with the fungal enzymes, as can be seen from the relatively low values for K_m/K_i (Table 3).

The yeast enzymes can be produced in high yields from whey fermentation[34] and are therefore relatively inexpensive.[29] They are also widely accepted as safe for use in foods. Their major disadvantage is that their optimum temperature (30–40°C) encourages microbial growth, so a relatively short hydrolysis period of 2–3 h is advisable. To achieve high levels of conversion in this time, with minimal oligosaccharide concentrations, requires relatively high enzyme levels and therefore higher process costs in a batch operation. In order to avoid these problems, it is often preferable to carry out hydrolysis for 16–24 h at 4–6°C, where microbial spoilage is minimized. The moderate temperature optima and rather poor thermostabilities of these enzymes have led to their use mainly in batch processes rather than in immobilized forms, with the notable exception of fibre-entrapment (see Section 8).

6.3. Bacterial Enzymes

Many bacteria produce β-galactosidases but relatively few bacterial species are regarded as safe sources. However, two sources are worthy of note: *Streptococcus thermophilus* and *Bacillus stearothermophilus*. The former is eminently suitable as a source organism from the safety viewpoint, because it is used as a starter in yoghurt and some cheese varieties, and has been consumed for several thousand years without ill-effect. The β-galactosidase from *S. thermophilus*, in common with those from most bacteria, is much larger than the corresponding enzymes from yeasts and fungi and almost certainly is made up of sub-units. It has a very low ratio K_m/K_i (Table 3), and is thus relatively insensitive to product inhibition. Its temperature optimum is 55°C, but it is quickly denatured in buffer at this temperature with a half-life of less than 10 min.[30] It is much more stable in milk,[31] but it is unlikely that it will work for long periods at temperatures exceeding 50°C.

The β-galactosidase from a selected variant of *B. stearothermophilus* is considerably more thermostable, with a half-life in buffer of 7·5 h at 60°C and 1·3 h at 65°C.[32] Thermostability in lactose solutions and milk has not been reported but should be even better, as is the thermostability of the enzyme in immobilized whole cells.[32] An enzyme with similar properties from a closely related thermophilic *Bacillus* is also available commercially in an immobilized form. All the known kinetic characteristics of the enzyme from the *B. stearothermophilus* variant are favourable, including low sensitivity to galactose inhibition during lactose hydrolysis as judged by the K_m/K_i ratio (Table 3). The only obvious disadvantage is the rather low reported yield of about 4000 ONPG (*o*-nitrophenol galactoside) units litre^{-1},[33] which is about 10 times lower than that attainable with yeast in batch culture.[34] However, this may be amenable to substantial improvement through optimization of growth conditions or through recombinant DNA technology.

6.4. Other Sources

Despite the large number of β-galactosidases already characterized, there will undoubtedly be a continuing search for enzymes from new sources so as to fit more ideally into a desired process scheme and to reduce costs. In recent years the thrust of these searches has been toward finding highly thermostable enzymes which can work at temperatures that minimize microbial growth in dairy substrates. A case can therefore be made for the need to discover β-galactosidases that work effectively (not very slowly) at low temperatures, e.g. 0–5°C. Unfortunately, most psychrophilic bacteria are not considered safe source organisms. Nonetheless, this may still be a worthwhile area of enquiry.

7. β-GALACTOSIDASE ACTIVITY AND STABILITY IN MILK AND WHEY

Most of the kinetic characterization of the β-galactosidases reported in the scientific literature has been performed in model systems, e.g. lactose or *o*-nitrophenyl β-D-galactopyranoside in buffer. While useful for comparison of enzymes, these data are often a poor guide to the behaviour of an individual enzyme in a dairy product such as milk or whey. The large differences observed in activity and stability reflect the complex nature of these substrates, particularly the ionic composition and the protein constituents.

The activity of some neutral-pH β-galactosidases in milk and whey is less than 30% of the activity in phosphate buffer.[27,31] This is mainly due to the high level of soluble calcium, which exerts a strong inhibitory effect.[27,28] It is also possible that the citrate anions in milk reduce activity compared to phosphate buffer, by complexing magnesium ions. The effect of milk proteins on the activity of β-galactosidase is less clear. There are several reports that heating milk and whey leads to an increase in enzyme activity,[35-39] and this has led to speculation that milk may contain a thermolabile lactase inhibitor.[36] Other reports are contradictory,[27,31] so the situation remains unresolved. The increase in activity had been thought to result from the thermal denaturation of some protein component in the milk, but it is also possible that heat treatments cause a shift in the soluble salt balance which accounts for the effect.

There are hardly any data on the stability of soluble β-galactosidase in dairy products, although it is widely accepted that the enzyme is stabilized to some (unknown) extent by the substrate. A recent study on the enzyme from *S. thermophilus* suggests that stability is strongly enhanced by the milk proteins, especially casein.[31]

8. PROCESSING OF DAIRY PRODUCTS WITH β-GALACTOSIDASE

The β-galactosidases already described can be used in a number of ways to hydrolyse lactose in milk, whey and whey permeate. The choice of process technology depends upon the nature of the substrate, the characteristics of the enzyme and the economics of production/storage/ marketing of the product. There are basically two different ways to use β-galactosidases: as a soluble enzyme and as an immobilized enzyme. The soluble enzyme is normally used for batch processes while the immobilized form lends itself to continuous operation.

8.1. Processing with Soluble β-Galactosidase

8.1.1. Processing of Milk by the Consumer
Perhaps the simplest way of getting low-lactose milk is for the consumer to make it himself by simply adding neutral yeast β-galactosidase to fresh

pasteurized milk. After overnight incubation, in the refrigerator to prevent microbial growth, the lactose content is reduced by 70% or more (depending upon the dosage). The enzyme (from *K. lactis*) is marketed as a liquid, stable for 18 months or more at room temperature, by Sugar-Lo Company in the United States and is sold through health food and drug stores.[40] The cost to the consumer is about 19 cents (US) per litre. A similar product, 'Kerulac', is produced by Gist Brocades NV, in the Netherlands.

8.1.2. *Processing of Milk at the Dairy*

Dairies can purchase neutral yeast β-galactosidase in bulk at considerably lower prices than individual users. Consequently, dairies can offer low-lactose milk for sale at prices that are considerably less than the cost of the home-produced product, while maintaining good added value.

Two basic process schemes are currently in operation. In the first of these, shown in Fig. 3, the milk is pasteurized and then cooled to the

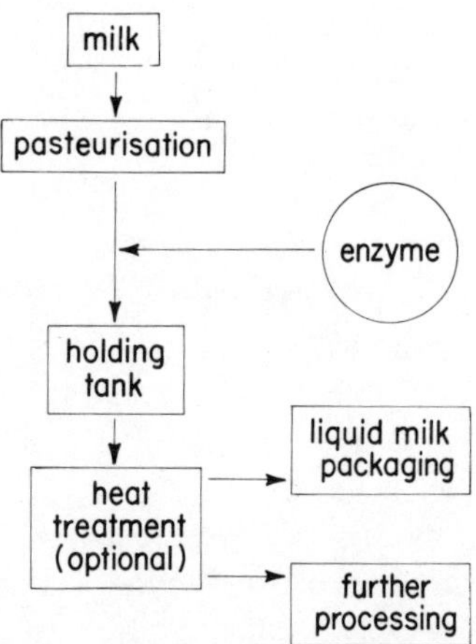

FIG. 3. Process scheme for batch hydrolysis of lactose in milk.

reaction temperature of the enzyme. The enzyme is then added and the milk is left in a holding tank where hydrolysis takes place. For overnight incubation, the milk is, typically, held at 4–6°C so as to prevent microbial growth. Alternatively, hydrolysis can be carried out at 35°C in 2–3 h, using a higher level of enzyme. When the desired degree of hydrolysis is reached the milk can be reheated to destroy the enzyme (if desired) and then packed or further processed, e.g. by evaporation and drying.

If the milk is to be concentrated, it is advisable to inactivate the enzyme by heat prior to evaporation, so as to limit oligosaccharide formation. Where the milk received by the dairy is fresh, with a low microbial load, hydrolysis at 4°C for 24 h can be carried out prior to normal pasteurization and packaging.

Low-lactose milk produced by this process and sold in supermarkets costs the consumer about 9 cents (US) per litre more than ordinary pasteurized milk.[40]

8.1.3. Milk Processing Using Very Small Quantities of Enzyme

The second process permits the use of a much smaller quantity of enzyme, which is then allowed to work for a longer period of time. The process, which was originally devised for use with sterilized milk,[41] is shown diagrammatically in Fig. 4. The enzyme (neutral-pH β-

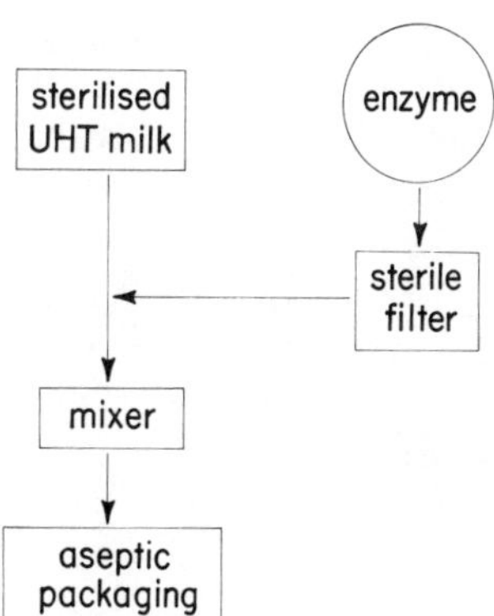

FIG. 4. Process scheme for hydrolysis of lactose in milk using minute amounts of enzyme.

galactosidase from *K. lactis*) is ultrafiltered to sterilize it and then mixed with UHT-sterilized milk immediately before the milk is aseptically packaged into paper containers. Only minute amounts of enzyme are needed (about 10 parts per million) to get almost complete hydrolysis in 7–10 days at room temperature.[42] However, the enzyme must be very pure, i.e. protease-free, or else the milk will deteriorate during storage. Even less enzyme can be used (5 parts per million) if the dairy is willing to delay sale of the milk for about 1 month after packaging. The resulting product has a shelf life of 4–6 months without refrigeration, which is about the same as that of ordinary sterilized milk.

This process has distinct advantages in that enzyme costs are greatly reduced and process costs are minimized. It is also possible to use pasteurized milk as a starting material,[40] but in this case the shelf life is much shorter and a larger enzyme dose is needed so as to complete hydrolysis before the milk reaches the consumer.

8.1.4. Batch Processing of Whey

Whey can be treated with β-galactosidase essentially by using the process scheme outlined for milk (Fig. 4). Sweet whey derived from cheesemaking is pasteurized to prevent further acidification by starter organisms, and cooled to the chosen reaction temperature. The pH is then adjusted towards neutrality with potassium hydroxide since sodium ions are inhibitory for the yeast enzymes. The enzyme is then added and the whey stored until hydrolysis is complete. The whey can then be concentrated by vacuum evaporation to give a sweet syrup. Where the substrate is acid whey, little or no pH adjustment is needed when employing the enzymes from *A. niger* or *A. oryzae*. This basic scheme can be modified, according to the desired end-product, by various operations such as deproteinizing the whey by ultrafiltration before hydrolysis, concentrating the whey before hydrolysis, and demineralizing the concentrated product by ion-exchange or electrodialysis.

A major cost element in batch processing is the enzyme, which is lost in the product. In the case of milk, this may be acceptable because of the value of the product. However, low-lactose syrups derived from whey face intense competition as food ingredients, so it is essential to minimize costs. This has led to development of processes in which the enzyme is recovered or used on a continuous basis.

8.1.5. Enzyme Recycling by Ultrafiltration

One way to reduce enzyme costs while retaining the kinetic advantages of using soluble enzymes is to recycle the enzyme using ultrafiltration. The basis for this approach is that lactose and its hydrolysis products readily permeate high-flux ultrafiltration membranes whereas the enzyme does not. A flow diagram for such a process is shown in Fig. 5.

The feed material (milk or whey) is first ultrafiltered to produce a permeate containing most of the lactose. The permeate is then treated with soluble β-galactosidase in a reactor until hydrolysis is considered sufficient. The optimum process temperature using the enzyme from *K. lactis* appears to be in the range 10–24°C.[43] The enzyme is then recovered by second-stage ultrafiltration and recycled back to the reactor. The permeate, containing monosaccharides and some residual lactose, may then be drawn off to make glucose–galactose syrups or

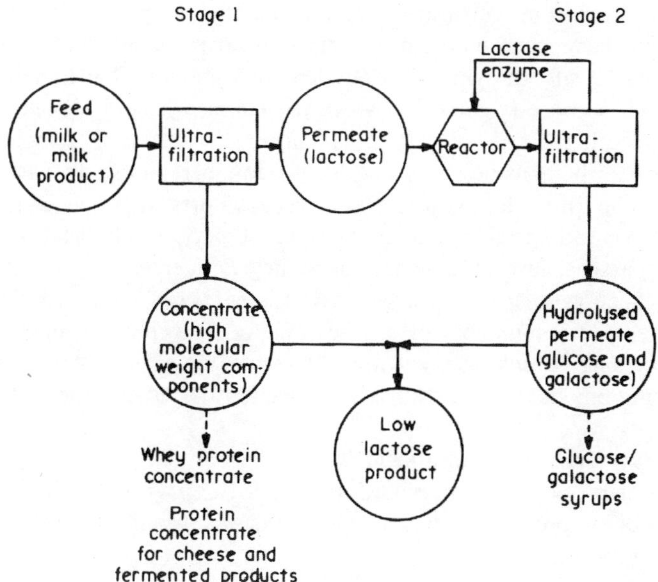

FIG. 5. Flow diagram for the enzyme recycle process using ultrafiltration. Reproduced with permission from reference (43), Food Technol. in Australia, 1980.

recombined with the retentate from the first stage to make a low-lactose product, e.g. low-lactose milk. The process may be operated continuously or batchwise, depending upon throughput volumes and ultrafiltration capacity.

Although attractive in some aspects, this process has not yet reached commercialization. The main problem seems to be prevention of microbial growth during continuous operation at ambient temperatures with non-sterile feed materials. Furthermore, continuous whey processing using immobilized enzymes (see below) is simpler in concept and requires less labour in operation.[44]

8.2. Processing with Immobilized β-Galactosidase

While enzyme recycling has potential (and also problems) the major effort in reducing enzyme costs had been directed toward using immobilized forms of β-galactosidase.

In essence, the enzyme can be immobilized on an extremely wide variety of supports employing all the primary immobilization techniques such as adsorption, entrapment and covalent linkage.[29,45,46] The supports that have been used range from natural polymers such as chitin, collagen and silk to synthetic organic resins and durable mineral particles such as porous glass beads or even stainless steel.[29] While all the systems work reasonably well at hydrolysing solutions of pure lactose, very few can be considered as suitable for commercial use. In part this is because the physical properties of the supports themselves are not suitable for industrial processes, because of poor mechanical strength, and in part because milk, whey and whey permeate are complex materials which pose special problems. At present, only three systems have been used for commercial or semi-commercial production: one system, for processing of milk, uses entrapped enzyme; the other two systems, for processing of whey, use adsorbed or covalently bound enzymes.

8.2.1. Hydrolysis of Lactose in Milk

Development of an immobilized enzyme reactor for continuous hydrolysis of lactose in milk has not proved easy. Among the difficulties to be overcome are:

1. The neutral pH of milk encourages microbial growth except at low and very high temperatures;
2. Milk proteins tend to adsorb onto the immobilized enzyme surface and foul the reactor;
3. The neutral-pH enzymes (from yeast) currently available are not very stable when immobilized by classical techniques such as adsorption or covalent linkage.

In order to overcome these difficulties, the neutral-pH enzyme from *K. lactis* has been immobilized by entrapping it in porous cellulose acetate fibres.[47,48] The porosity permits diffusion of low molecular weight compounds but does not allow the enzyme to escape. Since the enzyme exists in solution within the microcavities in the fibre, it maintains its natural configuration. Furthermore, there are no charge effects and, as a result, characteristics such as pH and temperature optima are not affected.

By entrapping the enzyme in the microcavities within the fibre, many of the problems associated with adsorbed or bound enzyme are avoided. When milk or whey are in contact with the fibres, only the low molecular weight substances, including lactose, can diffuse through and contact the enzyme. This is an advantage in that proteins which would otherwise tend to foul the enzyme surface are kept out, as are bacteria and proteases, which would damage the catalyst. Other advantages of this technique are that high levels of bound activity per gram of fibre can be obtained,[29] and the stability of the entrapped enzyme is very good under practical conditions. The stability of free β-galactosidase ($K.$ $lactis$) in solution at 25°C is rather poor, but when entrapped in fibres the half-life using lactose as substrate is about 100 days.[49] This increased stability has been attributed to isolation of the enzyme from proteases and microbial contamination, and the fact that the enzyme in the micro-cavities is highly concentrated in tiny water droplets and in this form the enzyme does not undergo dilution on repetitive use. It is also possible that the degree of hydration has been changed.[48]

The main disadvantage is that diffusion of substrate through the fibre to the enzyme surface is slow, despite the large surface-to-volume ratio. As a result, the measured activity is only a fraction of the free activity immobilized.[49] The effect of diffusional limitation is reflected in the K_m for the immobilized enzyme, which increases from 9·5 mM to 77 mM.[49]

A schematic diagram of the process in operation at Centrale del Latte in Milan is shown in Fig. 6. Skim milk is sterilized, cooled to 5°C and then continuously circulated through the enzyme reactor, which contains fibres in the form of skeins, packed vertically (parallel to the long axis) in a column. After about 20 h at 7°C, approximately 75% hydrolysis of lactose is achieved. At the end of each run, the fibres are washed with sterile buffer containing magnesium ions (to keep the enzyme active), and

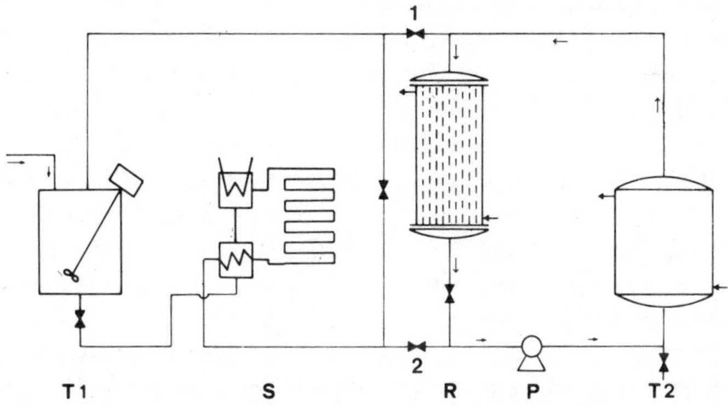

FIG. 6. Flow diagram for the hydrolysis of lactose in milk using fibre-entrapped enzyme: TI, stirred vessel for preparing washing solutions or for receiving milk; S, sterilizer; R, enzyme reactor; P, recycling pump; T2, reservoir; 1,2, valves. Reproduced with permission from reference (50), Academic Press (1976).

periodically treated with a quaternary ammonium compound to prevent build-up of psychrophiles on the fibre surfaces.[50]

Operational stability is considered good in that the loss in activity after 50 runs is less than 10%.[50] A semi-commercial batch system processing 10 000 litres of milk per day has been operating at Milan for several years. Despite the need to operate batchwise (to prevent microbial contamination and surface fouling of the fibres), this process represents the most advanced application of immobilized β-galactosidase technology to milk to-date.

8.2.2. Hydrolysis of Lactose in Whey

Processing whey poses fewer problems than with milk, particularly since the acidic pH restricts microbial growth. The two most successful systems both use immobilized β-galactosidase from *A. niger*.

The Corning process. Over the last decade, Corning Glass Co, Corning, New York, has developed and refined a process that employs β-galactosidase immobilized on porous glass beads. In essence, the controlled-pore glass beads are reacted with γ-aminopropyl-triethoxysilane, which places an alkylamino group on the bead surface. The 'silanized' beads are then reacted with glutaraldehyde to provide a terminal carbonyl group. This, in turn, is reacted with the enzyme, which is covalently coupled by an imine bond.[51] By this means, beads are prepared with bound activity of about 500 units per gram at 50°C and with a pH optimum in the range 3·2–4·3.[52]

Using this catalyst, two hydrolysis plants have been operated on a semi-industrial scale by the Milk Marketing Board in the UK and by Union Laitière Normande in France, to demonstrate the economics of the process. A flow diagram for the production of glucose–galactose syrups from whey, via whey permeate, is shown in Fig. 7. Raw whey is pasteurized and then passed to an ultrafiltration plant where the protein is removed. The permeate is demineralized using a twin-column ion-exchange system which removes 90% of the total ash. The demineralized permeate is bulked for storage and can, if necessary, be re-pasteurized. It is then transferred to the hydrolysis plant, where the pH is adjusted to 4·5 with HCl and the temperature adjusted accurately to the operating temperature of the enzyme reactor. The latter houses the beads packed into a cylindrical column. The column temperature is gradually increased (from 32 to 50°C) during long-term operation so that the degree of hydrolysis is maintained constant, at constant throughput, despite slow

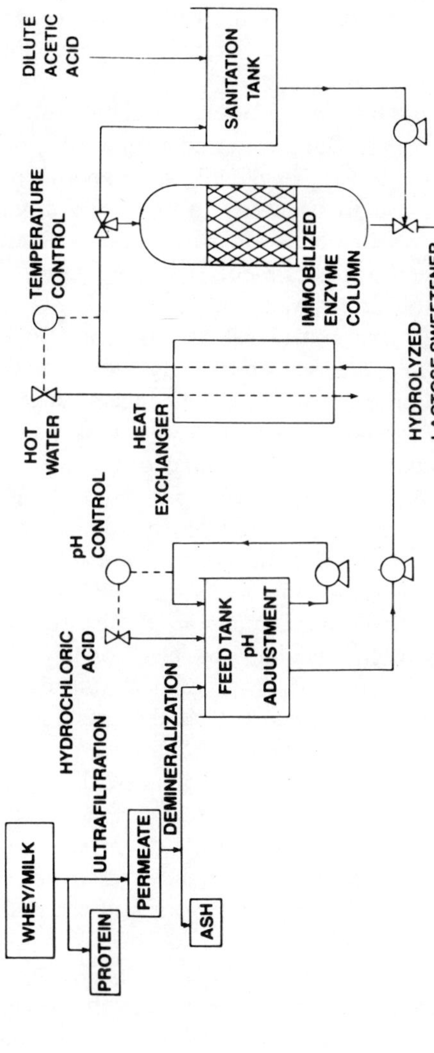

FIG. 7. The Corning process for hydrolysis of lactose in ultrafiltered milk or whey. Reproduced with permission from reference (53), Food Product Development, (1980).

loss of enzyme activity. The residence time in the column is 15–25 min and the emerging product is at least 80% lactose-hydrolysed.

With a column containing 23 kg of catalyst, each hydrolysis plant can process up to 9000 litres of permeate in a 20 h cycle. At the end of a run, the bead bed is cleaned by back-flushing with dilute acetic acid for half an hour. The life of the system is estimated at $1\frac{1}{2}$ years in continuous use.[53]

The hydrolysed material from the column is cooled to about 10°C by regeneration against the incoming feed and then stored, prior to evaporation to a syrup with 60–68% solids. Filtration and/or treatment with activated charcoal are usually necessary to produce a bright, sparkling, water-white product which corresponds to top-grade corn syrup in appearance, but these extra treatments are only needed for certain applications such as soft drinks and confectionery.

This immobilized enzyme system has also been developed to permit processing of whey. The process sequence is shown in Fig. 8 and is similar to that for permeate, except that demineralization takes place after hydrolysis, and a proprietary process step is needed after pH adjustment/pasteurization to prevent fouling of the enzyme reactor by whey proteins.[54] Cleaning of the enzyme bed requires circulation of a proprietary cleansing agent designed to remove protein deposits from the bed, followed by circulation of a bacteriocide and dilute acetic acid.[54] After hydrolysis, the whey syrup is cooled quickly to 10–15°C to minimize Maillard reactions, and demineralized. Finally, it is evaporated to give a light brown syrup containing 60–65% solids.

Hydrolysed whey and permeate syrups are often fully demineralized by ion-exchange because the minerals that are present can give rise to adverse flavours in some food products. However, the ion-exchange process is expensive. An alternative, which gives partial demineralization, is to use electrodialysis, which selectively removes more than 90% of the sodium, potassium and chloride ions when 70% of the ash is removed.[52] Removal of 50% of the ash by electrodialysis leaves an ash content of 2–3% in the final syrup, but the final product is still acceptable for many applications.

The cost of producing an 80% hydrolysed lactose syrup from permeate, in Western Europe, by the Corning process, has been estimated at 11–13 cents (US) per kg.[53] Concentration to 65% solids adds a further 11–13 cents, and demineralization costs range from 4–7 cents per kg for 50% de-ashing by electrodialysis to 13–20 cents for complete demineralization by ion exchange.[52,53] The final cost of 65% solids syrup is therefore in the range 26–46 cents per kg (all 1980 data).

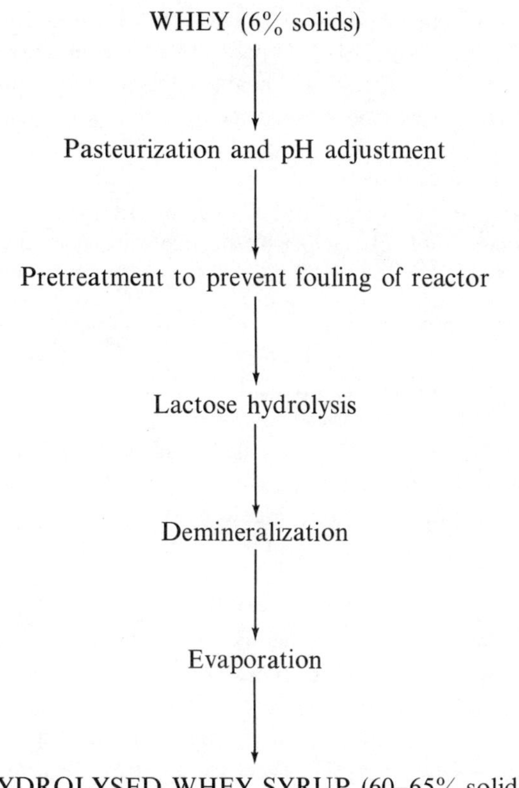

FIG. 8. Process scheme for the hydrolysis of lactose in whole whey. Adapted from reference (54), International Dairy Federation, 1981.

The Valio process. A different process for hydrolysing whey is operated commercially by the Valio Dairy in Finland. This process uses enzyme immobilized on a phenol–formaldehyde resin.[55] The enzyme is simply adsorbed onto the washed resin and then fixed in place by cross-linking with glutaraldehyde. This support system has good mechanical/physical properties and is capable of binding high levels of enzyme activity with high binding efficiency.[56] For hydrolysis, the resin is packed into a fixed-bed columnar reactor, which is operated in the plugged flow mode.[56]

The productivity of the process, i.e. total yield of hydrolysate over a given period of time, depends upon various operating parameters, each of which may influence the reaction rate and the stability of the enzyme. Highest productivity is obtained by:

1. Working at temperatures well below the optimum for the enzyme, e.g. 20–30°C; raising the temperature of the process in the range 10–50°C has a greater detrimental effect on immobilized enzyme half-life than it has on enhancement of the reaction rate.[56]
2. Pre-concentrating the feed; productivity increases with increasing lactose concentration.[56]
3. Operating at pH 3·5; productivity is essentially constant over the pH range 3·5–4·5, but at pH 3·5 no bacterial growth is observed in the reactor.[56]

In commercial practice, a typical production cycle takes 16–20 h for the hydrolysis of 20 000 litres of whey at 25–30°C and at a pH of about 3·5. Cleaning can be accomplished simply with water and typically takes no more than 3–4 h. Under these conditions, the half-life of the enzyme is in the range 150–200 days. The process has been in operation for several years and is quite suitable for treatment of whole whey, as well as permeate. Production costs are not readily available, but the simplicity of the system is obvious and appealing, especially in terms of catalyst fabrication.

Other processes for whey hydrolysis. β-Galactosidase entrapped in cellulosic fibres can be used to hydrolyse lactose in whey as well as in milk. A process has been described in which the enzyme reactor is packed with fibres wound radially around a perforated tube.[57] The reaction mixture flows from the outside through to the centre and is then recirculated until hydrolysis is complete. Either whole whey or permeate can be used as a feed material, but the rate of hydrolysis is dependent upon the nature of the feed and especially upon the ionic material present.[57]

Long-term enzyme stability in this process is very good, with more than 84% of the original activity remaining after 120 days of operation. However, this system has not attracted commercial interest, perhaps because it is not as efficient or economical as the other processes (above).

8.3. Transferase Activity Using Immobilized β-Galactosidase

Immobilization of an enzyme often leads to changes in the kinetic characteristics, the nature of which depends mostly upon the support and the method of operation. Among the most important changes caused is an increased diffusional resistance to mass transfer. This affects not only binding constants, such as K_m and K_i, but also the concentration of intermediate products such as the oligosaccharides produced by trans-

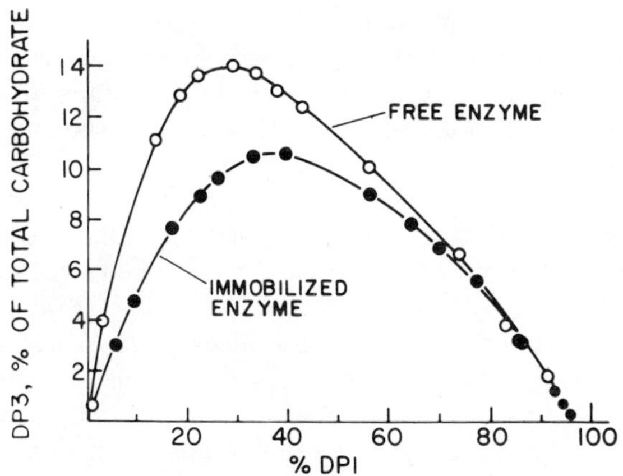

FIG. 9. Formation of trisaccharides during hydrolysis of 20% lactose with a thermophilic β-galactosidase from a *Bacillus* species. Reproduced with permission from reference (58), Humana Press, (1982).

ferase activity. An example of this is shown in Fig. 9, where trisaccharide production is shown as a function of degree of hydrolysis, using a commercial thermophilic β-galactosidase (*Bacillus* species). The diffusional resistance caused by immobilization produces concentration gradients of possible acceptors in the support particle and thereby lowers trisaccharide production compared to the free enzyme.[58] Lowered transferase activity is clearly an advantage of using immobilized enzyme in commercial practice.

8.4. Future Trends in Processing with Immobilized β-Galactosidase

The processes already developed for hydrolysing lactose in milk and whey will undoubtedly be further refined in order to reduce costs. Current research trends point toward two lines of development.

8.4.1. Use of More Thermostable Enzymes

Use of a thermostable neutral-pH β-galactosidase would permit processing of milk at much higher temperatures. Working at 60°C or higher would minimize microbial growth (probably the major problem in milk processing) as well as allowing faster catalysis. However, prolonged holding of milk at this temperature, e.g. in a process where the milk is recycled, may induce unacceptable flavour changes. Another potential problem in working at high temperatures is the increased likelihood of browning (Maillard) reactions, especially if the feed is concentrated.

8.4.2. Immobilization of Whole Cells Containing β-Galactosidase

The enzymes used in commercial processes are intracellular. Extraction, isolation and partial purification are costly procedures which can be eliminated by immobilization of whole cells of a suitable source organism. Whole cells of *L. bulgaricus*, *E. coli* and *K. lactis*,[59] and of a thermophilic *Bacillus* species,[32] have been immobilized in polyacrylamide gel. Cells of *K. fragilis* have been immobilized in cellulose derivatives,[44] while cells of *E. coli* have been immobilized in films and sponges cross-linked with glutaraldehyde.[60] A further advantage of using whole cells is that the β-galactosidase tends to be more stable in its natural environment than when isolated and immobilized by other techniques.[32] A potential drawback is the lower activity of whole cells, compared to isolated enzyme, due to diffusional restrictions. In a batch process, diffusional limitations can be largely eliminated by rapid stirring,[59] provided that the catalyst particles are sturdy enough. However, reduced activity is likely in a continuous process using a fixed-bed reactor. Another potential problem is that other enzymes, e.g. proteases or lipases, may still be active in the cells and may cause undesirable flavour changes.

9. APPLICATIONS

9.1. Lactose-hydrolysed Milk Products

While most of the lactose-hydrolysed milk (LHM) now produced is consumed directly, the changes brought about by conversion into glucose and galactose offer a number of opportunities for the product. These are based upon the increased solubility, sweetness and ease of fermentation of the sugars produced. In terms of perceived sweetness in liquid milk, hydrolysing 30%, 60% and 90% of the lactose present is equivalent to adding 0·3%, 0·6% and 0·9% sucrose, respectively.[61]

9.1.1. Flavoured Milk Products

Flavoured milk products, such as chocolate milk, are very appealing to children and are typically sweetened with sucrose. However, nutritionists are somewhat concerned about the effects of sucrose in providing unnecessary calories and contributing to dental caries formation. The amount of sucrose can be reduced by 20–40% without loss of sweetness by using LHM in the formulation of the product. A reduction of 10% in the energy from carbohydrates is also achieved.[62]

9.1.2. Sweetened Condensed Milk

Partial hydrolysis of lactose (25–30%) reduces the tendency for lactose crystallization so that seeding is unnecessary,[63] and also adds to the sweetness of the product.

9.1.3. Frozen Milk Products

The low solubility of lactose often leads to its crystallization during storage of frozen milk products. This phenomenon is often observed as 'sandiness' in ice-cream and can be prevented by the use of LHM.

Freezing milk as a 3:1 concentrate is an attractive way of preserving it since the flavour changes associated with heat-processed milks are largely avoided. However, in ordinary milk, crystallization of lactose during storage causes thickening and coagulation of the milk protein. Hydrolysis of the lactose (90%) before freezing reduces thickening and extends the storage life (Fig. 10). It provides a product that, after reconstitution, compares in flavour with untreated fresh milk that has been slightly sweetened.[61,64–66] Despite the advantages conferred by lactose hydrolysis, frozen concentrated milk has never been produced commercially (to the writer's knowledge). The reason is not entirely clear, but it may be that the product cannot compete in cost and convenience with instant skim-milk powder, to whose taste the public is well accustomed.

9.1.4. Fermented Milk Products

While most dairy starter organisms can ferment lactose, the actual hydrolysis is often the rate-limiting step. Pre-hydrolysis by β-galactosidase often stimulates the organisms and permits the use of strains that were previously considered unsuitable because they cannot grow when lactose is the sole source of carbohydrate.

Yoghurt. Use of LHM for yoghurt manufacture results in accelerated acid development by the starters. This leads to a reduction in the set time of 15–20%.[69,70] The increased rate of acid development is *apparently* due to the more rapid utilization of the available carbohydrate,[67,68] but the action of proteases, present as contaminants in commercial β-galactosidase preparations, may also play a role in stimulating acid production by the thermophilic streptococci used in yoghurt starters.[105] The final product reportedly has a higher viscosity, a longer shelf life and is sweeter.[71] The added sweetness is advantageous in making fruit-flavoured yoghurts since it permits lower levels of added sucrose, and

R. R. MAHONEY

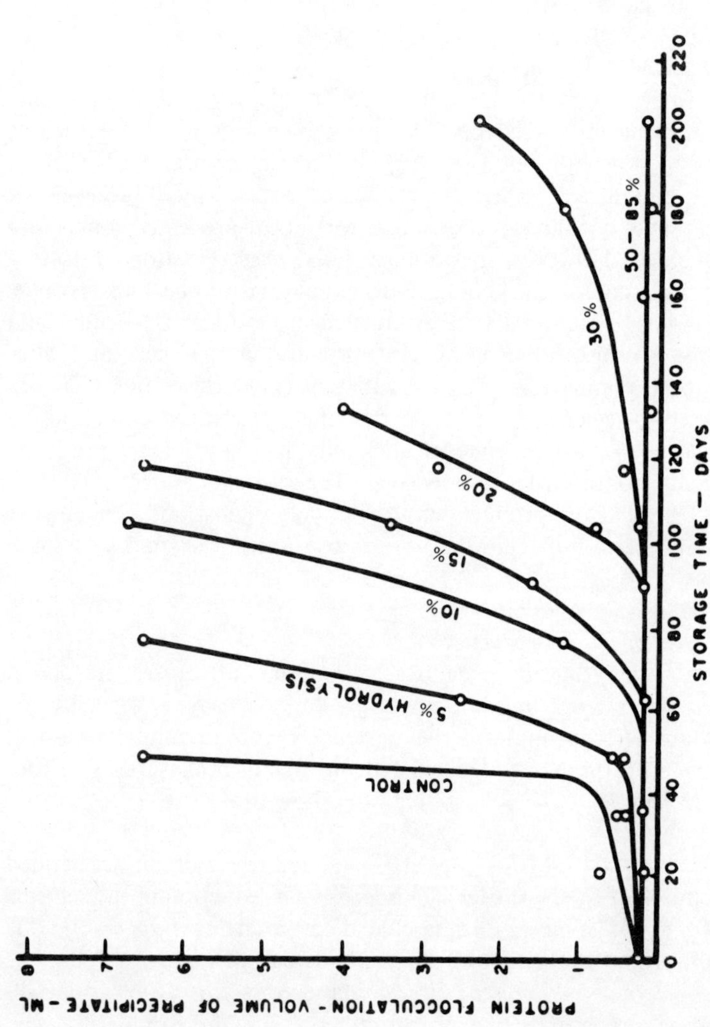

FIG. 10. Effect of lactose hydrolysis on protein flocculation in frozen concentrated milk (35% solids) stored at 15°F. Reproduced with permission from reference (64), J. Dairy Science, (1954).

also in plain yoghurts sold in some countries, such as the US, where the acid flavour is generally not well liked.[72]

Cheeses. Adding β-galactosidase to milk causes faster acidification by some of the starter organisms used in cheese manufacture.[76] In some cases this is accompanied by increased bacterial counts, but the results vary with strain and species of organism.[76,106]

When making fresh cheeses such as cottage cheese or quark, the net effects are a reduction in the set time and a firmer, more elastic curd with consequently smaller losses due to cheese fines.[73,74] This in turn can increase the yield by up to 10%.[74] However, it is arguable whether or not these advantages alone are worth the cost of enzyme processing.

Of greater interest is the use of milk treated with β-galactosidase preparations to make ripened or mature cheeses. Here a significant reduction in cost can be achieved by shortening the ripening process. For Cheddar cheese, the cost of ripening (ageing) was estimated at 2·9 cents (US) per kg per month in 1976,[75] and is probably still close to that figure at the time of writing (1983).

Adding β-galactosidase preparations to cheese-milk causes changes both in the process of cheese manufacture and in the ripening profile. The increased availability of monosaccharides, which are fermented more readily, leads to accelerated acid production by Cheddar cheese starter cultures and by individual strains of mesophilic lactic streptococci, of the order of 10–15%.[76] In contrast, lactose hydrolysis does not stimulate acid production by thermophilic lactobacilli and streptococci.[105]

Accelerated ripening has been reported for a number of cheese varieties. Studies on the manufacture of Cheddar cheese, using 70% and 90% LHM, showed that after 3–4 months the cheeses had the flavour, texture and body of 6–8 month-old cheeses made from ordinary milk.[75] In Camembert cheese made with LHM, mould development occurs earlier and is more prolific.[73] Blue cheese made with LHM ripens faster and has a more pronounced flavour than control cheese.[71] The greatest advantage of using LHM may be in the manufacture of very hard cheeses such as Parmesan, in which ripening may take up to 2 years, and in which flavour development does not begin until the proper body or texture has developed (5–6 months). For Cheddar cheese, the optimum degree of lactose hydrolysis is in the range 20–40%,[71,76] and the cost of achieving this is about 3 cents (US) per kg of cheese.[71]

Several studies have shown that rapid maturation of Cheddar cheese (and presumably other hard varieties) is caused by accelerated proteolysis, as indicated by increased soluble nitrogen[77] and by the development of free amino acids.[78] The subjective and chemical judgements regarding earlier ripening are supported by rheological data.[78] The origin of this increased proteolysis was originally presumed to be enhanced bacterial protease activity arising from accelerated growth of starter bacteria on the glucose in the LHM.[78] However, recent studies have shown that samples of the commercial β-galactosidase preparation used (Maxilact, derived from *Kluyveromyces lactis*) contained protease activity as a contaminant,[105–107] and that it was this protease activity, rather than β-galactosidase activity *per se*, that was responsible for the enhanced proteolysis.[106] Whether or not the protease activity can account for all of the acceleration of ripening and flavour development observed is not yet clear.[106] It may still be that β-galactosidase action makes some contribution by providing monosaccharides for fermentation by microorganisms involved in the ripening process.

Regardless of the cause of accelerated ripening, and whether or not it justifies the cost of using LHM, there is one indisputable advantage: the resulting whey contains partially hydrolysed lactose. It can therefore be used in a variety of products, as indicated below.

9.2. Lactose-hydrolysed Whey and Whey Permeate

Applications of syrups produced from whey by hydrolysis of lactose depend upon their chemical composition and functional properties. Typical data for the composition of 90%-hydrolysed syrups, concentrated to 65% solids and 90% demineralized, are shown in Table 4. Some of the properties of 90%-hydrolysed lactose, relative to lactose itself, are

TABLE 4

TYPICAL COMPOSITION OF HYDROLYSED-LACTOSE SYRUPS PRODUCED FROM
WHEY AND PERMEATE FEED AFTER 90% LACTOSE HYDROLYSIS[52,54]

	Whey feed (%)	Permeate feed (%)	Hydrolysed whey syrup (%)	Hydrolysed permeate syrup (%)
Total solids	6·1	4·7	65	65
Lactose	4·5	4·0	5·8	5·9
Glucose	—	—	26	29
Galactose	—	—	25	29
Nitrogen (× 6·38)	0·9	<0·1	7·3	0·5
Ash	0·7	0·6	0·6^a	0·5^a

^{a}After 90% of the ash is removed by ion-exchange.

TABLE 5
COMPARATIVE PROPERTIES OF LACTOSE AND 90% HYDROLYSED LACTOSE

	Lactose	Hydrolysed lactose	Ref.
Solubility (g per 100 ml) at 25°C	17	55	102
Viscosity (centipoise) of 50% solution at 25°C	17	105	103
Fermentability	Limited	Good	
Relative rate of browning	1	3·4	81
Sweetness of solution relative to sucrose	30–40[a]	65–90[a]	79, 104
Humectant properties	—	Comparable to sucrose	103

[a]A range of values depending upon concentration.

shown in Table 5. The most important differences are the increased solubility and sweetness.

The higher solubility of the monosaccharides makes it possible to prepare syrups with 55–65% total solids, which are microbiologically stable but not prone to crystallization. At higher concentrations, the galactose crystallizes out when the degree of hydrolysis is more than 75%.[79,80]

The sweetness of hydrolysed lactose syrups (relative to sucrose syrups of equal solids content) increases with concentration, as shown in Fig. 11. At high solids content, the hydrolysed lactose syrup is nearly equivalent in sweetness to sucrose[79] and is therefore sweeter than would be predicted by calculation using the relative sweetness values for the component sugars.[71] This synergistic effect of the mixed sugars on sweetness perception has also been observed in an ice-cream mix in which hydrolysed lactose was partially substituted for sucrose.[80]

Of the other changes observed, probably the most important with respect to processing applications is the increased reactivity of monosaccharides in browning and caramelization reactions. At neutral pH, glucose and galactose are 2·5 and 5 times more reactive than lactose,[81] while sucrose is essentially unreactive unless heated to high temperatures in concentrated form.

9.2.1. Lactose-hydrolysed Whey

Considered as a food ingredient, concentrated lactose-hydrolysed whey (LHW) is essentially a light brown, translucent sugar syrup containing

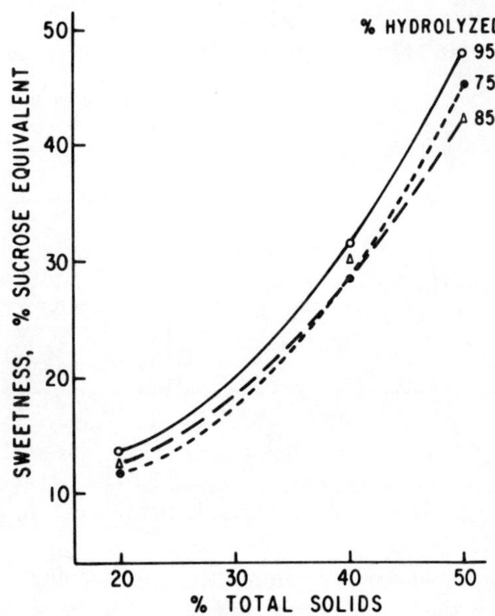

FIG. 11. Relative sweetness of hydrolysed lactose syrups and sucrose syrups. Reproduced with permission from reference (79), J. Dairy Science, (1978).

7–8% good-quality protein. Its potential uses include any product where protein (especially milk protein) is included together with sugar. Applications of LHW have been reported in a number of areas.

Ice-cream and frozen desserts. Hydrolysed whey syrup can be used to replace up to 25% of the serum solids in ice-cream, without flavour deterioration.[54,61,82] Hydrolysis of the lactose eliminates the problem of sandiness during prolonged storage and permits the use of higher levels of whey than would otherwise be possible. The result is a reduction in ingredient costs and a somewhat different texture due to the incorporation of whey proteins. There is a general tendency for ice-cream to become softer in texture as the level of hydrolysed whey is increased. This may or may not be advantageous depending upon the texture required. Studies so far reported do not agree on the extent to which hydrolysed whey must be demineralized for satisfactory incorporation into ice-cream,[54,82,83] but it appears that extensive demineralization (>70%) is not necessary in ice-cream made for the UK market.[54] Hydrolysed cottage cheese whey has also been used to make sherbets[84] and water-ices,[85] where it contributes to the acid requirement as well as providing sweetness.

Confectionery. Whey solids are already used in various confections such as toffees, caramels, fudge and Dulce de Leche, largely for the beneficial effects of the whey protein on texture.[86] Hydrolysis of the lactose permits the use of much higher levels of whey solids without risk of lactose crystallization.[87] LHW also prevents moisture loss from caramels during storage because of its properties as a humectant.[88]

Bakery products. Hydrolysed whey syrups contain native whey proteins which have good whipping properties. As a result, these syrups can partly replace egg protein, as well as replace sucrose in some bakery products.[89] Browning in baked goods is also enhanced owing to the presence of the highly reactive galactose. There are several reports that LHW can, with advantage, replace both ordinary whey solids and skim milk powder in bread making.[89-91] Among the benefits claimed are faster dough development and higher loaf volume, porosity and crumb compressibility, all due to the more readily fermentable sugar.

Fermentation products. LHW is readily fermented by yeasts (*K. fragilis* and *Saccharomyces cerevisae*) to produce ethanol,[92] but the yield and speed of fermentation are somewhat limited by the slow metabolism of galactose. Fermentation of LHW has also been used to prepare wine,[93] xanthan gum[94] and a flavouring material resembling soy sauce.[95] LHW is being used as a substrate for the growth of baker's yeast on a commercial scale in the US.[96]

9.2.2. Hydrolysed Whey Permeate

Hydrolysed whey permeate is basically a syrup of hydrolysed lactose. It is similar in sweetness to medium DE corn syrup but is less viscous. It can be used as a substitute for corn syrup in soft drinks, fermented beverages and confectionery products. Specific product applications reported to date include the following.

Fermented beverages. Hydrolysed permeate can replace corn syrup used as an adjunct in UK beers,[54] or can directly replace up to 50% of the wort in lager beer manufacture.[97,98] The permeate syrup can also be fermented directly to ethanol, using selected lactose-fermenting yeasts.

The 'beer' produced by fermentation may be distilled to provide high-quality alcohol for use in products such as vodka.[54]

Soft drinks and confectionery. Hydrolysed permeate can be used as a replacement for some or all of the corn syrup in soft drinks and some confections, but the syrup must be of the highest quality, i.e. minimum ash and protein contents, in order to avoid development of off-flavours and/or unwanted colouration. Satisfactory partial replacement of corn syrup in pectin jellies, fudge and jams has been reported,[54] and replacement of invert sugar as a canning syrup for fruits has also been successful.[99]

Some difficulties are encountered in confectionery, e.g. boiled sweets, because the syrup tends to colour rapidly during boiling owing to protein–sugar browning reactions.[81]

10. COMMERCIAL PROSPECTS

10.1. Lactose-hydrolysed Milk

The market for consumer-bought LHM is uncertain, but it is not likely to be very large. Lactose intolerance is not a major problem among the population of Western Europe, or among the white population of countries such as the US who are descended from Europeans. In most of the rest of the world, where hypolactasia is more prevalent, milk is an expensive item and the average consumer is unlikely to be willing to pay for the extra cost of LHM, even if it could be made available. The technology for making LHM is therefore likely to be used only in the highly developed countries, where the market is limited.

LHM is currently sold in Italy, and also in the US, where the major target population is black Americans who, because of their African negro ancestry, are more likely to exhibit lactose intolerance than their white compatriots. Sales of LHM in supermarkets in the US, where the product is often sold without extensive promotion, are reportedly slow but steady;[40] LHM typically accounts for about 1% of total sales of milk for the dairies that produce it.[40] Extensive promotion of the product, particularly to children who like the sweeter taste, will undoubtedly cause sales to increase. Some demand may also come from the growing geriatric population, which may be less tolerant to lactose than younger groups. Even so, it is unlikely that LHM will form a significant fraction of total liquid milk sales in the foreseeable future.

The major opportunity for LHM is as an alternative to ordinary milk in dairy foods such as cheese, ice-cream, yoghurt, etc. This substitution will occur only when the functional advantages of LHM are fully recognized and utilized, and when the cost is lowered to the point where it does not outweigh the benefits. It is the writer's opinion that this will happen, but only slowly. The rate-determining factor will be research into product applications.

10.2. Lactose-hydrolysed Whey and Permeate

The economics of producing lactose-hydrolysed syrups depend not only on the process technology but also on factors such as the volume of whey/permeate produced locally, the size of the processing plant and the location of the plant in relation to manufacturers who would want to use the syrups produced. Immobilized enzyme systems are designed for continuous operation, so a large volume of whey must be available. Large plants also lead to significant economies of scale.[100] The centralization of dairy production in Europe and the relatively small distances between major centres there help to make the process costs more attractive than elsewhere.

Hydrolysed whey and hydrolysed permeate syrups must compete directly with other sweeteners, especially corn syrup. The cost of producing hydrolysed permeate in Western Europe is in the range 31–44 cents (US) per kg, depending upon the degree of demineralization required.[53] Since corn syrup prices in Europe are in the range 44–55 cents (US) per kg,[53] the economics are attractive on this basis alone. Other advantages include improved whey utilization and savings on imports.

In the US, the outlook is not so promising since corn syrup is cheaper at 22–26 cents per kg,[53] and the costs for making hydrolysed lactose syrup are almost certainly higher than in Europe because of greater labour costs. However, there are other considerations. Whey protein has high nutritional quality and is an important food ingredient because of its functional properties. If this market grows then so will the volume of permeate, which has a high biological oxygen demand. Rather than paying disposal costs, dairy manufacturers may prefer to hydrolyse the permeate and use it as a sweetener in their own products. However, this will only be a feasible alternative for large dairy plants producing a range of products.

As with LHM, the stimulus for production of hydrolysed whey and permeate syrups will be research into product applications that exploit the special advantages of these materials.

In conclusion, the commercial outlook for hydrolysed whey and permeate syrups is quite good in Europe, where the economics are favourable, but very uncertain in the US because of competition from other sweeteners. There, as in Europe, it may well be that the most profitable use of hydrolysed syrup is fermentation to products such as ethanol.[101]

REFERENCES

1. COUGHLIN, J. R. and NICKERSON, T. A., *J. Dairy Sci.*, 1975, **58**, 169.
2. DEMAIMAY, M., LeHENAFF, Y. and PRINTEMPS, P., *Process Biochem.*, 1978, **13**(4), 3.
3. WALLENFELS, K. and WEIL, R., In: *The Enzymes*, 3rd edn, Vol. 7, P. D. Boyer (ed.), 1972, Academic Press, New York, p. 617.
4. BERGMEYER, H. U., *Principles of Enzymatic Analysis*, 1978, Verlag Chemie, New York.
5. WALLENFELS, K., SUND, H. and WEBER, K., *Biochem. Z.*, 1963, **338**, 714.
6. FOWLER, A. V. and ZABIN, I., *Proc. Nat. Acad. Sci. USA*, 1977, **74**(4), 1507.
7. WALLENFELS, K. and MALHOTRA, O. P., In: *The Enzymes*, 2nd edn, Vol. 4, P. D. Boyer, H. Lardy and K. Myrback (eds), 1960, Academic Press, New York, p. 409.
8. MAHONEY, R. R. and WHITAKER, J. R., *Food Biochem.*, 1977, **1**, 327.
9. LOOTIENS, F. G., WALLENFELS, K. and WEIL, R., *Eur. J. Biochem.*, 1970, **14**, 138.
10. HUBER, R. E., KURZ, G. and WALLENFELS, K., *Biochemistry*, 1976, **15**(9), 1994.
11. ROBERTS, H. R. and PETTINATI, J. D., *J. Agric. Food Chem.*, 1957, **5**, 130.
12. BURVALL, A., ASP, N–G. and DAHLQVIST, A., *Food Chem.*, 1979, **4**, 243.
13. GREENBERG, N. A. and MAHONEY, R. R., *Food Chem.*, 1983, **10**, 195.
14. TOBA, T. and ADACHI, S., *J. Dairy Sci.*, 1978, **61**, 33.
15. TOBA, T., YUKIHIRO, T., TAKATOSHI, I. and ADACHI, S., *J. Dairy Sci.*, 1981, **64**, 185.
16. WIERZBICKI, L. E. and KOSIKOWSKI, F. V., *J. Dairy Sci.*, 1973, **56**, 1400.
17. BURVALL, A., ASP, N–G. and DAHLQVIST, A., *Food Chem.*, 1980, **5**, 189.
18. NICKERSON, T. A., VUJICIC, I. F. and LIN, A. Y., *J. Dairy Sci.*, 1976, **59**, 386.
19. TAUBER, H. and KLEINER, I. S., *J. Biol. Chem.*, 1932, **99**, 249.
20. RICHMOND, M. L., BARFUS, D. L., HARTE, B. R., GRAY, J. I. and STINE, C. M., *J. Dairy Sci.*, 1982, **65**, 1394.
21. PIRISINO, J., *J. Food Sci.*, 1983, **48**, 742.
22. DEMAIMAY, M. and BARON, C., *Lait*, 1981, **61**, 261.
23. CHEN, S-L. Y., FRANK, F. J. and LOEWENSTEIN, M., *J. Assoc. Off. Anal. Chem.*, 1981, **64**(6), 1414.
24. ZARB, J. M. and HOURIGAN, J. A., *Aust. J. Dairy Technol.*, 1979, **34**, 184.
25. NIJPELS, H. H., EVERS, P. H., NOVAK, G. and RAMET, J. P., *J. Food Sci.*, 1980, **45**, 1684.

26. JEAN, I. J. and BASETTE, R., *J. Food Protect.*, 1982, **45**(1), 14.
27. MAHONEY, R. R. and ADAMCHUK, C., *J. Food Sci.*, 1980, **45**(4), 962.
28. GUY, E. J. and BINGHAM, E. W., *J. Dairy Sci.*, 1978, **61**, 147.
29. GREENBERG, N. A. and MAHONEY, R. R., *Process Biochem.*, 1981, **16**(2), 2.
30. GREENBERG, N. A. and MAHONEY, R. R., *J. Food Sci.*, 1982, **47**, 1824.
31. GREENBERG, N. A., PhD Thesis, University of Massachusetts, 1981.
32. GRIFFITHS, M. W. and MUIR, D. D., *J. Sci. Fd. Agric.*, 1978, **29**, 753.
33. GRIFFITHS, M. W., MUIR, D. D. and PHILIPS, J. D., US Patent 4332895, 1982.
34. MAHONEY, R. R., NICKERSON, T. A. and WHITAKER, J. R., *J. Dairy Sci.*, 1975, **58**, 1620.
35. AHN, J. K., LEE, S. H. and KIM, H. U., *Korean J. Animal Sci.*, 1981, **23**(6), 494.
36. WENDORFF, W. L., AMUNDSON, C. H., OLSON, N. F. and GARVER, J. C., *J. Milk Fd. Technol.*, 1971, **34**(6), 294.
37. WENDORFF, W. L., AMUNDSON, C. H. and OLSON, N. F., *J. Milk Fd. Technol.*, **34**(6), 377.
38. SFORTUNATO, T. and CONNORS, W. M., US Patent 2826502, 1958.
39. KOSIKOWSKI, F. V. and WIERZBICKI, L. E., *J. Dairy Sci.*, 1973, **56**(1), 146.
40. BANNAR, R., *Food Engineering*, 1980, **52**(7), 30.
41. DAHLQVIST, A., Asp, N–G. and RAUSING, H., *J. Dairy Res.*, 1977, **44**, 541.
42. ANON., *Process Biochem.*, 1981, **16**(2), 36.
43. MILLER, J. J. and BRAND, J. C., *Food Technol. in Australia*, 1980, **32**(3), 144.
44. WECKSTROM, L., LINKO, Y–Y. and LINKO, P., In: *Food Process Engineering*, Vol. 2, *Enzyme Engineering in Food Processing*, P. Linko and J. Larinkari (eds), 1980, Applied Science Publishers, London, p. 148.
45. FINOCCHARIO, T., OLSON, N. F. and RICHARDSON, T., *Adv. Biochem. Eng.*, 1980, **15**, 71.
46. WONDOLOWSKI, M. V., In: *Proceedings of 3rd International Biodegradation Symposium*, J. M. Sharpley and A. M. Kaplan (eds), 1976, Applied Science Publishers, London, p. 1033.
47. DINELLI, D., *Process Biochem.*, 1972, **7**(8), 9.
48. DINELLI, D., MARCONI, W. and MORISI, F., *Methods Enzymol.*, 1976, **44**, 227.
49. MORISI, F., PASTORE, M. and VIGLIA, A., *J. Dairy Sci.*, 1973, **56**, 1123.
50. PASTORE, M. and MORISI, F., *Methods Enzymol.*, 1976, **44**, 822.
51. WEETALL, H. H. and HAVEWALA, N. B., In: *Enzyme Engineering*, L. B. Wingard (ed.), 1972, Wiley, New York, p. 249.
52. DOHAN, L. A., BARET, J., PAIN, S. and DELALANDE, P., In: *Enzyme Engineering*, Vol. 5, H. H. Weetall and G. P. Royer (eds), Plenum, New York, p. 279.
53. MOORE, K., *Food Prod. Devel.*, 1980 **14**(1), 50.
54. COTON, S. G., POYNTON, T. R. and RYDER, D., Internat. Dairy Federation Document 147, Proceedings of Seminar on Dairy Ingredients, 1981, Brussels, p. 23.
55. OLSEN, A. C. and STANLEY, W. L., *J. Agric. Fd. Chem.*, 1973, **21**(3), 440.
56. HARJU, H., HEIKONEN, M., KREULA, M. and LINKO, M., In: *Food Process Engineering*, Vol. 2, *Enzyme Engineering in Food Processing*, P. Linko and J.

108   R. R. MAHONEY

Larinkari (eds), 1980, Applied Science Publishers, London, p. 133.
57. MARCONI, W., BARTOLI, F., MORISI, F. and MARIANI, A., In: *Enzyme Engineering*, Vol. 5, H. H. Weetall and G. P. Royer (eds), 1980, Plenum, New York, p. 269.
58. RUGH, S., *Appl. Biochem. Biotechnol.*, 1982, **7**, 27.
59. OHMIYA, K., HIROSHI, O., KABAYASHI, T. and SHIMIZU, S., *Appl. Environ. Microbiol.*, 1977, **33**, 137.
60. PETRE, D., NOEL, C. and THOMAS, D., *Biotechnol. Bioeng.*, 1978, **20**, 127.
61. GUY, E. J., TAMSMA, A., KONTSON, A. and HOLSINGER, V. H., *Food Prod. Devel.*, 1974, **8**(8), 50.
62. REIMERDES, E. H., Internat. Dairy Federation Document 147, Proceedings of Seminar on Dairy Ingredients, 1981, Brussels, p. 27.
63. DLUZEWSKI, M., WACSZKOWSKI, W. and BARANSKY, S., *Technoligia Rolno Spozwcza* (Warszawa), 1978, No. 12, 103 (*Food Sci. Technol. Abstr.*, 1980, 4P692).
64. TUMERMAN, L., FRAM, H. and CORNELY, K. W., *J. Dairy Sci.*, 1954, **37**, 830.
65. HOLSINGER, V. H. and ROBERTS, N. E., *Dairy and Ice Cream Field*, 1976, **159**(3), 30.
66. WOYCHIK, J. H. and HOLSINGER, V. H., In: *Enzymes in Food and Beverage Processing*, R. L. Ory and A. J. St. Angelo (eds), 1977, American Chemical Society, Washington, p. 67.
67. O'LEARY, V. S. and WOYCHIK, J. H., *J. Food Sci.*, 1976, **41**, 791.
68. O'LEARY, V. S. and WOYCHIK, J. H., *Appl. Environ. Microbiol.*, 1976, **32**, 89.
69. HILGENDORF, M. J., *Cultured Dairy Prod. J.*, 1981, **16**(1), 5.
70. HOLSINGER, V. H., *Food Technol.*, 1978, **32**(3), 35.
71. NIJPELS, H. H., In: *Developments in Food Carbohydrates*, Vol. 3, C. K. Lee and M. G. Lindley (eds), 1982, Applied Science Publishers, London, p. 23.
72. PERGAM, D. R. and PILGRIM, F. J., *Food Technol.*, 1957, **11**(9), insert 9–14.
73. Anon., *Dairy and Ice Cream Field*, March 1977, 66K.
74. GYURICSEK, D. M. and THOMPSON, M. P., *Cultured Dairy Prod. J.*, 1976, **12**, 13.
75. THOMPSON, M. P. and BROWER, D. P., *Cultured Dairy Prod. J.*, 1976, **11**, 22.
76. GILLIARD, S. E., SPECK, M. L. and WOODARD, Jr, J. R., *Appl. Microbiol.*, 1972, **23**(1), 21.
77. LABUSCHAGNE, J. H. and NIEWOUDT, J. A., *S. African J. Dairy Technol.*, 1978, **10**(22), 77.
78. WEAVER, J. C., KROGER, M. and THOMPSON, M. P., *J. Food Sci.*, 1978, **43**, 579.
79. GUY, E. J. and EDMONDSON, L. F., *J. Dairy Sci.*, 1978, **61**, 542.
80. SHAH, N. O. and NICKERSON, T. A., *J. Food Sci.*, 1978, **43**, 1085.
81. POMERANZ, Y., JOHNSON, J. A. and SHELLENBERGER, J. A., *J. Food Sci.*, 1962, **27**, 350.
82. LOEWENSTEIN, M., REDDY, M., WHITE, C., SPECK, S. and LUNSFORD, T., *Food Prod. Devel.*, 1975, **9**(9), 91.
83. HARJU, M. and KREULA, M., In: *Carbohydrate Sweeteners in Foods and Nutrition*, P. Koivistoinen and L. Hyvonen (eds), 1980, Academic Press, New York, p. 233.

84. WHITTIER, E. O. and WEBB, B. H., *Byproducts from Milk*, 1950, Reinhold, New York, p. 223.
85. GUY, E. J., VETTEL, H. E. and PALLANSCH, M. J., *J. Dairy Sci.*, 1966, **49**, 1156.
86. WEBB, B. H., *J. Dairy Sci.*, 1966, **49**, 1310.
87. CASTELAO, E., PERIN, O., REYNA, R., SBODIO, O., BACHETTA, R., ESCUDERO, J. and PAULETTI, M., *Ind. Lech.*, 1981, **61**, 10.
88. GUY, E. J., *J. Food Sci.*, 1978, **43**, 980.
89. Maxilact Technical Bulletin, G. B. Fermentation Industries, Des Plaines, Illinois, USA.
90. SHOGREN, M. D., POMERANZ, Y. and FINNEY, K. F., *Cereal Chem.*, 1979, **56**(5), 465.
91. DROBOT, V. I., KUDRYA, V. A. and SHCHELOKVA, I. F., *Pishchevaya Promyshlennost*, 1981, No. 4, 33 (*Food Sci. Technol. Abstr.*, 1983, 6M938).
92. O'LEARY, V. S., GREEN, R., SULLIVAN, B. C. and HOLSINGER, V. H., *Biotechnol. Bioeng.*, 1977, **19**, 1019.
93. ROLAND, J. F. and ALM, W. L., *Biotechnol. Bioeng.*, 1975, **17**, 1443.
94. COUGHLIN, R. W. and CHARLES, M., In: *Immobilised Enzymes for Food Processing*, W. H. Pitcher, Jr (ed.), 1980, CRC Press, Boca Raton, Florida, p. 153.
95. LUCAS, A. J., US Patent 3558328, 1971.
96. Anon., *Food Prod. Devel.*, 1982, **16**(1), 34.
97. HARJU, M., HEIKONEN, M., KREULA, M., PAJUNEN, E. and LINKO, M., *Karjantuote*, 1978, **61**(9), 4.
98. POZNANSKI, S., LEMAN, J., BEDNARSKI, W., SZMELICH, W., KOWALEWSKA, J., CHODKOWSKI, M. and WIELICZKO, R., *Nährung*, 1978, **22**, 275.
99. TWEEDIE, L. S. and MACBEAN, R. D., Proceedings, 20th Internat. Dairy Congress, 1978.
100. PITCHER, Jr, W. H., Proceedings of the Whey Products Conference, Chicago, 1974, USDA–ERRC, Philadelphia, p. 104.
101. HANSEN, R., *Nordeurop. Mej. Tidsskr.*, 1980, **46**(5), 119 (*Food Sci. Technol. Abstr.*, 1981, 9P1660).
102. TALLEY, E. A. and HUNTER, A. S., *J. Am. Chem. Soc.*, 1952, **74**, 2789.
103. SHAH, N. O. and NICKERSON, T. A., *J. Food Sci.*, 1978, **43**, 1081.
104. AMERINE, M. A., PANGBORN, R. M. and ROESSLER, R. B., *Principles of Sensory Evaluation of Foods*, 1965, Academic Press, New York, p. 90.
105. HEMME, D., VASSAL, L., FOYEN, H. and AUCLAIR, J., *Lait*, 1979, **59**, 589.
106. MARSCHKE, R. J., NICKERSON, D. E. J., JARRETT, W. D. and DULLEY, J. R., *Aust. J. Dairy Technol.*, 1980, **35**, 84.
107. GRIEVE, P. A., KITCHEN, B. J., DULLEY, J. R. and BARTLEY, J., *J. Dairy Res.*, 1983, **50**, 469.

Chapter 4

NUTRITIONAL SIGNIFICANCE OF LACTOSE: I. NUTRITIONAL ASPECTS OF LACTOSE DIGESTION

DAVID M. PAIGE and LENORA R. DAVIS

*Department of Maternal and Child Health,
School of Hygiene and Public Health,
Johns Hopkins University, Baltimore, Maryland, USA*

1. INTRODUCTION

Low levels of intestinal lactase activity occur in many otherwise healthy adults and children in various population groups. Current evidence indicates that these low levels are normal for most populations of the world, notable exceptions being peoples of northern European extraction: approximately two-thirds of the world's adult population are believed to be lactose malabsorbers. Individuals who have genetically acquired low lactase levels as adults were able to drink milk without ill-effects as infants.[1,2]

The onset of acquired lactose malabsorption depends on the population studied; in developing countries, as data from Peru indicate, 50% of the population is intolerant by age 3 years,[3] but in more technologically developed countries, such as the USA, 54% of the black population exhibits lactose malabsorption before the age of 10 years.[4,5] The difference in the rate of loss of enzyme activity between children in developing areas and blacks in the USA may reflect a relatively better environmental and/or nutritional status of the latter which may serve to delay the onset of lactose malabsorption.[6]

2. LACTOSE DIGESTION

Clinically identifiable lactose malabsorption evolves over time and may be classified into three distinct periods.

The initial period is characterized by a decline in lactase activity through childhood. This decline in lactase is measured by the use of a proxy indicator, i.e. peak blood glucose rise, which is observed to decrease progressively throughout childhood and adolescence, following a lactose challenge.

The second phase in this continuum is noted when the decline in lactase activity is sufficient to result in a level of undigested lactose quantitatively high enough to cause symptoms associated with the ingestion of a lactose load.

The third phase occurs when the level of lactase declines to the point that it is no longer sufficient to digest even the modest levels of lactose usually consumed. This may lead to milk rejection or milk intolerance, and/or the presence of symptoms after consumption of a glass of milk.

The events are staged and progressive with age. Therefore, as lactase levels continue to decline, individuals become symptomatic with increasing age as a result of the undigested lactose.[7] When the decline is available lactase activity with age reaches a clinical threshold, even higher levels of undigested lactose reach the colon, resulting in greater fermentation and the increased probability of symptom manifestation. Initially this occurs with a high lactose load, followed over time by an inability to digest usually consumed levels of lactose.[8] Symptoms in older children are significantly associated with decreasing lactase activity, incomplete lactose hydrolysis and the resultant higher levels of lactose reaching the colon.[7]

3. AETIOLOGY

Intestinal lactase activity is normally high in mammals during the suckling period but decreases to low levels following weaning. Although early studies on adults of north European origin suggested that humans are an exception to this pattern of mammalian development, subsequent work has shown that most groups of human adults have low intestinal lactase activity and, in fact, do conform to the usual mammalian ontogenetic pattern.

Despite difference in timing, primary lactose malabsorption is considered to be a genetically controlled non-sex-linked autosomal recessive event.[9] The genetic hypothesis has been supported by several recent studies of the prevalence of lactose absorption/malabsorption in racially mixed populations, and, more directly, from family pedigrees of diverse ethnic groups, including Finns, West Africans and American Indians. The family studies suggest that lactose absorption in the human adult, the unusual 'phenotype' on a world-wide basis, is inherited as a completely penetrant autosomal dominant characteristic. Johnson *et al.*,[10] reporting on Papago Indians in the southwestern United States, noted that the results of their family studies were consistent with this mode of inheritance; the observed and the expected proportion of lactose malabsorbers were strikingly similar.

There is little evidence to support the adaptive hypothesis in humans. The induction hypothesis, which postulates that lactase activity is regulated by the concentration of lactose in the diet, has not been substantiated. Nevertheless, adaptive regulation of lactase activity may occur to a small degree, at least in animals, but only over a relatively narrow, genetically determined range.

4. SYMPTOMS

Most individuals with a low lactase level are not capable of digesting a lactose load of 50 g per m^2 of body surface; almost all subjects will have a peak blood sugar rise of less than 20 or 25 mg dl^{-1}, and most will report symptoms when subjected to a lactose challenge (Fig. 1). Occasionally, individuals with normal jejunal lactase levels also report symptoms to a lactose challenge.[11] With a total lactose dose of 50 g, equivalent to approximately 1 litre of milk, a lesser number, perhaps 90–95%, will have a flat blood sugar curve and 85–90% will have symptoms. At this dosage there will be some individuals with low lactase levels who will not be symptomatic.[12] Breath hydrogen rise following a lactose challenge provides additional important information on lactose digestion. Utilizing a breath hydrogen rise of over 20 ppm as evidence that undigested lactose has reached the colon, 100% of persons with lactose intolerance are said to have a positive test.[13] When breath hydrogen and lactose tolerance tests (blood glucose rise) are done at the same time, there is generally a good correlation between the increase in breath hydrogen over 20 ppm and blood sugar rise under 20 mg dl^{-1}. It is worth noting, however, that some subjects who have a low blood sugar rise and show symptoms

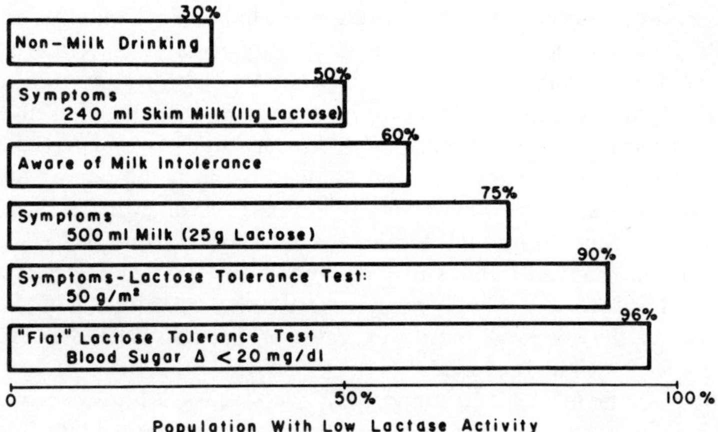

FIG. 1. The consequences of low intestinal lactase activity in adults, given as the frequency of responses to varying amounts of milk and lactose (from Ref. 63).

during the tolerance test may not experience a breath hydrogen rise over 20 ppm. These individuals, who total approximately 5% of those tested, have falsely negative breath hydrogen tests as a measure of lactose tolerance.

4.1. Symptoms in Children

The association of symptoms with non-digestion of lactose has also been studied in children.[5] Complete data on symptomatic response to a lactose load were recorded in 378 of 409 US black subjects on whom blood glucose rise information was available; 31 subjects on whom information was incomplete were excluded but these were not identifiably different from the remainder of the children. In children less than 4 years old there was no difference in peak blood glucose rise in the 16 subjects with symptoms and the 97 subjects without symptoms following a lactose challenge.[5] Similarly, no significant differences were found in the 27 symptomatic and 125 non-symptomatic subjects aged 4–7 years.

At 8 years of age, however, this pattern changed. Results indicated a significantly lower peak blood glucose rise ($13.6 \, \text{mg dl}^{-1}$) in the 40 symptomatic black subjects tested with a lactose load, compared to a blood glucose rise of $24.6 \, \text{mg dl}^{-1}$ in the 73 black children who were asymptomatic. The data suggest that the steady decline in available lactase reaches an identifiable clinical threshold by 8 years of age. These

TABLE 1

SYMPTOMS RELATED TO LEVEL OF BLOOD GLUCOSE RISE, BY AGE GROUPS, IN BLACK SUBJECTS[8]

Blood glucose rise (mg dl^{-1})	<4 years[a]		$\geqslant 8$ years[b]	
	No symptoms	Symptoms	No symptoms	Symptoms
<15	12	3	20	28
	(12%)	(19%)	(27%)	(70%)
15–25	14	3	23	7
	(14%)	(19%)	(32%)	(18%)
$\geqslant 26$	71	10	30	5
	(74%)	(62%)	(41%)	(12%)
Total subjects	97	16	73	40

[a] $X^2 = 0.81$; p is not significant.
[b] $X^2 = 19.77$; $p < 0.001$.

changes result in a pattern of increasing reports of symptoms with a lactose load (Table 1).

Despite the randomness in reported symptoms in children in the youngest age group, symptoms are significantly associated with a continued decline in blood glucose rise in older children. In 78 children $\geqslant 8$ years old, 35 (45%) reported symptoms with a blood glucose rise of <26 mg dl^{-1}, and 28 of the 35 children reporting symptoms (80%) had a peak blood glucose rise of <15 mg dl^{-1}. This significant association suggests a relationship between decreasing lactase activity, incomplete lactose digestion, and symptoms with increasing age (Table 2).

TABLE 2

PERCENTAGES OF BLACK CHILDREN REPORTING SYMPTOMS AFTER LACTOSE INGESTION, BY AGE[8]

	<4 years[a]	4–7 years[b]	$\geqslant 8$ years[c]	All ages
Malabsorbers	19% (32)	24% (63)	45% (78)	32% (173)
Absorbers	12% (81)	13% (89)	14% (35)	13% (205)

Figures in parentheses indicate the number of children in each group.
[a] $X^2 = 0.34$; p is not significant.
[b] $X^2 = 2.03$; p is not significant.
[c] $X^2 = 8.59$; $p < 0.001$.

5. MILK INTOLERANCE

Data on intolerance to commonly consumed amounts of milk have been reported. Studies using a 12 g dose of lactose are most relevant to clinical situations, since a 240 ml glass of whole or skim milk contains ~ 12 g of lactose. In one study,[12] 12 g of lactose in an electrolyte solution equivalent to the electrolyte concentration in milk caused symptoms in 15 of 20 adult subjects with confirmed (by biopsy) low lactase levels.

Bayless *et al.*[14] reported that 68% of 44 randomly selected lactose-intolerant male adults reported symptoms following a dose of 12 g of lactose in an electrolyte solution, while only 1 subject (2%) reported symptoms with an equivalent solution of the monosaccharide components of lactose, glucose and galactose. When an amount of skim milk containing 11 g of lactose was consumed, 59% of the lactose-intolerant men were symptomatic, while none of the 27 lactose-tolerant controls reported symptoms.[14] Gudmand-Hoyer and Simony,[15] who carried out similar studies on a double-blind basis, reported almost identical results; they gave a dose of reconstituted milk powder containing 10 g of lactose to 13 lactose-intolerant adults on a double-blind basis, 75% of whom were symptomatic. In a study by Jones *et al.*,[16] 69% of lactose-intolerant adult subjects were symptomatic to 15 g of lactose in skim milk. In the above studies the dominant symptoms noted by the lactose-intolerant subjects after drinking 240 ml of milk or 12 g of lactose in an electrolyte solution were flatulence, distension, or cramps; diarrhoea was not a prominent symptom at this lactose level. Some subjects were symptomatic to as little as 3 or 6 g of lactose (Fig. 2).[12]

Breath hydrogen excretion levels after ingestion provide an objective measure of inadequate lactose digestion.[17] In a study utilizing 240 ml of skim milk, containing 11 g of lactose, 40% of lactose-intolerant adults had a peak breath hydrogen rise of >20 ppm. If a rise of over 10 ppm is considered an indicator that unabsorbed lactose has reached the colon, then 58% of the subjects reacted positively to 240 ml of skim milk. Solomons *et al.*[18] also have reported that $>50\%$ of lactose-intolerant subjects tested had a rise of >20 ppm in breath hydrogen following ingestion of a glass of milk. Earlier work by Levitt and Donaldson[19] also had suggested that $\sim 50\%$ of the lactose-intolerant subjects tested had a rise in breath hydrogen after consuming 240 ml of milk. These results provide objective evidence of incomplete lactose digestion in usually consumed quantities of milk by 50% of lactose-intolerant adults.

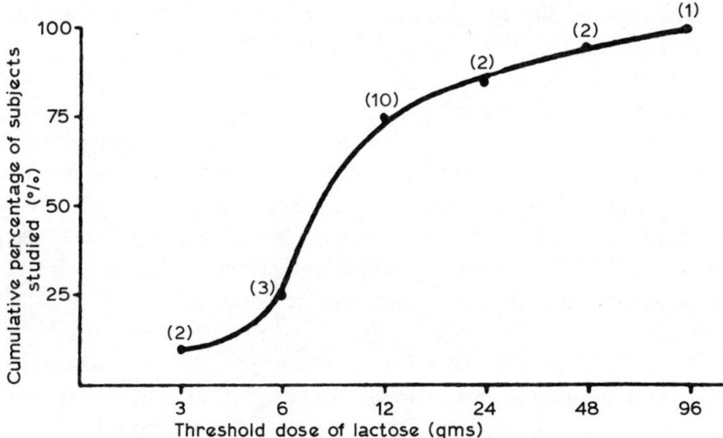

FIG. 2. Threshold doses of lactose in 20 subjects with low intestinal lactase levels (from Ref. 12).

5.1. Symptoms of Milk Intolerance

The level of ingested lactose that will cause symptoms in persons with low lactase levels varies. Some individuals may note symptoms of intolerance after one glass (240 ml) of milk, while others must drink three or four glasses before noting symptoms. It should be stressed that not everyone who has low lactase levels and develops symptoms after a lactose tolerance test load ($50\,\mathrm{g\,m}^{-2}$) will be aware of symptoms after drinking 240 ml of milk (12 g of lactose).

Distinctions must be made between lactose malabsorption, which is the incomplete hydrolysis of lactose without apparent symptoms, lactose intolerance, which is the presence of the symptoms described above along with malabsorption, and milk intolerance, which is the occurrence of the above symptoms following consumption of a moderate amount of milk, usually $\geqslant 240$ ml. It is possible to note one or all of the above manifestations in a single individual.

Incomplete absorption of some nutrients may take place despite the absence of overt symptoms while, conversely, total absorption of the nutrients in milk may occur with individuals experiencing symptoms. The occurrence of symptoms increases with age among both absorbers and malabsorbers. Data reported by Mitchell et al.[20] indicate that over 50% of 13 teenagers experiencing a flat tolerance curve and symptoms with a test dose of 50 g of lactose also experienced symptoms with 12 g of lactose in electrolyte solution or with 240 ml of milk.

5.2. Patterns of Milk Intake

A low level of milk consumption has been implicated as the cause of low lactase levels. Yet, milk drinking does not seem to affect lactase levels. Although some lactase-deficient individuals can gradually increase their milk intake, prolonged lactose or milk feeding does not increase lactase activity. On the other hand, prolonged abstinence from lactose by those individuals destined to have high lactase levels does not appear to diminish lactase activity; 9 white children with galactosaemia (7–17 years old) had normal lactose tolerance tests, despite their avoidance of all lactose-containing foods since early infancy.[61] Furthermore, in adults, complete lactose deprivation for 42 days was not associated with a significant change in lactase activity or in lactose tolerance.[20a]

Since there is a gradient of symptoms to lactose ingestion, the fact that some individuals are intolerant to a lactose load equivalent to 3–4 glasses of milk does not necessarily mean that they will experience symptoms to smaller quantities of milk. However, there is evidence that, even though intolerant subjects may not manifest symptoms with this smaller quantity of lactose, it is still not being digested. Subjects are less symptomatic if milk is taken with other foods rather than alone. The food may possibly dilute the milk or it may delay gastric emptying, thus permitting only small amounts of lactose to enter the small intestine at any one time.[63]

Simoons[21] supports the genetic explanation for lactose malabsorption, the roots of which lie in the 'traditional zone of non milk'. Those ethnic groups with a low incidence of lactose malabsorption, i.e. northwestern Europeans and certain East-African pastoral groups, have come from cultures that have traditionally consumed milk. Lactose tolerance has evolved as a result of dairying and significant changes in the diets of these groups. In the hunting and root-gathering stage prior to dairying, the total human population, like other mammals, probably experienced a decline in lactase at the time of weaning.

In dairying populations, those unique individuals who maintained a high level of lactase were able to drink milk comfortably during periods of restricted food supply. As a result of this alternative source of adequate nutrition, these individuals enjoyed an increased chance of survival. In a classical Mendelian way, the lactose absorbers would 'win out' over others. Estimates of gene selection intensities of 1–2% make this hypothesis entirely tenable. Flatz and Rotthauwe[22] suggest further that this selective pressure may have been at its maximum in northern Europe; this cloudy region would have had a high incidence of vitamin D deficiency and osteomalacia, and therefore the ability to drink milk and the enhanced effect in humans of dietary lactose hydrolysis on calcium absorption may have influenced selection.

5.3. Age of Onset

As a result of the age-specific decline in lactase levels, milk drinking patterns have been reviewed to determine if there is a change in milk consumption practices with age.[5] Information on levels of milk drinking by US black children was obtained by history from caretakers of 102 subjects below 5 years of age and by observation during the school lunch period on 160 low-income children at and above 5 years of age. A milk drinker was defined as an individual who consumed $>50\%$ of a 240 ml glass of milk per 24 h. This was determined by caretaker recall for children <5 years of age, and by direct observation and weighing of the milk carton at the end of the school lunch period on two independent occasions in subjects 5 years of age and above.

Approximately 10% of US black children <8 years old failed to meet the above criterion and were designated non-milk-drinkers. This level was similar whether information was obtained by recall or direct observation. At 8 years of age, over 20% of the observed black children were classified as non-milk-drinkers.[23] The pattern of milk consumption by 49 low-income US white children in the first three grades of the participating school differed from the pattern for black children. On one observation, only two white children (4%) failed to consume $>50\%$ of the 240 ml of milk offered. Two additional observations carried out on the same population of white children showed that only one child did not consume the designated amount of milk.

This overall proportion of black children rejecting milk parallels the level of milk rejection by lactose-intolerant school children reported earlier.[8] Furthermore, the number of children above 8 years of age who incompletely consume the 240 ml of milk provided in the school lunch approximates the expected number of malabsorbing children at this age with a maximum blood glucose rise of less than $15 \, \text{mg} \, \text{dl}^{-1}$. These data, when compared to previously reported prevalence data and milk rejection patterns, suggest an association between blood glucose rise following a lactose test and milk rejection.[24]

Johnson *et al.*[25] report an increasing prevalence of lactose malabsorption with age and an association with milk intolerance. Among full blooded Pima subjects, the proportions of lactose malabsorbers were 40, 71, 92 and 100% in the age groups 3–4, 4–5, 5–7 and >8 years, respectively. The incidence of symptoms associated with the test increased from 58% of subjects less than 12 years old to 74% of those in the 12–18 years age group, and to 91% of those 18 years or above. Recognition by malabsorbers of symptoms associated with the consumption of milk products also increased from 4% in subjects less than 12 years old to 7% in the 12–18 years group and 68% in subjects 18 years and older.

Data presented by Garza and Scrimshaw[26] support the inference that declining levels of lactase may have little or no clinical consequence in populations of healthy children less than 8 years old but have increasing clinical implications above 8 years of age. The authors report: (1) an increase in the age-specific prevalence of lactose intolerance, ranging from 11% in 4–5 year old blacks to 50% at 6–7 years of age and 72% in 8–9 year olds; (2) no differences between the milk intakes of 5–7 year old black and white children; (3) an increase in symptom response to graded amounts of lactose in peanut butter and whole cow's milk, with an increase in age; and (4) symptoms in 10% of blacks above 8 years old to 12 g of lactose when consumed with peanut butter.

Lisker *et al.*[27] report that, in rural and urban Mexican children aged 5–14 years, no less than 15% had gastrointestinal complaints after ingesting 240 ml of regular milk and that lactose-hydrolysed milk was tolerated significantly better than other types of milk. Among urban children, who had less chronic gastrointestinal problems and better general environmental conditions, 16·8% showed symptoms with regular milk and 3·9% with lactose-free milk. Rural children, who had a high incidence of poor nutrition, diarrhoea, and parasitosis, showed severe symptoms almost twice as often with regular milk as with lactose-free milk. Despite general agreement on the data reported above, only limited agreement has been found on the effect of lactose intolerance on milk drinking patterns.

5.4. Clinical Differences in Milk Intolerance

Reported differences in milk drinking patterns may be due in part to the fact that loss of lactase activity is not an all-or-none phenomenon. There appears to be a steady decline in available lactase activity with age, resulting in a decreasing ability to handle even the amounts of lactose usually consumed. This decline may be modified further by other foods consumed along with a 'usual lactose load'. While a smaller than usual lactose load may not provoke threshold-level symptoms in a subject, or the subject may not recognize the association, there are data that suggest that lactose is not completely hydrolysed and that carbohydrate absorption is altered in such cases.[18]

Furthermore, comparisons are made as if the age of the subjects were the same or similar, when in fact the age of subjects reported on are frequently quite disparate and preclude age-specific comparisons. Another factor contributing to the difficulty and confusion in interpreting the association between symptoms and milk drinking is the practice of attempting to quantify the reported symptoms. Many investigators have superimposed upon the patient's subjective reports their own subjective decisions regarding which symptoms, how many symptoms, and to what degree symptoms are considered sufficiently important to be included as present or absent. Attempting to quantify, grade, or score an individual's subjective feeling of discomfort by yet another set of subjective interpretations on the part of the investigator may distort further a very imprecise outcome measure. It may be better to report the individual's recognition of the presence or absence of a threshold level of discomfort following ingestion of the test product during or immediately following the test period. Attempts at quantitative or semi-quantitative statements regarding symptoms associated with lactose ingestion do not appear to increase the reliability or validity of such reports.

In addition, it would be useful if a specific standard were used for test products. Since the quality and quantity of test products differ, the literature remains quite confusing and it is almost impossible to compare results from two or more studies. Comparisons are made and inferences are drawn from reports in the literature as if the tests were similar, even when different milks, such as chocolate, skim or whole milk, were used in the studies being compared.[28]

An example of this confusion is illustrated by the reports of Kwon *et al.*[29] Inferences were drawn about levels of milk consumption and associated symptoms in teenagers although the 'milk' used was a synthetic chocolate-flavoured product. However, Bricker *et al.*[30] noted that college students experienced several side effects, such as nausea and cramps, following the consumption of 1 ounce of cocoa. This may help to explain why Kwon *et al.*[29] found no statistically significant differences in the incidence of symptoms reported by malabsorbers and absorbers after drinking 240 ml of a synthetic chocolate milk containing lactose.

6. NUTRIENT ABSORPTION

In addition to the inadequate rise in blood glucose, the absorption of other nutrients may also be affected by lactose ingestion by lactose malabsorbers. When the results of short-term balance studies on 4 lactose-intolerant and 2 lactose-tolerant Peruvian children were analysed, significant differences in the efficiency of absorption of selected nutrients were observed.[30a] The children were placed on a diet of casein, cottonseed oil and carbohydrate as sucrose or lactose. There was increased loss of water, nitrogen, potassium and sodium in the faeces when the lactose-intolerant subjects were switched from a 9-day period on sucrose to a 9-day period on lactose. Values for these parameters returned to the control levels when sucrose replaced lactose in the diet.

Leichter and Tolensky[31] conducted balance studies on post-weanling rats to assess the effects of dietary lactose on the absorption of protein, fat and calcium over a 10-day period. The experimental diets included either 10% or 30% of lactose, while the control diets contained equivalent amounts of sucrose. The authors report higher faecal losses of fat and nitrogen in the lactose groups (Tables 3 and 4). Faecal calcium, however, was lower in the lactose groups but significant differences were observed only in the case of 30% lactose diet (Table 5). Rats fed the 30% lactose diet were also observed to have significantly higher urinary calcium excretion. This may have resulted from the increased absorption of calcium in this group. The results suggest the possibility that high levels of dietary lactose in subjects with lactase deficiency may reduce the absorption of protein and fat, but not of calcium.

TABLE 3

FAECAL FAT EXCRETION OF RATS ON DIETS CONTAINING LACTOSE OR SUCROSE[31]

Dietary group	Faecal fat	p
10% Lactose	3.75 ± 0.94	
		< 0.01
10% Sucrose	2.30 ± 0.14	
30% Lactose	4.43 ± 0.72	
		< 0.05
30% Sucrose	3.51 ± 0.55	

Values refer to mean $\pm$ SD for six rats.

TABLE 4

FAECAL NITROGEN EXCRETION OF RATS ON DIETS
CONTAINING LACTOSE OR SUCROSE[31]

Dietary group	Faecal nitrogen	p
10% Lactose	5.02 ± 0.91	
		< 0.05
10% Sucrose	3.71 ± 0.55	
30% Lactose	6.11 ± 0.92	
		< 0.01
30% Sucrose	4.22 ± 0.48	

Values refer to mean $\pm$ SD for six rats.

TABLE 5

FAECAL CALCIUM EXCRETION OF RATS ON DIETS
CONTAINING LACTOSE OR SUCROSE[31]

Dietary group	Faecal calcium	p
10% Lactose	45.12 ± 13.15	
		> 0.05
10% Sucrose	51.42 ± 4.10	
30% Lactose	28.20 ± 6.10	
		< 0.001
30% Sucrose	56.07 ± 7.03	

Values refer to mean $\pm$ SD for six rats.

Since blood urea levels increase in proportion to the amount of protein in a meal,[32] attempts to determine protein absorption using the maximum rise of plasma urea levels as an index of protein absorption have been carried out in lactose-tolerant and -intolerant adults.[32a] Following an overnight fast, a drink of 50 g of lactose and 55 g of gelatin in 400 ml of water was given to 7 lactose-tolerant and 5 lactose-intolerant subjects. Venous blood samples were drawn at 0, 60, 120, 180 and 240 min after ingestion. The procedure was repeated 1 week later on the same subjects with 50 g of sucrose replacing the lactose in the drink. No significant differences in peak plasma urea levels in either the lactose-tolerant or -intolerant groups were reported (Table 6). It is of interest that none of the subjects reported symptoms when lactose was consumed with gelatin.

TABLE 6

MEAN MAXIMUM RISE IN PLASMA UREA AFTER GELATIN AND SUGAR INGESTION IN LACTOSE-TOLERANT AND -INTOLERANT SUBJECTS[32a]

	Plasma urea (mg/100 ml)		
	Lactose	Sucrose	p
Lactose-tolerant subjects $(N = 7)$	11.7 ± 5.57	10.3 ± 2.78	> 0.05
Lactose-intolerant subjects $(N = 5)$	9.8 ± 3.16	10.5 ± 5.03	> 0.05
p	> 0.05	> 0.05	

Subjects were given 55 g of gelatin and 25 000 IU of vitamin A with either 50 g of lactose or 50 g of sucrose.

The influence of lactose on the absorption of vitamin A and ascorbic acid in lactose-intolerant adults was studied following a similar protocol.[32a] No significant differences in the peak plasma ascorbic acid levels between the tolerant and intolerant adult subjects were observed whether ascorbic acid was ingested with lactose or sucrose.

7. LONG-TERM NUTRITIONAL CONSEQUENCES

In his review, Newcomer[33] notes that, despite attention to the problem of lactose intolerance and its immediate side-effects, potential long-term complications have received little attention. One effect described by Birge et al.[34] is that 9 of 19 osteoporotic subjects had lactase deficiency, whereas all 13 of the control subjects had 'normal' lactase levels. At the time this study was carried out, it was not fully appreciated that the prevalence of lactase deficiency varies widely among ethnic groups and hence the authors made no mention of ethnic background.

A similar study was conducted in 1978 at the Mayo Clinic on subjects of known ethnic background.[35] Lactase deficiency was determined by breath hydrogen analysis on 30 women with idiopathic post-menopausal osteoporosis and on 31 female control subjects with no evidence of metabolic bone disease. Results indicated that 8 subjects with osteoporosis were lactase-deficient compared to only one control subject

($p < 0.05$). Despite the fact that the lactase-deficient subjects were unaware of milk intolerance, their reported intakes of both calcium and lactose were significantly lower than those of the 'lactase normal' group. Additional reported relationships between lactase deficiency and osteoporosis provide further support. Kocian *et al.*[36] report that the cortical thickness of the clavicle in subjects after gastric resection was significantly less in subjects with lactase deficiency than in those with sufficient lactase.

Nevertheless, investigations into the role of lactose on the absorption of calcium have resulted in conflicting reports, both in animals and in man. Debongnie *et al.*[37] observed that 4 subjects with low levels of lactase did not absorb calcium contained in regular milk as well as calcium in lactose-free milk. It has been observed that blacks, who are known to have low levels of lactase, have a low prevalence of osteoporosis which appears to argue against a simple cause-and-effect relationship between hypolactasia and osteoporosis.[38] However, the greater bone density in blacks, with thicker cortices at maturity, may afford protection from osteoporosis later in life, despite the presence of a high prevalence of low lactase activity.[38] A number of questions in this important area of study remain to be resolved.

Many people with hypolactasia who show symptoms after consuming physiologic amounts of milk have unconsciously recognized the relationship between milk intake and intestinal symptoms and have decreased milk intake to the tolerance level. An additional area of interest is how commonly occult lactase deficiency accounts for chronic non-specific abdominal complaints. It is not clear how many individuals have unrecognized lactase deficiency and secondary intestinal symptoms which mimic irritable bowel syndrome. A review of available information does not provide an answer to this question. It has been reported that 20–95% of individuals complaining of irritable bowel-type symptoms are lactase-deficient, with many of these patients reporting improvement following lactose restriction.[15,39–44] However, other studies[45,46] report that only 5–6% of patients with symptoms of an irritable bowel have lactase deficiency as the underlying cause. Early studies do not include data on ethnic background and many of the studies provide no information on control groups. Additionally, almost all such studies include subjects who already were aware of intolerance to milk, while many of the diet studies which showed improvement on reducing lactose levels were not double-blind, which further confounds the issue.

8. LACTOSE HYDROLYSIS

Enzymatic modification of the lactose in dairy products has been recognized since the early 1950s. The development of effective lactases isolated from microbial sources had made this process commercially feasible. Currently, lactose-modified beverage milk may be purchased in the US at local supermarkets and some dairy stores.

Holsinger[47] notes that the lactases of commercial importance are those derived from the yeast *Saccharomyces lactis* and the fungus *Aspergillus niger*. These lactases vary considerably in their properties, particularly in pH and temperature optima. Lack of stability below pH 6·0 precludes the use of *S. lactis* lactase in treating acid whey, whereas *A. niger* lactase is suitable.[48] The literature on lactases and their applications are reviewed in Chapter 3 of this book.

Some attention must be given to the organoleptic qualities of lactase-treated milk. Slight difficulty has been observed because hydrolysis of lactose to its constituent monosaccharides, glucose and galactose, results in a marked increase in the intensity of sweetness of milk. Hydrolysis of 30%, 60% and 90% of the lactose in milk has almost the same effects on flavour as adding 0·3, 0·6 or 0·9% of sucrose, respectively, to the milk.[49] The acceptability of these changes in sweetness was studied at Johns Hopkins University.[50] It was observed that black adolescents, a target population for low-lactose milk, found milk in which 90% of the lactose has been hydrolysed acceptable to drink even though 56% of those reporting thought it was sweeter than untreated control. The mean maximum blood glucose rise in the 22 lactose-intolerant subjects after drinking 8 ounces ($\sim$240 ml) of untreated whole milk, 50% lactose-hydrolysed milk, or 90% lactose-hydrolysed milk were 4·4, 8·8 and 14·5 mg dl^{-1}, respectively ($p < 0.001$). In the 10 lactose-tolerant subjects, the mean peak blood glucose rise was not appreciably different whether they consumed untreated whole milk or lactose-hydrolysed milk.

8.1. Lactose-hydrolysed Milk Products

Other applications of modified milk to be considered are frozen 3:1 concentrates, dried products and cultured products. Frozen 3:1 concentrate is an attractive way of preserving milk with minimal flavour change, but unfortunately the concentrates thicken and coagulate during storage owing to lactose crystallization. Early research[51a] demonstrated improved physical stability of concentrated milks during frozen storage by hydrolysing the lactose, but preheat treatments are necessary in addition to lactose hydrolysis to give adequate storage stability.[51]

Cultured products have been of interest to lactase-deficient individuals because of the fermentation of lactose. Some of the products manufactured from lactase-treated milk include yoghurt, buttermilk, cottage cheese and Cheddar-type cheese. Consumers found the increased sweetness of cultured buttermilk prepared from lactose-hydrolysed milk objectionable.[52]

Yoghurt is enjoying a very rapidly increasing popularity, particularly the fruit-flavoured types. Yoghurt generally has been thought to contain low levels of lactose because of the utilization of lactose during the fermentation process. However, only about 15–20% of the lactose is utilized during fermentation, and the practice of fortifying yoghurt with non-fat milk solids results in appreciable amounts of added lactose. Lactose values of 3·3–5·7% have been reported in commercial yoghurts, i.e. amounts comparable to that in fluid milk.[53] Kolars *et al.*[54] have shown, on the basis of breath hydrogen excretion, that lactase-deficient persons absorb lactose better from yoghurt than from milk, and that this improved absorption appears to result from the intraluminal action of bacterial lactase, which substitutes for the patient's endogenous lactase.[55] However, another recent study[56] demonstrated that lactase-deficient persons do not tolerate unfermented acidophilus milk, which contains the lactase-producing organism *Lactobacillus acidophilus* ($<10^9$ organisms in 240 ml), any better than unaltered milk.

8.2. Whey

Nutritional aspects of lactose cannot be discussed adequately without some consideration of the applications of whey from cheese or casein manufacture. Although whey contains small amounts of high quality protein, equivalent to that of egg, fluid whey is essentially a crude solution of lactose. Development of economic methods of whey disposal has been of major concern to the dairy industry for many years. Modification of the whey by lactase treatment offers considerable potential in this regard.

The increased sweetness of lactose-hydrolysed whey suggests that it might be useful as an ingredient in sweet foods. In addition, the increase in solubility brought about by hydrolysis permits the manufacture of non-crystallizing high-solids syrups for use in beverages, ice creams, and other foods.[51] Applications of whey and its fractions for beverage manufacture include wines, beers and snack drinks, milk analogues, and liquid breakfasts with protein contents ranging from less than 0·5% to more than 3·5%.[57]

A lactase-hydrolysed whey permeate has been used to develop a prototype snack-type soft drink, 'Lactofruit', projected to be cheaper to produce and sell than other whey beverages because of the unique process employed.[58] The process involves ultrafiltration of the whey, followed by lactose hydrolysis by enzymatic electrocatalysis. Enzymatic electrocatalysis is advantageous in that it is able to counterbalance the loss of enzyme activity with time; the current is regulated continuously so that local pH variations are compensated for and enzyme activity is maintained at a constant level. The basic composition of Lactofruit shows that, in spite of lactase treatment, only 50% lactose hydrolysis is attained. However, since the drink is not fermented, the taste quality is sweet and different from that of most other whey beverages.[47]

A high protein milk analogue, based on a whey–soy mix, was formulated to feed children in developing countries.[47] Whey solids represented 41% of dry matter and the finished product contained ∼50% of carbohydrate, mostly lactose. Since a high incidence of lactose malabsorption occurs in children in developing countries[3] it seemed advisable to modify the lactose. Trial production runs and storage tests on whey with 90% of its lactose hydrolysed indicated that no production difficulties existed. An additional benefit occurred from lactase treatment, since consumers preferred the hydrolysed lactose product because of its noticeably sweeter taste.[51]

Apart from the benefits accruing to lactose-intolerant individuals, the use of lactase-treated whey as an ingredient in ice cream permits a reduction of at least 10% in the concentration of the sucrose without loss of quality.[59] Unfortunately, reduction of the milk-solids-not-fat concomitant with increasing levels of whey caused a 30% increase in the ash and a 20% reduction in the protein content of an ice cream containing 11% of whey solids. Sherbet and water ices on a stick are other possiblities for utilization of lactase-treated whey.[47,60,61]

9. CONCLUSIONS

Most African, Asian, and South American populations that have low lactase levels drink little, if any, milk after infancy; the milk products that they do consume are usually in a fermented form (e.g. yoghurt or cheese), in which some or all of the lactose has been hydrolysed and should be well tolerated by even a lactase-deficient person. Therefore, lactose intolerance is not a problem in most areas of the world as long as these individuals are permitted to follow their normal dietary habits.[61] Milk is an important food with a potential for providing much-needed protein and other nutrients to disadvantaged populations. It is therefore useful that lactose-intolerant individuals should be capable of both consuming milk without symptoms and utilizing its nutrients without interference and loss.[61]

The absorption of fat, protein, and calcium from physiologic amounts of milk seems to be normal in adults with lactase deficiency. The importance of lactase deficiency as a cause of recurrent abdominal pain in children is debatable; when appropriate control groups are used along with double-blind dietary studies, lactase deficiency does not seem to be a significant mechanism.[55]

Researchers have been divided in their recommendations in regard to the consumption of lactose by lactose-intolerant individuals; some have suggested continued milk consumption despite the occurrence of known intolerance when symptoms are absent, while others have argued for the removal of milk from the diet. It is clear that, for those individuals incapable of drinking milk, alternatives to milk should be identified; one such alternative exists through the pre-hydrolysis of lactose.[61]

Lactose-hydrolysed milk is a viable product of enzyme technology, but its manufacture must be conducted optimally in terms of adaptation to milk processing and lactase efficiency; the latter is dependent on product inhibition, which appears to be a common problem for lactases.[62] Possibilities for widespread production of lactose-modified dairy products have been opened up by the development of commercial sources of the enzyme lactase. Low-lactose fluid milk is commercially available. Such milk, when condensed, may be used to manufacture frozen 3:1 concentrates that do not thicken and gel during storage. Lactose-modified non-fat dry milk may be manufactured for use as a food ingredient, and cultured products such as yoghurt may be prepared successfully from lactase-treated milk. Lactose-modified whey may be condensed to high-solids non-crystallizing syrups for use in soft drinks, confections and ice cream. Use of lactase-treated products may lead to production efficiencies and permit monetary savings as well as provide nutritional advantages to lactase-deficient populations.[47]

Nutrition planners should be sensitive to the fact that a portion of the population is lactose-intolerant, and individuals responsible for public nutrition programmes also should recognize this fact. In this manner, the nutritional needs of milk-intolerant groups will be fully recognized and addressed.[61]

REFERENCES

1. JOHNSON, J. D., KRETCHMER, N. and SIMOONS, F. J., *Adv. Pediatr.*, 1974, **21**, 197.
2. SIMOONS, F. J., JOHNSON, J. D. and KRETCHMER, N., *Pediatrics*, 1977, **59**, 98.
3. PAIGE, D. M., LEONARDO, E., CORDANO, A., NAKASHIMA, J., ADRIANZEN, T. B. and GRAHAM, G. G., *Am. J. Clin. Nutr.*, 1972, **25**, 297.
4. PAIGE, D. M., BAYLESS, T. M., HUANG, S. S. and WEXLER, R., *Am. J. Clin. Nutr.*, 1979, **28**, 818.
5. PAIGE, D. M., BAYLESS, T. M., MELLITS, E. D. and DAVIS, L., *Am. J. Clin. Nutr.*, 1977, **30**, 1018.
6. JONES, D. V. and LATHAM, M. C., *Am. J. Clin. Nutr.*, 1974, **27**, 547.
7. PAIGE, D. M., BAYLESS, T. M., MELLITS, E. D., DAVIS, L., DELLINGER, W. S. and KREITNER, M., *Agric. Food Chem.*, 1979, **27**, 677.
8. PAIGE, D. M., In: *Lactose Digestion and Absorption: Clinical and Nutritional Implications*, D. M. Paige and T. M. Bayless (eds), 1981, Johns Hopkins University Press, Baltimore, p. 151.
9. KRETCHMER, N., RANSOME-KUTI, O., HURWITZ, R., DUNGY, C. and ALKIJA, W., *Lancet*, 1971, **ii**, 392.
10. JOHNSON, J. D., SIMOONS, F. J., HURWITZ, R., GRANGE, A., SINATRA, F. R., SUNSHINE, P., ROBERTSON, W. V., BENNETT, P. H. and KRETCHMER, N., *Am. J. Clin. Nutr.*, 1978, **31**, 381.
11. BAYLESS, T. M. and ROSENSWEIG, N. S., *J. Am. Med. Assoc.*, 1966, **65**, 735.
12. BEDINE, M. S. and BAYLESS, T. M., *Gastroenterology*, 1973, **65**, 735.
13. METZ, G., JENKINS, D. J. A., PETERS, T. J., NEWMAN, A. and BLENDIS, L. M., *Lancet*, 1975, **i**, 1155.
14. BAYLESS, T. M., ROTHFELD, B., MASSA, C., WISE, L., PAIGE, D. M. and BEDINE, M. S., *N. Engl. J. Med.*, 1975, **292**, 1156.
15. GUDMAND-HOYER, E. and SIMONY, K., *Am. J. Dig. Dis.*, 1977, **22**, 177.
16. JONES, D. V., LATHAM, M. C., KOSIKOWSKI, F. V. and WOODWARD, G., *Am. J. Clin. Nutr.*, 1976, **29**, 633.
17. SOLOMONS, N. W., VITERI, F. E. and HAMILTON, L. H., *J. Lab. Clin. Med.*, 1977, **90**, 856.
18. SOLOMONS, N. W., GARCIA-IBANEZ, R. and VITERI, F. E., *Am. J. Clin. Nutr.*, 1980, **33**, 545.
19. LEVITT, M. D. and DONALDSON, R. M., *J. Lab. Clin. Med.*, 1970, **75**, 937.
20. MITCHELL, K. J., BAYLESS, T. M. and PAIGE, D. M., *Pediatrics*, 1975, **56**, 718.
20a. COOK, G. C. and KAJUBI, S. K., *Lancet*, 1966, **i**, 725.
21. SIMOONS, F. J., *Am. J. Dig. Dis.*, 1978, **23**, 963.
22. FLATZ, G. and ROTTHAUWE, H. W., *Progr. Med. Genet.*, 1977, **2**, 207.
23. BAYLESS, T. M. and PAIGE, D. M., In: *Nutritional Management of Genetic Disorders*, M. Winick (ed.), 1979, John Wiley, New York, p. 79.
24. LISKER, R., AQUILAR, L. and ZAVALA, C., *Am. J. Clin. Nutr.*, 1978, **31**, 1499.
25. JOHNSON, J. D., SIMOONS, F. J., HURWITZ, R., GRANGE, A., MITCHELL, C. H., SINATRA, F. R., SUNSHINE, P., ROBERTSON, W. V., BENNETT, P. H. and KRETCHMER, N., *Gastroenterology*, 1977, **13**, 1299.

26. GARZA, C. and SCRIMSHAW, N. J., *Am. J. Clin. Nutr.*, 1976, **29**, 192.
27. LISKER, R., AQUILAR, L., LARES, I. and CRAVIOTO, J., *Am. J. Clin. Nutr.*, 1980, **33**, 1049.
28. GUTHRIE, H. A., *J. Am. Diet Assoc.*, 1977, **71**, 35.
29. KWON, P. H. JR, RORICK, M. H. and SCRIMSHAW, N. S., *Am. J. Clin. Nutr.*, 1980, **33**, 22.
30. BRICKER, M. L., SMITH, J. M., HAMILTON, T. S. and MITCHELL, H. H., *J. Nutr.*, 1949, **39**, 445.
30a. PAIGE, D. M. and GRAHAM, G. G., *Pediatr. Res.*, 1972, **6**, 329.
31. LEICHTER, J. and TOLENSKY, A. F., *Am. J. Clin. Nutr.*, 1975, **28**, 238.
32. EGGUM, B. O., *Br. J. Nutr.*, 1970, **24**, 983.
32a. LEICHTER, J., In: *Lactose Digestion: Clinical and Nutritional Implications*, D. M. Paige and T. M. Bayless (eds), 1981, Johns Hopkins University Press, Baltimore, p. 142.
33. NEWCOMER, A. D., In: *Lactose Digestion: Clinical and Nutritional Implications*, D. M. Paige and T. M. Bayless (eds), 1981, Johns Hopkins University Press, Baltimore, p. 124.
34. BIRGE, S. J. JR, KENTMANN, H. T., CUATRECASAS, P. and WHEDON, G. D., *N. Engl. J. Med.*, 1967, **276**, 445.
35. NEWCOMER, A. D., HODGSON, S. F., McGILL, D. B. and THOMAS, P. J., *Ann. Intern. Med.*, 1978, **89**, 218.
36. KOCIAN, J., VULTERINOVA, M., BEJBLOVA, O. and SKALA, I., *Digestion*, 1973, **8**, 324.
37. DEBONGNIE, J. C., NEWCOMER, A. D., McGILL, D. B. and PHILLIPS, S. F., *Dig. Dis. Sci.*, 1979, **24**, 225.
38. TROTTER, M., BROMAN, G. E. and PETERSON, R. R., *J. Bone Joint Surg.*, 1960, **42**, 50.
39. McMICHAEL, H. B., WEBB, J. and DAWSON, A. M., *Lancet*, 1965, **i**, 717.
40. WESER, E., RUBIN, W., ROSS, L. and SLEISENGER, M. H., *N. Engl. J. Med.*, 1965, **273**, 1070.
41. JUSSILA, J., LAUNIALA, K. and GORBATOW, O., *Acta Med. Scand.*, 1969, **186**, 217.
42. FUNG, W. P. and KHO, K. M., *Aust. N. Z. J. Med.*, 1971, **4**, 374.
43. GUDMAND-HOYER, E., RIIS, P. and WULFF, H. R., *Scand. J. Gastroenterol.*, 1973, **8**, 273.
44. ARVANITAKIS, C., CHEN, G.-H., FOLSCROFT, J. and KLOTZ, A. P., *Am. J. Clin. Nutr.*, 1977, **30**, 1597.
45. PENA, A. S. and TRUELOVE, S. C., *Scand. J. Gastroenterol.*, 1972, **7**, 433.
46. FAIRMAN, M., SCOTT, B. and LOSOWSKY, M., *Br. Med. J.*, 1975, **26**, 227.
47. HOLSINGER, V. H., In: *Lactose Digestion: Clinical and Nutritional Implications*, D. M. Paige and T. M. Bayless (eds), 1981, Johns Hopkins University Press, Baltimore, p. 231.
48. WOYCHIK, J. H., WONDOLOSKI, M. V. and DAHL, K. J., In: *Immobilized Enzymes in Food and Microbial Processes*, A. C. Olson and C. L. Cooney (eds), 1974, Plenum Press, New York, p. 41.
49. GUY, E. J., TAMSMA, A., KONTSON, A. and HOLSINGER, V. H., *Food Prod. Devel.*, 1974, **8**(8), 50.
50. PAIGE, D. M., BAYLESS, T. M., HUANG, S. S. and WEXLER, R., In:

Physiological Effects of Food Carbohydrates, A. James and J. Hods (eds), 1975, American Chemical Society Symposium Series No. 15, Washington, p. 191.

51. HOLSINGER, V. H. and ROBERTS, N. E., *Dairy and Ice Cream Field*, 1976, **159**(3), 30.

51a. TUMERMAN, L., FRAM, H. and CORNELY, K. W., *J. Dairy Sci.*, 1954, **37**, 830.

52. GYURISEK, D. M. and THOMPSON, M. P., *Cultured Dairy Prod. J.*, 1976, **11**, 12.

53. GOODENOUGH, E. PhD Dissertation, Rutgers University, New Brunswick, NJ, 1975.

54. KOLARS, J. C., JAVIANO, D. A., AOUJI, M. and LEVITT, M. D., *Gastroenterology*, 1983, **84**, 1211.

55. NEWCOMER, A. D. and McGILL, D. B., *Clin. Nutr. Suppl.*, 1984, **3**(2), 53–8.

56. NEWCOMER, A. D., PARK, H. S., O'BRIEN, P. C. and McGILL, D. B., *Am. J. Clin. Nutr.*, 1983, **38**, 257.

57. HOLSINGER, V. H., POSATI, L. P. and DE VILBISS, E. D., *J. Dairy Sci.*, 1974, **57**, 849.

58. FESNEL, J. M. and MOORE, K. K., *Food Prod. Devel.*, 1978, **12**, 45.

59. GUY, E. J., *J. Food Sci.*, 1980, **45**, 129.

60. GUY, E. J., VETTEL, H. E. and PALLANSCH, M. J., *J. Dairy Sci.*, 1966, **49**, 1156.

61. PAIGE, D. M., In: *Animal Products in Human Nutrition*, D. C. Beitz and R. Hanson (eds), 1982, Academic Press, New York, p. 373.

62. RAND, A. G. JR, In: *Lactose Digestion: Clinical and Nutritional Implications*, D. M. Paige and T. M. Bayless (eds), 1981, Johns Hopkins University Press, Baltimore, p. 219.

63. BAYLESS, T. M., In: *Lactose Digestion: Clinical and Nutritional Implications*, D. M. Paige and T. M. Bayless (eds), 1981, Johns Hopkins University Press, Baltimore, p. 117.

Chapter 5

NUTRITIONAL SIGNIFICANCE OF LACTOSE:
II. METABOLISM AND TOXICITY OF GALACTOSE

ALBERT FLYNN

*Department of Nutrition,
University College, Cork, Republic of Ireland*

1. INTRODUCTION

The principal dietary sources of galactose are milk and certain milk products in which it occurs mainly as a component of lactose. Lactose is hydrolysed by β-galactosidase (lactase) in the brush border of the small intestine to its component monosaccharides, galactose and glucose, which are both absorbed by an active transport mechanism across the gut wall into the blood.[1]

The brush border of the intestinal mucosa contains a sodium-dependent carrier which is shared by glucose and galactose, and interaction between the carrier and a localized sodium pump allows the monosaccharides to be actively concentrated within the cell.[2] Although the maximum rate of galactose absorption is higher than that of glucose, the affinity of the carrier for glucose is higher than for galactose.[3] This has important implications for the rate of galactose absorption from the intestine; following ingestion of equimolar quantities of glucose and galactose, the former is absorbed more rapidly in humans,[4] rats[5] and dogs.[6]

Hydrolysis of lactose by intestinal lactase is believed not to limit the rate of galactose absorption in lactose-tolerant humans.[7,8] However, the absorption of both galactose and glucose is slower when ingested as lactose than when ingested as a monosaccharide mixture.[9] It is not clear why this should be so. In lactose-intolerant subjects the absorption of galactose and glucose is much slower when ingested as lactose than when ingested as hydrolysed lactose because lactose hydrolysis by intestinal lactase is rate-limiting.

2. PATHWAYS OF GALACTOSE METABOLISM

The main route of metabolism of absorbed galactose is the Leloir pathway (Fig. 1) which involves four enzymes and results in the overall conversion of galactose into glucose-1-phosphate.[10-12]

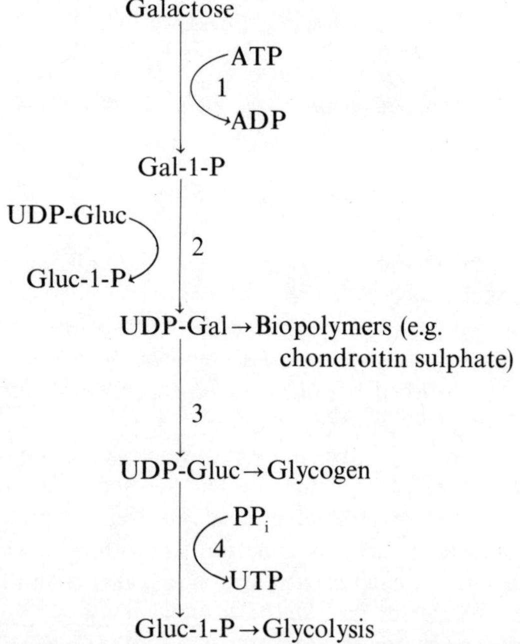

1, Galactokinase; 2, Gal-1-P:uridyl transferase; 3, UDP-Gal-4-epimerase; 4, UDP-Glu-pyrophosphorylase.

FIG. 1. The Leloir pathway for galactose metabolism.

The liver is the primary site of the Leloir pathway but it is also present in other tissues, including kidney and erythrocytes. The congenital absence of either one of the first two enzymes of this pathway results in the clinical condition galactosaemia which will be discussed later.

There are three minor pathways for galactose metabolism in humans. The Isselbacher pathway involves conversion of Gal-1-P into UDP-Gal by the enzyme UDP-galactose pyrophosphorylase, and it is believed that development of this pathway allows increased galactose metabolism in older galactosaemic patients.[13]

$$\text{Gal-1-P} \xrightarrow{\qquad\qquad} \text{UDP-Gal}$$
$$\text{UTP} \quad \text{PP}_i$$

Reduction of galactose to the sugar alcohol galactitol is catalysed by aldose reductase; this reaction is important in the development of cataract.[14]

$$\text{Galactose} \xrightarrow{\qquad\qquad\qquad} \text{Galactitol}$$
$$\text{NADPH} \quad \text{NADP}$$

Human tissues have only a very limited capacity for further metabolism of galactitol.[15]

Dehydrogenation of galactose produces galactonic acid which can be detected in the urine of both galactosaemic and non-galactosaemic subjects after a galactose load.[16]

3. FACTORS AFFECTING GALACTOSE ABSORPTION AND METABOLISM

The galactose concentration in the serum of normal fasting individuals is below 4·3 mg per 100 ml.[17] Serum galactose concentration increases after ingestion of galactose, and the time required for serum galactose concentration to reach a peak and the height of that peak both increase with increasing galactose load.[18] The increase in serum galactose also varies among individuals.[19]

In humans, blood galactose levels can be reduced either by its uptake into the tissues or its excretion in the urine. The liver is the major site of galactose uptake in humans and, while the erythrocyte and kidney are also capable of metabolizing galactose, the extent of their contributions to the handling of ingested galactose is unknown.[11] In normal humans, the maximum rate of galactose uptake by the liver is less than 500 mg min^{-1}, and this is reached at a plasma galactose concentration of 750 mg $litre^{-1}$.[20] The capacity of the liver to metabolize galactose is smaller in females than in males and decreases slowly with age.[21] Excretion of galactose in urine is very low compared to liver uptake at plasma galactose concentrations below 500 mg litre^{-1},[22] but it increases rapidly with rising blood galactose above 800 mg litre^{-1}.[18]

Oral ingestion of galactose results in increased blood glucose concentration within 20–30 min,[18,19,23] owing to the high capacity of liver for the uptake of galactose and its rapid conversion into glucose-1-P,[24] some of which is converted into glucose and excreted into blood while the remainder is metabolized. Orally ingested galactose produces a slight increase in serum insulin levels in humans,[23] which is a response to the increase in blood glucose resulting from galactose intake[25] rather than a direct effect of galactose.

In normal humans, the simultaneous ingestion of glucose with a galactose test meal reduces the subsequent serum galactose response[18] to an extent proportional to the quantity of glucose ingested.[19] It appears that this is due to an increased rate of removal of galactose from blood by the liver, mediated by increased blood glucose and insulin, rather than to the inhibition of galactose absorption by glucose.[19] A similar effect has been observed when galactose and glucose are given in the form of lactose to normal subjects[26] and when a glucose–galactose monosaccharide mixture is given to lactose-intolerant subjects.[27] Indeed, the serum galactose response to orally ingested glucose–galactose monosaccharide mixtures is the same in lactose-tolerant and lactose-intolerant subjects.[27] This suggests that lactose-intolerant individuals in whom the amount of galactose normally available for metabolism is minimal (owing to impaired lactose digestion) have retained their capacity for galactose metabolism and have normal galactose tolerance.[27]

In humans, simultaneous ingestion of ethanol and galactose results in a larger blood galactose response than that seen after galactose alone because ethanol inhibits the clearance of galactose from the blood.[28] Oxidation of ethanol in the liver increases the $NADH/NAD^+$ ratio which inhibits UDP-galactose-4-epimerase in the Leloir pathway[29] and thus retards galactose metabolism and uptake.

4. GALACTOSE TOXICITY IN MAN: GALACTOSAEMIA

The potentially hazardous nature of galactose and its metabolites to certain susceptible individuals has been highlighted by studies of galactosaemia and galactose-induced cataract. There are two clinically recognized forms of galactosaemia: classical galactosaemia and galactokinase-deficient galactosaemia. Both occur as inborn errors of metabolism in which an enzyme concerned with the normal metabolism of galactose is deficient. Tolerance of galactose is reduced and the ingestion of galactose in either condition results in a chronically elevated blood galactose level with the subsequent accumulation and/or excretion of galactose and galactose derivatives. Both conditions can be controlled by means of a galactose-free diet.[30]

Classical galactosaemia is due to a deficiency of galactose-1-P : uridyl-transferase;[31] ingested galactose can be phosphorylated to galactose-1-P but not metabolized further, and this results in the accumulation of galactose and galactose-1-P in the tissues.[32,33] At the cellular level, galactose-1-P interferes with the metabolism of glucose and the synthesis of glycoproteins and glycolipids[15,34] and reduces the level of ATP in the cell.[35] Galactosaemic infants appear normal at birth but subsequent to milk ingestion they develop symptoms including liver malfunction, cataract, mental retardation and failure to thrive. Early introduction of a galactose-free diet restores normal growth and can prevent permanent liver damage and halt (and sometimes reverse) cataract development. However, mental retardation is irreversible if a galactose-free diet is not introduced within 2–3 months. Classical galactosaemia is a rare condition occurring with a frequency of about 1 in 55 000.

Galactokinase-deficient galactosaemia, which occurs with a frequency of about 1 in 40 000, is characterized by cataracts which occur in the first or second decade of life but the subject is otherwise normal.[36] In such patients, ingested galactose remains unphosphorylated and may be converted into galactitol which causes damage to the lens fibres of the eye. The marked differences between the two conditions in the severity and diversity of symptoms and in the time scales within which they occur demonstrate that galactose-1-P is much more toxic than galactitol. The mechanism by which cataract develops in galactosaemia has been elucidated by studies on experimental animals, mainly the rat.

5. GALACTOSE-INDUCED CATARACT IN THE RAT

Mitchel and Dodge[37] reported that feeding diets rich in lactose (70%) to young rats resulted in rapid development of cataract, i.e. opacity of the lens of the eye, which was subsequently attributed to the galactose component of lactose. This experimental animal model of lactose/galactose-induced cataract has been widely studied.

In a study of the dose dependence of galactose-induced cataract in young rats, the lowest dietary level of galactose that produced cataract was about 15%.[38] Older rats are less susceptible to galactose-induced cataract than younger rats.[39,40]

The mechanism by which galactose causes cataract in the rat has been described by Kinoshita *et al.*[14] and van Heyningen.[41] The primary cause of cataract is the prolonged elevation of plasma galactose levels resulting in a high concentration of galactose in the lens of the eye. Galactitol, formed in the lens from galactose by the action of aldose reductase, accumulates to high concentrations and this exerts a strong osmotic effect. Sugar alcohols do not diffuse readily through biological membranes, so the retention of galactitol within the lens leads to the imbibition of water to maintain osmotic equilibrium. The development of galactose-induced cataract in galactosaemic humans is thought to occur by the same mechanism as described for the rat.

Richter and Duke[39] reported that rats fed exclusively on a diet of yoghurt develop cataracts. This observation is not surprising given the sensitivity of the rat to galactose, together with the fact that yoghurt may contain in excess of 30% galactose on a dry weight basis.[42] Hydrolysed lactose is more cataractogenic to rats than lactose;[43] when fed hydrolysed lactose at 40% of the diet for 160 days, 14 out of 15 rats developed cataract while those rats fed lactose at 40% of the diet were unaffected. This observation can be explained by the different rates of galactose absorption from the two diets. Absorption of galactose from ingested lactose first requires intestinal digestion of the lactose; since this is slow in weanling rats, owing to the low level of intestinal lactase, the rate of release of galactose from lactose determines the rate at which it can be absorbed. This slow rate of galactose absorption does not exceed the capacity of the rat liver to metabolize galactose, and thus plasma galactose does not reach high levels. On the other hand, ingestion of hydrolysed lactose leads to rapid absorption of galactose, faster than the liver can clear it, and plasma galactose levels rise, resulting in cataract.

6. DOES GALACTOSE CAUSE CATARACT IN NORMAL (NON-GALACTOSAEMIC) HUMANS?

While galactose has been clearly implicated in cataract development in the laboratory rat and in galactosaemic humans, there is no reliable evidence that it contributes to the cataract that occurs in normal non-galactosaemic humans. Bhat and Gopolan[44] reported on a study in India in which 7 out of 15 adults with cataract had an impaired tolerance to galactose. However, this study was poorly conducted and the authors did not exclude the possibility that these individuals may have been suffering from galactosaemia. The cause(s) of cataract in non-galactosaemic humans are not known at present.[45]

There are several strands of evidence which, taken together, suggest that it is unlikely that dietary galactose causes cataract in non-galactosaemic humans. First, the weanling rat is uniquely sensitive to dietary galactose among animal species. It is therefore not possible to make a quantitative extrapolation of the effects of galactose on the rat to its effect on humans. In addition, the lowest dietary level of galactose that has been shown to induce cataract in the rat (15% by weight of the diet) is equivalent to a daily galactose intake of $\sim 15\,g$ per kg of body weight. This is equivalent to the ingestion of 3 litres of cow's milk per day for a 5 kg infant or 30 litres per day for a 50 kg adult. Normally, the maximum daily galactose intake by infants rarely exceeds $3.7\,g\,kg^{-1}$ (~ 1 litre of breast milk per day for a 5 kg infant) and for adults about $1\,g\,kg^{-1}$ (equivalent to 2 litres of cow's milk per day for a 50 kg adult). The low incidence of cataract in infants indicates that normal levels of dietary galactose are well tolerated.

Galactose is normally ingested with at least an equal quantity of glucose (e.g. as lactose or hydrolysed lactose). This has been shown to enhance considerably the rate of galactose clearance by the liver[19] and thus prevents plasma galactose concentrations from rising to high levels. Thus, the quantities of lactose and galactose consumed in the usual human diet do not appear to exceed the capacity of the normal individual to avoid hazardous excesses.[45,46]

The development of lactose-hydrolysed products has led to renewed investigations of the absorption and metabolism of galactose when ingested as the monosaccharide, particularly by lactose-intolerant individuals. In these people, both the rate of galactose absorption and the quantity absorbed would be greater from a lactose-hydrolysed product than from a lactose-containing product since absorption would no longer be limited by intestinal lactase levels. Would such individuals be able to tolerate levels of galactose to which they have not been exposed since infancy? Williams and MacDonald[27] showed that the serum galactose response to orally ingested glucose–galactose mixtures is the same in lactose-intolerant and lactose-tolerant subjects. This suggests that lactose-intolerant individuals, in whom the amount of galactose normally available for metabolism is minimal, retain their capacity for galactose metabolism and have normal galactose tolerance.

REFERENCES

1. MacDonald, I., *J. Soc. Dairy Technol.*, 1978, **31**, 196.
2. Crane, R. K., *Internat. Rev. Physiol.*, 1977, **12**, 325.
3. Cook, G. C., *Scand. J. Gastroent.*, 1977, **12**, 733.
4. Holdsworth, C. D. and Dawson, A. M., *Clin. Sci.*, 1964, **27**, 371.
5. Cori, C. F., *Proc. Soc. Exp. Biol. Med.*, 1926, **23**, 290.
6. Annegers, J. H., *Proc. Soc. Exp. Biol. Med.*, 1968, **127**, 1071.
7. McMichael, H. B., Webb, J. and Dawson, A. M., *Clin. Sci.*, 1967, **33**, 133.
8. McMichael, H. B., *Acta Hepato-Gastroent.*, 1972, **19**, 281.
9. Gray, G. M. and Santiago, N. A., *Gastroenterology*, 1966, **51**, 489.
10. Herman, R. H. and Zakim, D., *Am. J. Clin. Nutr.*, 1968, **21**, 127.
11. Cohn, R. M. and Segal, S., *Metabolism*, 1973, **22**, 627.
12. Hansen, R. G. and Gitzelman, R., Amer. Chem. Soc. Symposium Series, No. 15, 1975, p. 100.
13. Isselbacher, K. J., *Science*, 1957, **126**, 652.
14. Kinoshita, J. H., Merola, L. O., Satoh, K. and Dikmik, E., *Nature* (London), 1962, **194**, 1085.
15. Gitzelman, R. and Hansen, R. G., In: *Inherited Disorders of Carbohydrate Metabolism*, D. Burman, J. B. Holton and C. A. Pennoch (eds), 1980, MTP Press, London, p. 61.
16. Bergren, W. R., Ng, W. G., Donnell, G. N. and Markey, S. P., *Science*, 1972, **176**, 683.
17. Rommel, K., Bernt, E. and Schnitz, F., *Klin. Wschr.*, 1968, **46**, 936.
18. Stenstam, T., *Acta Med. Scand.*, 1946, Suppl. 177.
19. Williams, C. A., PhD Thesis, University of London, 1981.
20. Waldstein, S. S., Greenburg, L. A., Biggs, A. D. and Corn, L., *J. Lab. Clin. Med.*, 1960, **55**, 462.
21. Tygstrup, N., *Acta Med. Scand.*, 1964, **175**, 281.
22. Tygstrup, N., *Acta Physiol. Scand.*, 1961, **51**, 263.
23. Ganda, O. P., Soeldner, J. S., Gleason, R. E., Cleator, I. G. and Reynolds, C., *J. Clin. Endocr. Metab.*, 1979, **49**, 616.
24. Segal, S. and Blair, A., *J. Clin. Invest.*, 1961, **40**, 2016.
25. Gitzelman, R. and Illig, R., *Diabetologia*, 1969, **5**, 143.
26. Williams, C. A., Phillips, T. and MacDonald, I. A., *Metabolism*, 1983, **32**, 250.
27. Williams, C. A. and MacDonald, I., *Human Nutrition: Clinical Nutrition*, 1982, **36C**, 149.
28. Salaspuro, M. P. and Kesaniemi, Y. A., *Scand. J. Gastroent.*, 1973, **8**, 681.
29. Isselbacher, K. J. and Krane, S. M., *J. Biol. Chem.*, 1961, **236**, 2394.
30. Clothier, C. M. and Davidson, D. C., *Human Nutrition: Applied Nutrition*, 1983, **37A**, 483.
31. Kalckar, H. M., Anderson, E. P. and Isselbacher, K. J., *Biochim. Biophys. Acta*, 1956, **20**, 262.
32. Schwarz, V., Goldberg, L., Komrower, G. M. and Holzel, A., *Biochem. J.*, 1956, **62**, 34.
33. Anderson, E., Kalckar, H. M. and Isselbacher, K. J., *Science*, 1957, **125**, 113.

34. SCHWARZ, V., *Biochem. Soc. Trans.*, 1975, **3**, 234.
35. RENNERT, O. M., *Ann. Clin. Lab. Sci.*, 1977, **7**, 443.
36. GITZELMAN, R., *Paediat. Res.*, 1967, **1**, 14.
37. MITCHEL, H. S. and DODGE, W. M., *J. Nutr.*, 1935, **9**, 37.
38. KEIDING, S. and MELLENGAARD, G., *Acta Ophthalmol.*, 1972, **50**, 174.
39. RICHTER, C. P. and DUKE, J. R., *Science*, 1970, **168**, 1372.
40. LERMAN, S., *Physiol. Rev.*, 1965, **45**, 98.
41. VAN HEYNINGEN, R., *Exp. Eye Res.*, 1971, **11**, 415.
42. PAUL, A. A. and SOUTHGATE, D. A. T., In: McCance and Widdowson's *The Composition of Foods*, 1978, HMSO, London.
43. POIFFAIT, A., FRANGNE, R., DAVID, C., MANENT, P. J. and ADRIAN, J., *Internat. J. Vit. Nutr. Res.*, 1982, **52**, 456.
44. BHAT, K. S. and GOPOLAN, C., *Nutr. Metab.*, 1974, **17**, 1.
45. BUNCE, G. E., *Nutr. Rev.*, 1979, **37**, 11.
46. WILLIAMS, C. A. and MACDONALD, I., *Wld. Rev. Nutr. Diet.*, 1982, **39**, 23.

Chapter 6

THE MILK SALTS: THEIR SECRETION, CONCENTRATIONS AND PHYSICAL CHEMISTRY

C. HOLT

Hannah Research Institute, Ayr, Scotland, UK

1. INTRODUCTION

It is over 20 years since the physical chemistry of milk salts was comprehensively reviewed by Pyne,[1] and progress in the interim has been somewhat uneven. While much has been learned about the secretion of milk salts, our knowledge of their behaviour during the manufacture of milk products has not advanced to the same degree. Little interest has been shown in the nutritional manipulation of milk salt composition in spite of its potential importance as a means of influencing processing characteristics. Possibly this is because [lactose], [Na], [K] and [Cl] are not readily altered by diet[2,3] and subtle changes in the equilibria of multivalent ions may go unnoticed. In recent years there has been a resurgence of interest in the milk salts following the application of a range of physicochemical techniques, one consequence of which has been a renewed debate on the nature of the micellar calcium phosphate—a question once thought to have been settled.[4]

Accordingly, the emphasis in this chapter differs from that in Pyne's review. A section is devoted to a summary of current ideas on how the milk salts are secreted, to give some insight into how the composition of milk is established and might be manipulated. Many factors influence the concentrations of milk salts to produce complex patterns of change; the trends with stage of lactation, for example, have been well studied.[5] Nevertheless, there are patterns of variation in composition that apply to

many different circumstances, arising from constraints imposed by the
secretory mechanisms and the physicochemical equilibria, particularly
amongst the multivalent ions. These patterns and principles are discussed
in Section 3. Two principles of milk salt composition can be identified
with confidence apart from the (by no means trivial) requirement of
overall electrical neutrality. The first is that, whatever the composition,
the milk shall have an osmotic pressure close to that of blood, and to
accommodate this constraint a family of correlations exists involving
principally [lactose], [Na], [K] and [Cl]. The second is that casein
micelles must form and be stable, so pH and free $[Ca^{2+}]$ are constrained
to certain domains, and the calcium phosphate must form a complex
with the casein. Following from this are two further families of correlations, one involving the colloidal and the other the diffusible multivalent salts.

Skim milk can be considered a two-phase system of calcium phosphate
in quasi-equilibrium with an aqueous solution of salts and proteins. The
phase boundary, however, is not well defined because of the intimate
association of the calcium phosphate with the casein phosphoproteins.
The nature of this calcium phosphate and the interactions of salts and
proteins in the aqueous solution are the subjects of Section 4. When raw
milk is heated or concentrated, or when citrate or phosphate salts are
added, the quasi-equilibrium state of the milk salts may be perturbed,
and the kinetics of the resulting transformations can be as important as
the thermodynamics. Neither type of information is currently available at
a sufficiently fundamental level to be transferable from one manufacturing application to another.

2. SECRETION OF MILK SALTS

2.1. Historical

The mechanisms of secretion of all the milk salts are interrelated and
cannot be considered without discussing also the synthesis and secretion
of lactose and casein. Nevertheless, in the historical development of ideas
in this field, progress was made by considering Na, K, Cl and lactose
alone. Only recently has secretion of multivalent ions been a frequent
subject for experiment and informed theoretical speculation (but see
Wright[6]).

For many years it has been known that milk is isosmotic with blood;[7]
its freezing point normally varies only within close limits and hence there

is a general inverse relation between the activities of univalent ions and lactose. Rook and Wood[8] noted that, while the [lactose]/[K] ratio in milk of an individual healthy cow in mid-lactation was a constant, these constituents varied inversely between such animals. However, during mastitic infections, at the start or end of lactation or on treatment with oxytocin there may be a direct relation between these constituents. To explain these or related observations, various theories have been proposed in which milk is supposed to be formed by mixing two fluids of very different salt and lactose compositions.[8-12] For example, Barry and Rowland[11] proposed that a primary secretion, or 'true milk', is formed and diluted to a variable extent with a fluid having a salt composition similar to that of extracellular fluid.

Linzell and Peaker[13-15] advanced an alternative scheme for univalent ion secretion, based mainly on measurements of intracellular ion concentrations, epithelial permeabilities and electrical potential differences in the goat and guinea pig mammary gland, which has been widely accepted in its essential features. Intracellular [Na] and [K] are established by a Na/K-activated ATPase on the basolateral surface of the secretory cell, and there is a dynamic electrochemical equilibrium of these ions across the apical membrane. The salt composition established in the alveolus by this mechanism corresponds to the 'true milk' of Barry and Rowland.[11] However, exchange of solutes between extracellular fluid and the primary secretion can occur if the junctions between cells are not fully sealed.

It is now recognized that a fat-free primary secretion is formed inside vesicles derived from the Golgi apparatus of the secretory cell, and the salt concentrations in mature vesicles are essentially the same as in skim milk. Furthermore, there is some understanding of how the synthesis of lactose in vesicles is related to transmembrane ion transport and swelling of vesicles,[16,17] allowing a more detailed discussion of how milk composition is related to secretory cell functions than has hitherto been possible.

2.2. The Golgi Vesicle Route

In this pathway, membranous vesicles bleb-off the distal face of Golgi dicytosomes and pass through the cytoplasm to the apical membrane where exocytosis occurs. Vesicles contain casein, the galactosyl transferase of lactose synthesis and α-lactalbumin. There is good evidence that lactose synthesis takes place in Golgi vesicles,[18,19] and it has been estimated, in the rat, that mature vesicles contain the same lactose concentration as milk.[20] Fully formed casein micelles can be seen in vesicles and they appear to be at the same concentration as in the alveoli.[21] Furthermore, Wooding and Morgan[22] localized Ca in swollen vesicles by a pyroantimonate staining method. In so far as it has been determined, therefore, the intravesicular fluid resembles fat-free milk, at least in mature vesicles immediately prior to exocytosis.

Although Na^+ and K^+ can permeate the apical membrane, permeation by Ca, P_i and citrate does not occur in the forms in which they are found in milk and intracellular fluid. Active transport of Ca has, however, not been ruled out.[23] The transit times from blood to milk of radiolabelled Ca, P_i and citrate are consistent with a Golgi vesicle route of secretion in the goat.[23,24] Rat mammary gland fractions enriched in Golgi membranes contain a Ca-stimulated Mg-ATPase[25] and can accumulate Ca, particularly in the presence of anions that could form intravesicular precipitates.[26] Preliminary evidence has been given for citrate transport by Golgi membranes from cow mammary gland,[27] but no study has been reported on Mg transport. Accumulation of P_i has not been studied in great detail but the evidence of Kuhn and White[16] on the rat is that P_i is formed intravesicularly during lactose synthesis. They have established that a uridine nucleotide cycle is involved in the formation of lactose from UDP-galactose and glucose. Within the vesicle these precursors form lactose and UDP which cannot cross the membrane. The UDP is hydrolysed to UMP and P_i, both of which can re-enter the cytosol, avoiding product inhibition of the lactose synthetase.

For lactose synthesis to begin, it is probably necessary for site II of galactosyl transferase to be occupied by Ca^{2+},[28] which in turn requires a concentration of free Ca^{2+} within the vesicle of the order of mM. Since the cytosol concentration is probably μM,[29] concentration by a factor of 10^3 is required, possibly against an electrical potential gradient. In the scheme illustrated in Fig. 1, the Ca/Mg-ATPase fulfils this role but, since the free ion can form complexes with citrate, casein and P_i, Ca accumulation can greatly exceed that established by the free ion concentration alone. Indeed, once the solubility product of the micellar calcium phosphate salt is reached, phase separation can occur with the concomitant formation of casein micelles. Some hypotheses regarding these and other transcellular secretory mechanisms are summarized in Fig. 1 but, whatever the details, some general principles can be discerned regarding the relations between secretion and composition. To the extent that the paracellular processes are of secondary importance in established lactation in the cow and goat,[30] the concentrations of some milk salt

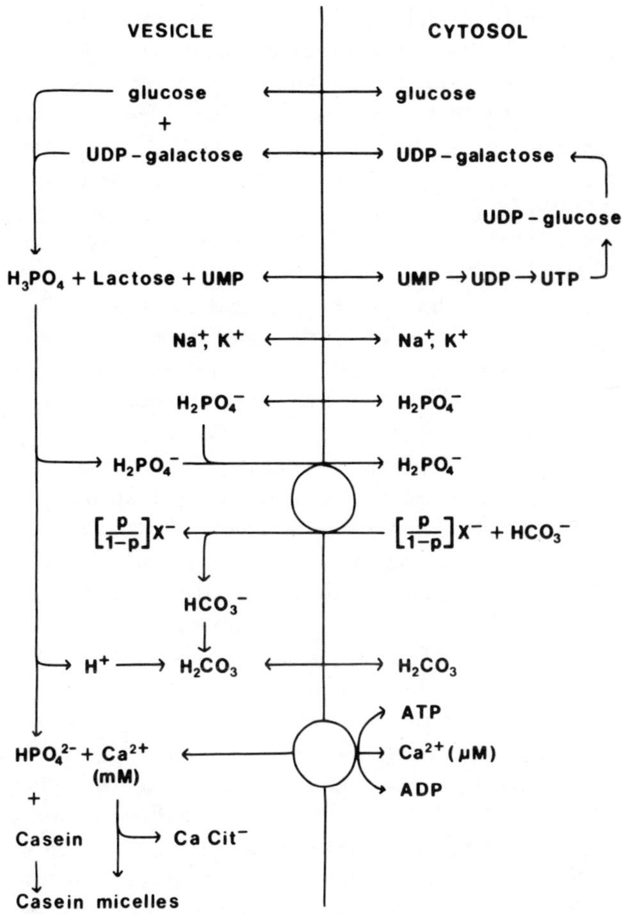

FIG. 1. Summary of some proposed transport mechanisms for milk salts between the cytosol of the secretory cell and the inside of Golgi vesicles.[13–17,25] Stimulation of the Ca-ATPase results in the establishment of a large concentration gradient in free $[Ca^{2+}]$ and activation of lactose synthetase. The Ca^{2+} may combine with HPO_4^{2-} and casein to form the micellar calcium phosphate and drive the uridine nucleotide cycle to synthesize more lactose and transport more P_i. The cations Na^+ and K^+ are free to permeate the membrane but, because of restricted transport of anions, an electrical potential gradient develops to satisfy the condition of overall electrical neutrality. The membrane is assumed to be freely permeable to water, so swelling of vesicles occurs through the net accumulation of lactose and salts. Lactose synthesis is possibly coupled to salt transport, firstly through Ca^{2+} transport and secondly through the neutralization of H^+, though these proposals are largely hypothetical.[17,19]

components (largely the diffusible Na, K and possibly citrate, Cl and P_i) reflect cellular levels. Others, such as colloidal Ca, Mg and P_i, are a consequence of the amount accumulated in Golgi vesicles and the osmotic dilution resulting from lactose synthesis. The interplay of these factors is reflected in the interrelationships of milk salt concentrations and the patterns of change in milk composition, some of which are considered in Section 3.

2.3. The Paracellular Route

In the cow, the paracellular route is particularly important during the onset and cessation of lactation, during mastitic infection and on treatment with non-physiological doses of oxytocin. However, it is probably never completely absent,[30] though the amount of exchange via this route and the constituents able to exchange may alter with time. The exchange is between extracellular fluid and alveolar milk, i.e. between a fluid higher in Na, Cl and pH and lower in K and lactose than milk, and could involve only a simple diffusive exchange of solutes. Exchange of the diffusible Ca, P_i and citrate of milk with extracellular fluid should occur since the concentration gradients are appreciable.[31] Little work has been reported in this area; Tallamy and Randolph[32] found the expected decreases in [Ca] and [P_i] in centrifuged-milk serum, but [Mg] increased in mastitic quarters compared to healthy controls. Mammary lymph [Mg] is about 2 mM whereas diffusible [Mg] in milk is about 3 mM. It is conceivable, though probably of secondary importance, that the rates of exchange on the two sides of the epithelium will differ, tending to generate an osmotic gradient and hence a flow of water. If, for example, lactose exchanges more slowly than the ions, the flow of water will dilute the alveolar fluid, but little is known about the mechanisms or kinetics of the processes, and no firm conclusions can be made.

3. SALT CONCENTRATIONS IN MILK

Some average concentrations of salts in bulk milks are given in Table 1, though it is stressed that appreciable variation occurs, even in bulked samples from many cows, due to factors associated with stage of lactation, mastitis, diet and season.

3.1. Variation with Stage of Lactation

It is convenient to discuss the changes in milk composition associated with stage of lactation while realizing that these are the resultant of a number of causal factors, and it is seldom if ever possible to quantify the effects of each one separately. The literature on lactational changes is voluminous, so full coverage has not been attempted. A thorough study of the post-partum changes in total and diffusible salt concentrations in Ayrshire cows was made by White and Davies;[5] their work remains the single most reliable source of data. Rook and Campling[33] observed lactational variations in total Na, K, Ca and P in experiments with controlled feeding and management, and Keogh et al.[34] have conducted a survey of seasonal and lactational changes in the salt composition of Irish milk. The largest changes in composition occur at around parturition and near the end of lactation; for the greater part of lactation in the healthy cow, milk composition is relatively constant. Some typical concentrations in early, middle and late lactation are summarized in Table 1.

TABLE 1

TOTAL AND DIFFUSIBLE CONCENTRATIONS (mM) IN SKIM MILK FROM AYRSHIRE COWS[a]

	Bulk	*Early*[b]	*Middle*	*Late*	*Mastitic*[c]
Total Ca	30·1	33·2	29·4	32·1	29·4
Diffusible Ca	9.5	12·8	9·4	8·0	9·1
Total Mg	5·1	5·7	5·0	5·4	4·9
Diffusible Mg	3·3	3·9	3·2	3·3	3·2
Total P_i	20·9	19·4	20·9	18·4	19·0
Diffusible P_i	11.2	10·5	11·7	7·0	9·2
Total Cit	9·8	9·8	9·1	8·5	8·8
Diffusible Cit	9·2	9·1	8·9	7·8	8·3
Total Na	25·5	29·7	24·8	48·8	34·5
Total K	36·8	41·8	40·3	26·9	36·1
Total Cl	30·3	36·7	29·7	46·5	40·5
Milk pH	6·72	6·53	6·73	6·98	6·87

[a] Data taken from White and Davies[5] but with citrate values increased by 4%,[35] and diffusible citrate increased by a further 5%.[36]
[b] Colostrum samples.
[c] Subclinical mastitis.

Compared to mid-lactation values, in early lactation average concentrations of casein, total and diffusible Ca and Mg, free Ca^{2+}, total and diffusible citrate, diffusible phosphate esters and total Na and Cl are higher whereas lactose and milk pH are lower. In late lactation average concentrations of casein, total Ca, Na and Cl, colloidal P_i and milk pH are higher whereas lactose, diffusible Ca, free Ca^{2+}, diffusible P_i and total K are lower. Milk citrate concentration is sensitive to diet and other factors but generally passes through a maximum early in lactation.[37-41] The presence of a paracellular flux in early and late lactation can explain, but only in part, some of the lactational changes. Thus, the raised [Na] and [Cl] and reduced [lactose] and [K] in early and late lactation would be expected from such a flux, but the low pH in early lactation is not explained. Lactational trends in total concentrations of multivalent ions are often more complex because of separate trends in diffusible and colloidal concentrations. Colloidal Ca, Mg and citrate are associated predominantly with casein, and their concentrations are usually closely correlated. However, in early lactation, casein micelles are not as mineralized as later in lactation, so colloidal [Ca], for example, is not as high then as might otherwise be expected. Diffusible ion concentrations, in the absence of paracellular exchange, may be related to cellular levels through secretory mechanisms such as those illustrated in Fig. 1, but satisfactory determinations of multivalent ion concentrations in the cell have not yet been made.

3.2. Variation Through Mastitis

The effects of mastitis on milk composition have been reviewed in detail by Kitchen.[42] Quantitative differences in effect are observed with the cause and severity of infection, such that in severe cases the secreted fluid may differ radically from normal milk. Some characteristic changes in salt composition associated with sub-clinical mastitis are summarized in Table 1. The pH and [Na] and [Cl] increase while [K] and [lactose] decrease, indicating paracellular exchange processes.[30] The [citrate] may also fall,[12] and in severe infections other salt concentrations change appreciably with lowered levels of total and diffusible $[P_i]$ being usually observed.[32,43-45] Total [Ca] and [Mg] may decrease[12,32,43] or increase,[44] and from limited studies[12,32] it appears that diffusible [Ca] may be depressed. An increased total [Ca] may result from some degree of mammary gland involution in which milk casein and colloidal Ca concentrations increase, compensating for the drop in diffusible [Ca].

3.3. Effects of Diet and Variation with Season

Apart from starvation or underfeeding, nutrition, with few exceptions, appears to have little effect on [lactose], [Na], [K] and [Cl], although milk yield may be altered.[2] In contrast, multivalent ion concentrations are, by and large, subject to change when milk yield is manipulated, and they may also alter even when yield is relatively constant. However, there have been relatively few controlled experiments designed to manipulate milk salt concentrations by dietary means in spite of the importance of these components in milk processing. Some, at least, of the reported effects are indirect, mediated by hormonal changes, rather than a direct response in milk output to diet. Thus, the elevation of [lactose] and depression in [Na], [K] and [Cl] levels produced by feeding iodinated casein are almost certainly mediated by thyroxine.[46–49]

In an experiment with two levels of hay intake and either oat or barley based concentrate, Fisher $et\ al.$[50] found that milk [Ca], but not [Mg], [Na] or [K], was significantly greater with the barley based concentrate. On a diet deficient in minerals, supplementation with calcium carbonate or phosphate can produce a general improvement in milk yield and composition, including increased [Ca] and $[P_i]$.[51–56] Other observations, however, have revealed no effect of mineral supplementation.[57–61] Nevertheless, even if no large change in concentration is obtained, the effects on the physicochemical properties of the milk can be significant. Thus, in the experiments of Sommer and Binney[62] and Labuschagne and Landrey,[63] although only a small increase in milk [Ca] was produced, ethanol stability was reduced by the mineral supplement. Similarly, in the work of Atramentova and Iopa[59] the coagulation properties of the milk in cheesemaking were improved by calcium phosphate supplements, yet there were no significant differences in total [Ca], [P] or [protein]. Golovnin and Molchanov[52] also found a reduction in chymosin clotting time when there was an excess of Ca in the cows' ration compared to controls. Destabilization of milk to ethanol can result from cows grazing forages rich in Ca,[64,65] suggesting that an increase in free $[Ca^{2+}]$ is produced.[66] Furthermore, the observations of Labuschagne and Landrey,[63] that milk from cows receiving a mineral supplement contains lower diffusible [citrate] and $[P_i]$ and is unstable to ethanol, are also consistent with a raised free $[Ca^{2+}]$. Milks that exhibit the 'Utrecht abnormality' have increased sensitivity to ethanol and heat and a higher than normal free $[Ca^{2+}]$. The abnormality can be corrected by administration of citrate to the cows or by supplementary feeding.[67,68] In recent work on the 'Utrecht abnormality' in Japan, Yoshida[69] has shown that the cows show signs of chronic Mg deficiency and the condition can be treated by supplementary feeding. Thus, the effects of mineral supplementation on milk composition may be small, provided that diets are not deficient in any way, but more subtle influences on salt equilibria may still occur with consequent effects on milk processing characteristics.

In a study of seasonal variations in creamery milk composition in south-west Scotland, Holt and Muir[70] found that diffusible [Ca] and [citrate] were temporarily depressed shortly after the cows went out to grass, and there were associated changes in the heat stability of the skim milk and concentrated skim milk.[71,72] There are many other reports of seasonal changes in milk [citrate],[37,73–78] but it is difficult to separate the effects of diet from the effects associated with stage of lactation (see above), environmental temperature,[79] mastitis[12] or other possible factors such as pregnancy, ketosis or oestrus. Controlled experiments, however, have shown that milk [citrate] can be varied by feeding.[80–82] Diets with low roughage and high concentrate content produced consistent reductions in milk [fat], [citrate] and diffusible [Ca], whereas supplementary fat increased milk [fat], [citrate] and diffusible [Ca]. A rationale for these observations and others, such as the effect of starvation, which increases [citrate], and the lactational trend, has been offered by Ormrod, Faulkner and co-workers.[72,82–84] Milk [citrate] is postulated to reflect, in some way, that in the cytosol of the mammary secretory cell. A major pathway of citrate metabolism in the cytosol is its oxidative carboxylation via isocitrate to 2-oxoglutarate, a reaction that supplies much of the NADPH required for *de novo* fatty acid synthesis. Feeding high-fat diets or starvation depresses fat synthesis, and the suggestion is that a build-up of [NADPH] occurs which inhibits the oxidative decarboxylation and raises cytosolic and milk [citrate]. On high concentrate diets, acetate, 2-hydroxybutyrate and long-chain fatty acid uptake by the mammary gland is depressed, and it is proposed that this results in reduced synthesis of short-chain fatty acids and citrate, and a reduced conversion of citrate into isocitrate, 2-oxoglutarate and malate. Further details are given in the review by Faulkner and Peaker[85] and the doctoral thesis of Ormrod.[83]

3.4. Interrelationships of Salt Concentrations

Three families of correlations of milk salt concentrations can be identified. The first, involving [lactose], [K], [Na] and [Cl] (Fig. 2a), has been recognized for many years[2,86] and results from the requirement that milk be isosmotic with blood (e.g. [lactose] negatively correlated with [K]) and the operation of a paracellular pathway (e.g. [Na] positively correlated with [Cl]). Correlations of various constituents with pH were tabulated by White and Davies,[5] and the physicochemical basis of some of these has since been more clearly defined. Thus, a second family of correlations can now be identified (Fig. 2b,c,d) involving (i) diffusible [Ca], [Mg] and [citrate], and (ii) $[Ca^{2+}]$, $[HPO_4^{2-}]$ and pH. The third family, recognized for some time, involves colloidal [Ca], [Mg], $[P_i]$ and [citrate].

At constant pH, diffusible [Ca] and [Mg] are closely and directly correlated with [citrate] (Fig. 2b), and slightly less well correlated if pH varies.[70,87] This is because, in whatever way milk [citrate] is established during secretion, the strength of the soluble $CaCit^-$ and $MgCit^-$ complexes, compared to those of $CaHPO_4$, $MgHPO_4$ and others, ensures that diffusible [Ca], [Mg] and [citrate] are closely related.[70] The negative correlation of free $[Ca^{2+}]$ and $[HPO_4^{2-}]$ arises from the solubility product type equation for the micellar calcium phosphate (Fig. 2d). The HPO_4^{2-} and $H_2PO_4^-$ ions form only weak complexes in solution with milk cations, so most of the diffusible P_i is present in these forms (see Table 3) with the ratio $[HPO_4^{2-}]/[H_2PO_4^-]$ being strongly dependent on pH. With these interrelationships in mind, it is instructive to consider the lactational trends discussed in Section 3.1. In early lactation, the free $[Ca^{2+}]$ is high at 2–4 mM, so the $[HPO_4^{2-}]$ must be low to satisfy the solubility product condition. However, pH is low at about 6·4–6·6, so most of the diffusible P_i is $H_2PO_4^-$ rather than HPO_4^{2-} and its concentration is, on average, similar to that in mid-lactation where higher pH and lower free Ca^{2+} concentrations prevail. In late lactation free $[Ca^{2+}]$ is lower,[5] so higher concentrations of HPO_4^{2-} are allowed. Nevertheless, much of the diffusible P_i is HPO_4^{2-} rather than $H_2PO_4^-$ because pH is also raised, so its concentration is reduced compared to mid- or early lactation values.

Correlations of colloidal milk salts arise largely from the nature of the micellar calcium phosphate which contains Ca, P_i, Mg and citrate in fairly fixed proportions.[5,88,89] Concentrations of these constituents are normally closely related to casein content, although in early lactation milk the degree of mineralization of the casein is less than at other times.[5] If phosphoserine residues are necessary for the formation of micellar calcium phosphate (see Section 4.4) then an upper limit may be placed on the degree of mineralization of casein micelles, imposed by the content of phosphorylated amino acids. Accumulation of Ca in Golgi vesicles by the active transport mechanism discussed in Section 2.2 will thus be limited by casein content. Casein content, in turn, is thought to be determined by its rate of synthesis and secretion in relation to osmotic dilution through lactose synthesis.

Holt[17] has given a theoretical analysis of the relations between milk

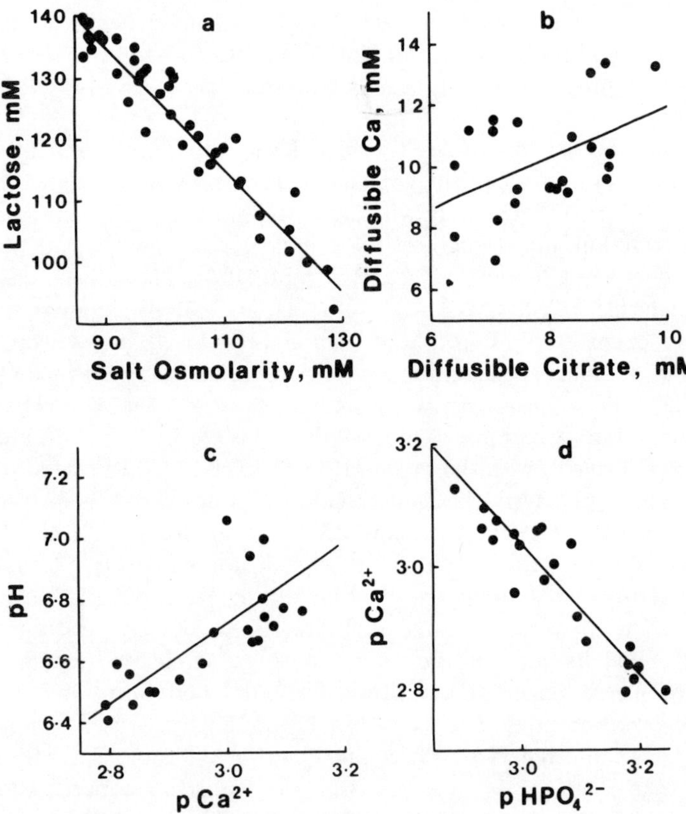

FIG. 2. Some interrelationships of diffusible salts and lactose in milks from cows in different stages of lactation.

(a) The negative correlation of salt osmolarity, $\{S\}$, with lactose concentration, $[L]$, is in accord with a theoretical expression for the total osmolarity, $\Pi = \{L\} + \{U\} + \{S\}$, in which Π and the osmolarity of miscellaneous uncharged species, $\{U\}$, are constant. The linear regression equation is:

$$[L] = 226 \cdot 4(\pm 4 \cdot 6) - 1 \cdot 01(\pm 0 \cdot 045)\{S\},$$
$$r^2 = 0 \cdot 94; \; p < 0 \cdot 001.$$

(b) The positive correlation of diffusible [Ca] and [citrate] is expected because of the formation of the strong, soluble, complex $CaCit^-$ at milk pH. The scatter of points is almost entirely due to the differences in free $[Ca^{2+}]$ between milks. Diffusible [Mg] and [citrate] are similarly correlated.

(c) The positive correlation of $p[Ca^{2+}]$ with pH reflects the facts that milks are relatively constant in their degree of saturation with respect to micellar calcium phosphate, and $[H_2PO_4^-]$ is relatively invariant. The linear regression equation composition and the known mechanisms of secretion. It was proposed that the swelling of Golgi vesicles through lactose synthesis is controlled by a coupling to anion transport characterized by a constant, Ω, such that:

$$[\text{Cas}]_m = [\text{Cas}]_g (1 - \Omega[\text{L}]_m/\Pi) \tag{1}$$

where $[\text{Cas}]_m$ and $[\text{Cas}]_g$ are the casein concentrations in the mature vesicle and Golgi apparatus respectively, Π is the total osmolarity and $[\text{L}]_m$ is the final lactose concentration. According to the analysis, the main source of interspecific differences in casein concentration is the variation in lactose synthesis, but for intraspecific differences the situation is less clear and variation in $[\text{Cas}]_g$, associated with changes in the flux of vesicles, could be responsible for some of the observed variability. Superimposed on the effects of vesicle flux and vesicle swelling are changes induced by the operation of the paracellular route, so the relation between [casein] and [lactose] predicted by eqn (1) will not always be apparent in the skim milk.

In conclusion, it can be seen that quite different factors influence diffusible and colloidal concentrations of milk salts, which may explain why patterns of change in total [Ca], [Mg] and [citrate] have been less readily discerned than those in [Na], [K], [Cl] and [lactose]. In this regard, the value of determining the partition of milk salts is obvious.

4. PHYSICAL CHEMISTRY OF MILK SALTS

4.1. The Partition of Milk Salts
An equilibrium diffusate is prepared by dialysis of a volume of water against a very much larger volume of skim milk, allowing the diffusible ions to distribute across the semipermeable membrane and reach

---

is:

$$\text{pH} = 2.82(\pm 0.62) + 1.30(\pm 0.21)\text{p}[\text{Ca}^{2+}],$$
$$r^2 = 0.66, \ p < 0.001.$$

(d) The negative correlation of free $[\text{Ca}^{2+}]$ with $[\text{HPO}_4^{2-}]$ is a consequence of the solubility product type equation for the micellar calcium phosphate. For brushite the equation is $\text{p}[\text{Ca}^{2+}] = 6.58 - \text{p}[\text{HPO}_4^{2-}]$ whereas the linear regression equation is:

$$\text{p}[\text{Ca}^{2+}] = 6.27(\pm 0.31) - 1.08(\pm 0.10)\text{p}[\text{HPO}_4^{2-}],$$
$$r^2 = 0.85; \ p < 0.001.$$

electrochemical equilibrium. An ultrafiltrate with a composition closely similar to an equilibrium diffusate may be prepared by pressure filtration of skim milk through a suitable membrane,[36] provided that the flow is low enough to avoid any sieving effect. Estimation of the diffusible (or ultrafilterable) salt concentrations is achieved by calculation from the composition of the diffusate (or ultrafiltrate), and a colloidal concentration is calculated as the total skim milk concentration less the diffusible concentration.

Diffusible ions are sometimes said to be 'soluble' but this term is avoided here because of the implication that colloidal components are therefore insoluble. While it can be argued that the micellar calcium phosphate is a separate phase and hence insoluble, other non-diffusible components (e.g. proteins) are soluble in the thermodynamic sense and are not in a separate thermodynamic phase. The term 'the aqueous phase of milk' is sometimes used to denote the non-fat secretion of the mammary gland and sometimes it is applied to diffusible constituents only. For this reason and because of the semantic difficulty with the word 'phase', its use is avoided here. There is no satisfactory brief way of describing the calcium-phosphate-magnesium-citrate salt associated with casein micelles. It is designated micellar calcium phosphate (MCP) in this chapter to indicate its uniqueness and the fact of its intimate association with the casein proteins. The term colloidal will be used in preference to the more clumsy non-diffusible and non-ultrafilterable.

To convert an ion concentration in diffusate to the diffusible ion concentration in the skim milk requires two corrections. There is a Donnan potential arising from the charge on colloidal components which increases diffusate pH by about 0·02 over that in the skim milk. The activity in milk of a diffusible univalent ion differs from the diffusate activity by about 1%, and for a divalent ion the difference is about 2%. A second correction is required for the excluded-volume effect of co-solutes. White and Davies[5,36] considered that diffusate concentrations should be multiplied by the ratio of the weight of water in a given weight of skim milk to that in diffusate, to obtain diffusible ion concentrations in the skim milk; this is, in effect, an excluded-volume correction that neglects the possible non-solvent properties of water of hydration. The excluded-volume correction is typically about 4% and is additive with the Donnan effect for anions but compensates for the Donnan effect on cations. Davies and White[36] considered the Donnan effect small enough to be neglected and found no colloidal lactose or Cl, small colloidal concentrations of Na, K and citrate, and larger colloidal concentrations of Ca, Mg and P_i. To compute the Donnan correction for the multivalent ions, account must be taken of the formation of complexes since the effect on $CaCit^-$ or $MgCit^-$ is opposite in sign but smaller in magnitude to that on Ca^{2+} or Mg^{2+}. Calculation of the concentrations of such complexes can now be made routinely,[90,91] so Donnan corrections are feasible. Moreover, it is possible to estimate the Donnan effect by a determination of the pH difference between skim milk and diffusate. Care is required because slow changes in pH can occur through loss of CO_2, temperature and pressure dependent re-equilibrations, and post-secretory precipitation of calcium phosphate. Differences in pH of 0·04 or greater are unlikely to be due to a Donnan effect in cow's milk.

A representative determination and calculation of the partition of salts from the compositions of skim milk and diffusate is given in Table 2. The diffusate composition is the same as that in Table 3 and similar to the synthetic milk serum of Jenness and Koops.[92]

TABLE 2
CALCULATION OF THE PARTITION OF MILK SALTS

	Concentration (mM)			
	Skim milk	*Diffusate*	*Diffusible*	*Colloidal*
Ca	30·3	10·2	10·0	20·3
Mg	5·2	3·4	3·3	1·9
Cit	9·5	9·4	9·1	1·0
P_i	21·4	12·4	11·9	9·5
Na	22·8	22·0	21·8	1·0
K	39·7	38·0	37·5	2·2
Cl	31·5	32·3	31·5	0·0

Excluded-volume factor = 0·965.
Skim milk pH = 6·68; diffusate pH = 6·70.

About two-thirds of the Ca, one-third of the Mg and one-half of the P_i are colloidal in a typical milk. It is presumed that all colloidal P_i is present in the micellar calcium phosphate, and it appears that all the colloidal citrate is similarly located,[88,89] whereas colloidal Ca and Mg are partly incorporated in the MCP and partly bound in a more direct manner to casein. A small proportion of colloidal Ca is bound to α-lactalbumin;[93–95] β-lactoglobulin can also bind Ca and Mg,[96] and there are other, minor, calcium-binding proteins.[97] Nearly all the Na, K and Cl is diffusible.

4.2. Ion Equilibria in Milk Diffusate

Whereas the univalent ions Na^+, K^+ and Cl^- exist in milk largely as free ions, the proton and multivalent ions such as Ca^{2+}, Mg^{2+}, PO_4^{3-} and Cit^{3-} form complexes such as HPO_4^{2-}, $CaCit^-$ and $MgHPO_4$. The free ion concentrations (e.g. $[Ca^{2+}]$, $[Mg^{2+}]$) may therefore be very much less than the total present in the diffusate, and few of these free ion concentrations have been determined experimentally. One exception is $[Ca^{2+}]$ which has been determined colorimetrically,[5,90,98–100] by ion-selective electrode,[90,101–103] by ion-exchange resin methods,[90,102,104–107] and by calculation.[90,91,98,99,108,109] Free $[Mg^{2+}]$ has also been determined by ion-exchange resin methods[90,104–107] and by calculation.[90,91,98,99] Results obtained by the various methods are in agreement to the extent that all give about 20–30% of Ca and Mg as free ions. However, where comparisons have been made on the same samples, the results of determining $[Ca^{2+}]$ by the ion-exchange resin method of Christianson et al.[105] agree with the ion-selective electrode results[102] or are, on average, slightly greater,[90] the more so in early lactation milks. On the other hand, the colorimetric determination of $[Ca^{2+}]$ appears to give results that are higher by about 0·8 mM than those obtained by the three other techniques. Calculated $[Mg^{2+}]$ were found to be similar to results obtained by an ion-exchange method.[90] The $[Ca^{2+}]$ and $[Mg^{2+}]$ are typically in the ranges 1–3 mM and 0·4–1·3 mM, respectively, with values tending to be higher in early lactation milks of lower pH.[5,90] Concentrations of anions such as HPO_4^{2-} and Cit^{3-} can, in principle, be determined by resin exchange methods or by nuclear magnetic resonance spectroscopy but no experimental work has been reported; available values are by calculation only.

The calculation of the ion equilibria in milk or diffusate requires a model describing the complexes assumed to be important. Early attempts at building suitable models[98,99,108,109] required, for ease of computation, some or all of the following approximations: Mg was treated as equivalent to Ca; complex formation by P_i was neglected; complex formation by minor constituents such as phosphate esters was neglected; complex formation by univalent ions such as Na^+, K^+ and Cl^- was neglected; differences of ionic strength were not taken into account. Computer programs are now available[90,91] that allow more exact calculations to be made, and calculated free Ca^{2+} and Mg^{2+} concentrations are in reasonable agreement with experiment.

Central to the calculation is the construction of a mass balance equation for each of the thermodynamic components in the system. (If thermodynamic components are defined as ions rather than as neutral substances, the number of degrees of freedom is unchanged because of the requirement for charge balance.) The model must include all important components and all important forms in which these components exist in solution. In both published programs, only 1:1 complexes of metal ions with anions are considered, a good approximation for the major cations in milk. Furthermore, in a chemical analysis of milk diffusate it is not considered feasible to identify and quantify all the minor constituents, particularly anions. In the program of Lyster,[91] $[M^+]$ (where $[M^+]=[Na^+]+[K^+]$) or $[Cl^-]$ is adjusted to achieve charge balance, effectively treating the minor anions as having the same ion-binding properties as Cl^-. In the program of Holt *et al.*,[90] allowance is made for sulphate, carbonate and phosphate esters, and a charge balance is achieved by varying the concentration of a monocarboxylic

TABLE 3

CALCULATED CONCENTRATIONS (mM) OF IONS AND COMPLEXES IN A TYPICAL MILK DIFFUSATE[90]

Anion	Free ion	Cation complex			
		Ca^{2+}	Mg^{2+}	Na^+	K^+
H_2Cit^-	+	+	+	+	+
$HCit^{2-}$	0·04	0·01	+	+	+
Cit^{3-}	0·26	6·96	2·02	0·03	0·04
$H_2PO_4^-$	7·50	0·07	0·04	0·10	0·18
HPO_4^{2-}	2·65	0·59	0·34	0·39	0·52
PO_4^{3-}	+	0·01	+	+	+
Glc-1-PH^-	0·50	+	+	0·01	0·01
Glc-1-P^{2-}	1·59	0·17	0·07	0·10	0·14
H_2CO_3	0·11	−	−	−	−
HCO_3^-	0·32	0·01	+	+	+
CO_3^{2-}	+	+	+	+	+
Cl^-	30·90	0·26	0·07	0·39	0·68
HSO_4^-	+	+	+	+	+
SO_4^{2-}	0·96	0·07	0·03	0·04	0·10
$RCOOH$	0·02	−	−	−	−
$RCOO^-$	2·98	0·03	0·02	0·02	0·04
Free ion		2·00	0·81	20·92	36·29

$[Ca]=10·2$; $[Mg]=3·4$; $[Na]=22·0$; $[K]=38·0$; $[Cl]=32·3$; $[Cit]=9·4$; $[P_i]=12·4$; $[Glc$-1-$P]=2·6$; $[SO_4]=1·2$; $[CO_2]=0·44$; $[RCOOH]=3·1$; $pH=6·70$; $\Gamma/2=0·073$ (calculated).
Concentrations shown as: (+) <0·005; (−) not estimated.

acid fraction. The mass balance equations are solved by efficient algorithms, details of which are provided,[90,110] to give calculated concentrations of all ions and complexes assumed to be present in the original model. In solving the mass balance equations, conversion of activities to concentrations is required and a set of association constants is employed. Particular care is needed in selecting association constants from the literature since the conditions of their determination must be consistent with the method of calculation employed. Holt *et al.*[90] used constants for the most important equilibria that were corrected for complex formation by the background electrolyte, and hence computed the interactions in diffusate in the same way. Lyster[91] employed a set of constants which gave satisfactory results in predicting the pH titration of Ca and Mg solutions with H_3PO_4.

There is a semi-quantitative agreement between results calculated by the two ion equilibria programs, and such differences as there are appear to be explicable by the choice of association constants rather than the method of calculation. Most of the Ca and Mg and virtually all the citrate form $CaCit^-$ and $MgCit^-$ ions whereas nearly all the P_i exists as HPO_4^{2-} and $H_2PO_4^-$ ions. The ion concentrations in the typical milk diffusate used in Table 2 for the salt partition calculation are given in Table 3.

4.3. Mineral Phosphates of Calcium

Before considering the nature of MCP it is appropriate to review briefly the mineral phosphates of calcium that may occur in milk. The phase relationships of calcium phosphates are described in detail by Brown,[111] and a broad perspective on the occurrence of calcium phosphates in biological systems can be obtained from the monograph edited by Nancollas.[112]

There are several calcium phosphate salts that may be either thermodynamically or kinetically stable under physiological conditions, ranging from the relatively acid dicalcium salts to the basic mineral hydroxyapatite. Dicalcium phosphate (DCP) occurs in the form $CaHPO_4$ (monetite) and $CaHPO_4.2H_2O$ (brushite), but only the latter has been identified in bone and a number of other calcifying systems.[113,114] Brushite is the most soluble of the calcium phosphates considered here, but because of favourable crystal growth kinetics it precipitates from solution in circumstances where it is metastable with respect to more basic salts. Octacalcium phosphate (OCP, $Ca_4H(PO_4)_3.2\cdot5H_2O$) can appear as a precursor phase in the nucleation and growth of hydroxyapatite; it occurs, for example, in an epitaxial relationship with hydroxyapatite in dental enamel.[115] The mineral whitlockite (β-TCP, $Ca_3(PO_4)_2$) is second only to hydroxyapatite in its insolubility above about pH 6, and cations such as Fe^{3+} and Mg^{2+} seem either to make it more stable or to facilitate its formation from aqueous media. The apparently amorphous calcium phosphates (ACP) formed by rapid precipitation from aqueous solution at alkaline pH can transform, in the absence of inhibitory substances such as Mg^{2+}, ATP and pyrophosphate, into more basic salts. The first-formed crystalline phase is, however, OCP in the pH range 7–8·6 but this hydrolyses at a pH-dependent rate to an apatitic phase with TCP stoichiometry.[116] Formation of crystalline, stoichiometric apatite (HA, $Ca_5OH(PO_4)_3$) normally occurs only slowly, and many systems will remain metastable with respect to HA indefinitely.

Phase separation of calcium phosphates from solution can proceed by either homogeneous or heterogeneous nucleation mechanisms; the latter, because of its faster rate, is commonly observed at lower degrees of supersaturation than the former. How effective a substance is at acting as a heterogeneous nucleating agent depends in part on the similarity of its structure to that of the forming phase (the lattice disregistry parameter). The theory is based on the assumption that the interfacial energy is at a minimum when the lattice of the forming crystal matches that of the nucleating catalyst.

4.4. Micellar Calcium Phosphate

That milk contains a calcium phosphate salt has been recognized for more than a century, yet there is a continuing controversy over its physical form and the nature of its linkage to casein. In brief, the evidence that milk contains a calcium phosphate salt is as follows: (a) The total concentrations of Ca and P_i in milk are in excess of the solubilities of calcium phosphates; indeed, milk diffusates appear to be saturated or supersaturated solutions but contain only a fraction of the total Ca and P_i concentrations. (b) Addition to milk of Ca-chelating agents or cation exchange resins produces an increase of diffusible P_i and, similarly, chelation of P_i appears to solubilize Ca. (c) The X-ray absorption spectrum of casein micelles reveals a calcium phosphate phase resembling brushite[117] although the crystallites appear small by electron microscopy.[118] This evidence also supports the contention that an intimate linkage exists between casein micelles and the MCP. Other evidence bearing on this point is that it is impossible by differential centrifugation or column fractionation techniques to obtain MCP and casein micelles as separate fractions.[119,120] Further, on dissolving MCP the micelles dissociate into much smaller protein aggregates or monomers. The MCP appears to be identifiable as distinct regions of higher electron density in electron micrographs of thin, unstained, sections of native micelles.[121–123]

Prior to the work of Pyne in the 1930s it was disputed whether MCP was a di- or tri-calcic salt (reviewed by Pyne[124]), but the oxalate titration method of Pyne and Ryan[125] was accepted, until recently, as having resolved the issue in favour of the more basic salt.[88] In other work concerned with determining the stoichiometry of MCP, a difficulty has arisen in determining what fraction of the colloidal Ca is directly bound to casein (caseinate Ca) and what fraction is part of the MCP. Pyne and McGann[88] and Morr et $al.$[126] attempted to solve this problem by preparing a milk free of colloidal phosphate and dialysing it against raw milk. The non-diffusible Ca was then taken to be the same as the original caseinate Ca. By such means Pyne and McGann[88] found $Ca/P_i \approx 1.6$ in the MCP compared to $Ca/P_i \approx 1.4$ by the oxalate titration method. The difference was partly accounted for by their finding that citrate was bound to the colloidal P_i, which would lead to an underestimation of Ca/P_i by the oxalate titration method by about 0·1. Others[4,127,128] have computed caseinate Ca by less direct means. Thus, Schmidt[4] used the Ca-binding isotherms of Dickson and Perkins[129] to calculate that there was approximately as much caseinate Ca as caseinate P (about 6 mol per mol of whole casein) which is higher than the 4·0 typically obtained by the Pyne and McGann[88] or Morr et $al.$[126] procedures, and this has the effect of reducing the calculated Ca/P_i to about 1·5. A similar ratio was calculated by Alekseeva et $al.$,[130] again using Ca-binding data. Later work[89] has, however, largely confirmed the findings of Pyne and McGann[88] in giving caseinate $Ca \simeq 4$ mol per mol of whole casein and $Ca/P_i = 1.6$, albeit by a similar experimental approach. It was also found that colloidal citrate was associated exclusively with the MCP in the ratio $Cit/P_i = 0.097$ whereas Mg, like Ca, was partly bound directly to casein (typically 1 mol per mol of whole casein) and partly associated with MCP ($Mg/P_i = 0.044$).

Milk diffusates appear to be saturated or supersaturated with respect to calcium phosphates, and the ion equilibria programs make it particularly easy to calculate the degree of saturation of a diffusate with respect to any solid phase.[89,131] This provides an indirect means of determining the nature of the MCP. It is invariably found that diffusates are highly supersaturated with respect to HA, to a lesser degree with respect to β-TCP and OCP, marginally supersaturated with respect to brushite and monetite, and undersaturated with respect to the calcium monophosphates. Diffusates are saturated or undersaturated in tricalcium citrate, and far from saturated with respect to calcium hydroxide, calcium carbonates and magnesium hydroxide, carbonates, phosphates and citrate. Spontaneous precipitation of calcium phosphates can occur in diffusates prepared from raw milk, particularly if the temperature is raised, but some diffusates remain in a metastable state for many days. Presumably this is due to the slow kinetics of the spontaneous precipitation from solution of HA and β-TCP compared to the faster-forming, more-acidic phases such as brushite and OCP.

Holt[89] calculated the ion equilibria in over 80 ultrafiltrates in the natural pH range 6·4–7·4 and found an ion activity product independent of pH for a molecular formula $Ca(HPO_4)_{0.7}(PO_4)_{0.2}$. The significance of this is that the form of the solubility product equation is dictated by the stoichiometry of the equilibrium calcium phosphate phase. Ion activity products for DCP, OCP, TCP (or ACP) and HA were calculated also but that for OCP and the more basic phosphates had a clear dependence on pH. The solubility product evidence therefore indicates that MCP is a calcium phosphate salt only slightly more basic than DCP (see Fig. 2d). Doi and Niki[132] found earlier that, by the criterion of the ease with which MCP dissolves following addition of sodium citrate to milk, it resembles DCP rather than TCP.

Although Rose and Colvin[121] obtained a diffuse X-ray diffraction pattern of two or three rings from freeze-dried casein micelles, these may have been artefacts since, in other investigations on unheated milk systems[118,122,133] by X-ray and electron diffraction methods, no evidence has been found of an extensive crystal lattice. High-resolution electron microscopy of an enzymically deproteinated MCP preparation also revealed no regular pattern of interference fringes corresponding to a crystalline lattice that persists over distances greater than about 1·5 nm.[118] From a low-angle X-ray scattering study on a hydrazine deproteinated MCP preparation, McGann et al.[134] derived the radial distribution function for atoms in their sample and showed that it resembled that of a Mg-containing ACP. The decay in correlations with distance indicates that order does not persist for more than about 1 nm but it is not clear whether the short-range correlations are unique to ACP or could also be exhibited by other microcrystalline calcium phosphates; the first two peaks correspond to nearest neighbour atomic separations and are common to all calcium phosphate radial distribution functions, while the subsequent peaks are broad and strongly damped. Nevertheless, the interpretation of the radial distribution function by McGann et al.[134] constitutes the strongest challenge to the proposal that MCP has a brushite-like structure.[117,118]

With the development of high intensity photon sources from synchrotron storage rings, routine measurement of X-ray spectra at high resolution has become feasible. Extended X-ray absorption fine structure (EXAFS) spectroscopy and X-ray absorption near-edge structure (XANES) spectroscopy have proved useful in characterizing the short-range structure in MCP. Holt et al.[117] compared the X-ray absorption near the Ca edge of enzymically deproteinated MCP preparations[121] and freeze-dried casein micelles with those of brushite, monetite, β-TCP, crystalline HA and an ACP preparation. A clear similarity of MCP to brushite and, to a lesser extent, monetite, was revealed; the more basic calcium phosphates were distinctly different. The EXAFS and XANES measurements are thus in agreement with the solubility behaviour in indicating a DCP-like phase whereas chemical analysis gives a Ca/P_i ratio indicating a basic salt. As a resolution of this apparent paradox, Holt et al.[117] suggested that the phosphate moiety of phosphoserine residues in the caseins can substitute in surface sites in a brushite-type lattice. For example, with the colloidal concentrations shown in Table 2 and 1 mM casein—concentrations that are typical of cow's milk—casein phosphate groups (P_o) amount to about 6 mM and caseinate Ca is about 4 mol per mol of whole casein. Thus the ratio Ca/P_i in the MCP is 1·6, typical of apatites, but the ratio $Ca/(P_i + P_o)$ is 1·0, as required for a DCP.

Holt et al.[135] have responded to the interpretation by McGann et al.[134] of the X-ray radial distribution function by measuring the EXAFS spectrum of a hydrazine deproteinated MCP preparation and again found a brushite-like spectrum, indistinguishable from those obtained previously.

One difficulty in interpreting the structure of MCP as a brushite-like salt attaches to the oxalate[125] and related[136] pH titrations. These have led to the conclusion that P_i in the MCP exists mostly as PO_4^{3-} ions with only about 20% HPO_4^{2-}. There is in the interpretation of these experiments an unstated assumption that the pH titration of the casein is unaffected by solution of the MCP. Clearly this is not so if phosphoserine residues, as RPO_4^{2-} groups, are part of the MCP lattice, for they will tend to form RPO_4H^- groups at milk pH in the absence of colloidal P_i, and hence the fraction of P_i as HPO_4^{2-} will be underestimated.

Similarly, any other amino acid side-chain that becomes exposed and protonated after dissolution of the MCP will have the same consequence, and calculations indicate that about 4 mol per mol of whole casein of such other groups are required to explain the titration results with a brushite-like MCP model.

If it is accepted that MCP has a brushite-like structure then, by analogy with the mineral calcium phosphates, some further conclusions can be made that may be relevant to the long-term storage of milk and some milk products. First, while milk diffusates are saturated with respect to MCP they are supersaturated with respect to the less soluble and more basic calcium phosphates, and possibly also to monetite. Even if formation of HA is too slow to be observed, β-TCP may form at the natural pH and above. Second, a brushite-like MCP may hydrolyse to form an OCP-like and, subsequently, an apatitic phase, though again the kinetics of the transformation may be slow. Third, more stable phases of calcium phosphate may form independently of casein micelles and may grow at the expense of MCP, causing micelles to dissociate. Fourth, even if new phases are not formed, growth of larger crystallites of MCP at the expense of smaller ones (Ostwald ripening) may be favoured thermodynamically. Whether or not ripening is favoured depends on how the interfacial free energy affects the growth of MCP within the protein matrix of the micelle. The caveat to these arguments is that the analogy with brushite should not be taken too far; the interaction with the caseins makes MCP a substance unique to milk, or at least to phosphoprotein–calcium phosphate complexes, and this may confer correspondingly unique physical properties.

4.5. Calcium and Magnesium Ion Binding to Casein

A considerable number of studies have been made of the ion-binding properties of isolated individual caseins, but problems remain in applying the results to an understanding of the caseinate Ca and Mg of milk.

It has been established that isolated whole casein binds Ca^{2+} to an increasing extent as pH and the free $[Ca^{2+}]$ increase (reviewed by Waugh et al.[137]). Dickson and Perkins[129] found that, at a given pH, Mg^{2+} was bound more strongly than Ca^{2+}, and the binding capacities of the caseins were in the order $\alpha_{s1}->\beta->\kappa$-casein which is in the order of their phosphoserine contents. Phosphorylation increased and dephosphorylation or esterification decreased binding capacity.[129,138] This, and changes in the infrared spectrum of casein produced by Ca^{2+},[139–141] implicate phosphoserine residues in the binding, possibly as parts of the stronger binding sites. Waugh *et al.*[142] studied Ca^{2+} binding to α_s- and β-casein at 37°C, pH 6·6 and various ionic strengths. Imade *et al.*[143] measured Ca^{2+} binding to α_{s1}-, β-, κ- and whole casein in the pH range 6·0–9·0, and found the binding capacities at a given pH to be in the order α_{s1}-> whole > β- > κ-casein. Dalgleish and Parker[144,145] studied the effect of temperature, pH and ionic strength on Ca^{2+} binding by α_{s1}- and β-casein and considered also earlier work.[129,142] They represented binding isotherms by a two-parameter equation involving a dissociation constant K^* for the first ion bound, and a substitution parameter N ($0 \leqslant N \leqslant 1$) which allowed for a decrease in binding strength as more Ca^{2+} ions were bound. Both K^* and N increased with increase of pH and temperature and decreased with increase in ionic strength. The K^* for β-casein was the same as that for α_{s1}-casein under comparable conditions, but N was less and hence responsible for the generally lower binding capacity of β- compared to α_{s1}-casein. Upton and Holt[146] have applied the two-parameter model[144,145] to describe both the separate and the simultaneous binding of Ca^{2+} and Mg^{2+} to whole casein in the pH range 6·0–7·4, ionic strength 0·08 and 40°C. Both K^* and N for Ca^{2+} binding were less than or equal to the values for Mg^{2+} and generally similar to the parameters of β-casein. The effects of the binding of Mg^{2+} on K^* and N for Ca^{2+} binding were complex in that the binding of the first Mg^{2+} appeared to increase K^* whereas subsequent binding decreased K^* to below its original value. At the typical free $[Ca^{2+}]$ and $[Mg^{2+}]$ of milk (Table 3) it was calculated that 2·5 mol of Ca^{2+} and 1·3 mol of Mg^{2+} bind per mol of whole casein. It is presumed, but not established, that the Ca and Mg caseinate of native micelles are comparable to the complexes formed by Ca^{2+} and Mg^{2+} with isolated caseins. However, if the phosphoserine residues in native micelles are mostly part of the MCP,[117] they cannot function in Ca^{2+} binding sites.

4.6. Distribution of Ions Around the Casein Micelle

The surface region of micelles exerts a dominant influence on their clotting and aggregation reactions, so the distribution of ions near the surface is of considerable importance for micelle structure and stability.

Casein micelles have a negative surface potential and are surrounded by an ion atmosphere of thickness about 1 nm. Thus, if the surface were smooth and flat, an electrophoretic mobility of $-1\cdot3 \times 10^{-8}\,m^2\,s^{-1}\,V^{-1}$, typical of experimental values,[147,148] would correspond to a ζ-potential of about $-17\,mV$. The concentrations of divalent cations and univalent anions near the plane of shear are changed by factors of 3·7 and 0·5 respectively compared to their values at an infinite distance from the surface. There is evidence, however, for a hairy layer on the outside of micelles, formed at least in part from the macropeptide of κ-casein.[149–152] Such a hairy layer provides a powerful electrosteric stabilizing mechanism for micelles,[153] an important component of which arises from the free energy of interaction of the distributed double layers. Calculations of the distribution of ions near to a hairy micelle surface depend on factors that include the density of hairs, their length and the number of fixed charges. For the particular model considered in Fig. 3, the potential at the inner surface of the hairy layer, about $-13\,$mV, decreases to a plateau of about $-5\,$mV over the greater part of the hairy layer, begins to decrease again within 3 nm of the outer boundary, and is

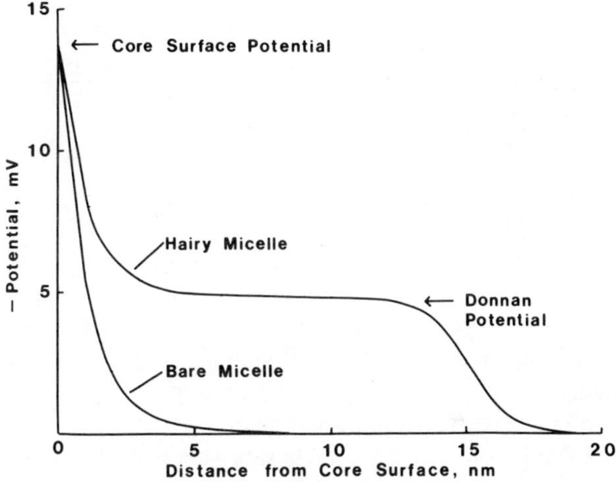

FIG. 3. Variation of the electrical potential near the surface of a bare and a hairy casein micelle. The parameters in the hairy micelle model were chosen to fit experimental results as follows. (a) An electrophoretic mobility of $-1.3 \times 10^{-8}\,\text{m}^2\,\text{s}^{-1}\,\text{V}^{-1}$, calculated from the theory of Donath and Pastushenko.[154] (b) A hydrodynamic voluminosity of about $3\,\text{ml}\,\text{g}^{-1}$.[150,155–157] (c) A change in hydrodynamic radius of $\approx 5\,$nm (calculated from the theory of McCammon et al.[158]) when 90% of the hairs are removed by chymosin,[159] with a concomitant decrease in electrophoretic mobility to $-0.9 \times 10^{-8}\,\text{m}^2\,\text{s}^{-1}\,\text{V}^{-1}$. These constraints require that all the κ-casein is at the surface, forming a hairy layer of about 15 nm thickness. Potential calculations were made with the theory of Hesselink[160] using a linearized Poisson–Boltzmann equation, but within the plateau region of constant potential the Donnan approximation ($\nabla^2 \Psi = 0$) is valid.

close to zero about 3 nm beyond this boundary. The Donnan approximation gives a potential close to that in the plateau, indicating that only small departures from electrical neutrality occur. Thus, although the potentials are less negative than close to a smooth particle of equal surface charge, the diffuse double layer is deeper and interacting micelles generate a repulsive force at large distances of separation. Compared to their concentrations at an infinite distance from the surface, mobile divalent cations and univalent anion concentrations within the plateau region are changed by factors of 1·40 and 0·85, respectively.

5. EFFECTS OF SOME PROCESSING VARIABLES ON THE ION EQUILIBRIA AND PARTITION OF SALTS

In the manufacture of dairy products a number of operations are commonly performed, each of which can affect the physicochemical equilibria of the salts. These operations include the use of permitted additives, concentration by a variety of means, heating, cooling and prolonged storage.

5.1. Effects of Additives

Various salts, acids and alkalis are added to effect desired changes in the processing characteristics of milk and milk products. These include Ca salts to improve coagulation in cheesemaking, various phosphate and citrate salts to improve the heat stability and storage life of concentrated milk products and to prevent deposit formation on heat exchanger and various membrane surfaces, and polyphosphates and metaphosphates to alter a wide range of properties including the viscoelastic and melting behaviour of cheese. A qualitative understanding of the effects of some of these additives on the ion equilibria and partition of milk salts has existed for many years and was reviewed by Pyne.[1] A quantitative description of the consequences of salt additions in kinetic and thermodynamic terms is still, however, some way from being realized. The principles that determine the thermodynamic equilibria have been described in the previous section: chelation of Ca^{2+} or HPO_4^{2-} or protonation of HPO_4^{2-} reduces the ion activity product of the MCP which then dissolves; increasing free $[Ca^{2+}]$ or $[HPO_4^{2-}]$ tends to precipitate calcium phosphate; Ca^{2+} binds to casein to an increasing extent as pH and free $[Ca^{2+}]$ increase. Other Ca and Mg salts (e.g. calcium carbonate, pyrophosphate, lactate or citrate, or more stable phases of calcium phosphate) may precipitate if their solubility product is exceeded.

Acidification of milk, either by bacterial souring or by addition of acid, dissolves progressively the MCP and reduces the binding of Ca^{2+} and Mg^{2+} to casein. Re-equilibration is slow and normally requires more than the 1–48 h allowed in most scientific investigations.[161] Pyne and McGann[88] were able to prepare milk free of calcium phosphate by acidification to pH 4·9 in the cold, and Davies and White[36] found that about 88% of the total $[P_i]$ was diffusible at 20°C when the pH was reduced to 5·60. At its natural pH, milk is highly supersaturated with respect to the more basic calcium phosphates, marginally supersaturated with respect to the dicalcium phosphates, and saturated or under-saturated with respect to tricalcium citrate.[109] On lowering the pH to below about 6, however, the situation is reversed in that milk can become saturated or supersaturated with respect to tricalcium citrate and undersaturated in the dicalcium phosphates. Only when the pH moves to below about 5 does the natural milk system become undersaturated with respect to all phases.[131] Acidification with citric acid will raise the degree of supersaturation in tricalcium citrate compared to that obtained by other means of reducing pH. On raising pH, $H_2PO_4^-$ is converted into HPO_4^{2-} and calcium phosphate tends to precipitate, causing free $[Ca^{2+}]$ to fall. Furthermore, if Na_2HPO_4 is used to raise pH, the degree of supersaturation produced is correspondingly greater and the free $[Ca^{2+}]$ at equilibrium smaller.[99,162] Addition of trisodium citrate also raises pH but Ca is chelated by citrate so the MCP dissolves and the fall in free $[Ca^{2+}]$ is not as large as with an equimolar addition of Na_2HPO_4.[70,99,162–164] Little effect on diffusible $[Ca]$ or free $[Ca^{2+}]$ and $[Mg^{2+}]$ occurs when NaH_2PO_4 is added, and pH is reduced only slightly.[162] Likewise, the salt partition is perturbed only slightly by additions of $MgCl_2$ for, while free $[Ca^{2+}]$ is raised by displacement from its citrate complex, $[HPO_4^{2-}]$ is reduced by conversion into $H_2PO_4^-$ at the lower pH. Rarely, a milk or concentrated milk is found to be stabilized to heat by small additions of $CaCl_2$; the effect is thought to be due to the drop in pH induced largely by precipitation of calcium phosphate, for free $[Ca^{2+}]$ and $[Mg^{2+}]$ are raised.[99,107,165,166]

Of particular interest in milk processing are the effects of additions of pyrophosphate, polyphosphates and metaphosphates, but there are qualitative and quantitative differences in their action, compared to those of orthophosphates and other milk salts, which are by no means well understood. These substances increase the viscosity of milk, decrease its whiteness, and inhibit gelation or coagulation on heating or during storage of concentrated milk products.[162,163,167-171] Other instances can be given of their use in a wide range of food manufacturing processes.[167,171-175] On their addition to milk, pH is raised to a variable extent, hardly at all with sodium hexametaphosphate and most with sodium pyrophosphate,[162] and, because of their Ca-chelating properties,[176-178] free $[Ca^{2+}]$ is reduced. However, unlike the orthophosphates, additions of which can cause milks to gel,[171,179-182] combination of hexametaphosphate and some other polyphosphates with casein can lead to dispersion of casein micelles and re-peptisation of coagulated protein.[163,164,167] Odagiri and Nickerson[163] found that, when hexametaphosphate was added to skim milk to a final concentration of 30 mM, colloidal P_i was almost completely dissolved and diffusible [Ca] increased. Most of the hexametaphosphate remained diffusible but about 2 mM was colloidal. In contrast, Vujičić et al.[162] added various polyphosphates at a level equivalent to 22·6 mM P and found diffusible [Ca] appreciably reduced, particularly with pyrophosphate, and it is clear from their data that most of the added P also became colloidal, possibly through formation of insoluble Ca salts or complexes with protein. After heat treatment or on long-term storage, polyphosphates in milk products are hydrolysed to shorter chain lengths, mainly mono- and di-phosphate, and the rate may be enhanced by enzymic reaction.[183-186] Jones et al.[173] have described a whey protein precipitant formed by mixing Fe^{3+} with polyphosphates; its precipitating ability was enhanced by ageing for 2 weeks.

5.2. Effects of Concentration

The method used to concentrate milk influences the final partition and equilibria of milk salts. Ultrafiltration allows diffusible components through the membrane so their concentrations are hardly changed while concentrations of colloidal salts are increased by the volume concentration factor. Filtration by the reverse osmosis procedure and evaporative concentration retain most, or all, of the milk salts and precipitation of diffusible Ca and P_i occurs at a temperature-dependent rate, to reach equilibrium or quasi-equilibrium values on storage for long times.

Changes with time in the ion equilibria and partition of salts in evaporative concentrates do not follow directly from the effect of concentration since the effect of the heat treatment also exerts an influence. Discussion of this point is deferred to the following section.

Evaporative concentration by a factor of about 3 produces a pH decrease of 0·2–0·7, mainly due to concentration and acid production through heating, rather than through liberation of H^+ in calcium phosphate formation, though up to 5 mM P_i becomes colloidal. The free $[Ca^{2+}]$ and $[Mg^{2+}]$ increase, but by less than the concentration factor.[99,104,187,188] The increased ionic strength reduces ion activity coefficients, so larger concentrations of diffusible P_i and Ca are allowed, at a given pH, while still satisfying the ion activity product for the MCP.[99] That some degree of redistribution of salts occurs on prolonged storage is demonstrated by the occasional formation of inorganic deposits in evaporated or sweetened condensed milks. Long-term storage at ambient temperatures and at a pH of about 6 results occasionally in crystals of almost pure (98%) tricalcium citrate.[189,190] As would be expected, crystal formation is promoted, and its rate is enhanced by additions of $CaCl_2$ and citric acid and retarded by additions of disodium phosphate prior to sterilization.[191] Storage at low temperatures inhibits the formation of crystals of tricalcium citrate; instead, granules of material rich in calcium phosphate can occasionally form.[192] Deposits with an infrared spectrum very similar to that of the calcium phosphate rich granules in evaporated milk were obtained by Fox *et al.*[192] on storage of a milk diffusate at 4°C for 1 month, and both had a calcium to total phosphorus ratio close to 1. Their interpretation of the infrared spectra was that they were similar to that of brushite but not identical with it.

Concentration by ultrafiltration usually involves only a mild heat treatment, and the pressure perturbs the equilibria only slightly, so the small changes in diffusible salt concentrations observed can be explained by the increase in the Donnan effect as a result of concentrating the proteins.[193–195] The findings of Brulé *et al.*[194] on the effect of pH adjustment of milk on diffusible salt composition were in qualitative agreement with the studies of Davies and White:[36] at pH 5·6, the lowest pH studied, ultrafiltrate [Ca] and [P] were increased by factors of about 2 and 1·4, respectively. With up to 6% added NaCl there was a notable increase in ultrafiltrate [Ca], a lesser increase in [Mg], and some dissociation of casein micelles. Brulé and Fauquant[196] acidified retentates, and analysis of the ultrafiltrates prepared therefrom some 30 min later showed that the colloidal Ca concentration per gram of protein was greater at higher degrees of concentration. As in milk, addition to retentate of trisodium citrate increased pH and diffusible Ca; most of the changes occurred within 1 h but these were still not complete after 4 h. On adding $CaCl_2$ to retentates, the proportion of Ca becoming colloidal was greater the higher the degree of concentration, suggesting that the casein was able to bind more Ca^{2+}.

5.3. Effects of Temperature

The effects of temperature change on the milk salts were reviewed by Pyne[1] and there has been only a little subsequent progress in our understanding of the phenomena involved. On cooling, some of the colloidal salts dissolve, and the previous partition may be re-established on re-warming. Likewise, when milk is gently warmed and then cooled, the original state is re-established. Prolonged storage of frozen milk or severe heating, however, can cause changes in salt partition that are irreversible.

Davies and White[36] cooled skim milk to 3°C and showed that almost 0·5 mM more P_i and Ca became diffusible within 24 h. A similar change in Ca partition was observed in milks concentrated by ultrafiltration,[196] with some evidence that the loss per unit weight of casein was less in more highly concentrated suspensions.[197]

The destabilization of frozen milk and milk products during storage is caused partly by a slow change in the salt partition, as measured after thawing and re-equilibration.[1] Chen and Yamauchi[198] reported the changes in salt partition in skim milk during frozen storage at $-7°C$ for up to 210 days. In one experiment, ultrafiltrate [Ca] decreased progressively up to about 120 days and then levelled off; the amount of Ca found at any stage increased with the time allowed for equilibration after thawing. In contrast, ultrafiltrate $[P_i]$ was constant at its initial value for about the first 15 days, decreased rapidly up to 60 days and then remained fairly constant at about 5 mM up to 150 days, after which a further decrease occurred. Equilibration time after thawing had a relatively small effect on ultrafiltrate $[P_i]$ in the plateau region from 90 to 150 days. In a second experiment with a different milk, qualitatively similar changes were seen but the decrease in ultrafiltrate $[P_i]$ was delayed somewhat more. Precipitation of calcium-phosphate-citrate was observed on prolonged storage of milk diffusate but the characterization of this and the skim milk precipitate has proceeded no further.[199]

Heating of milk leads to many changes, including a decrease in diffusible Ca and P_i which may reverse to some extent on cooling.[200–204] Moreover, if the heated milk is dialysed against a large volume of raw milk, the normal diffusible concentrations are restored and little if any of the calcium phosphate precipitated by heating can then redissolve.[202] Free $[Ca^{2+}]$ also declines and can recover on cooling,[99,101,102,205] though not completely if the heat treatment is severe. The pH of milk also declines on heating, partly through expulsion of CO_2,[206] adjustments in other ion equilibria, precipitation of calcium phosphate and, to an increasing extent as temperature is raised above 90°C, degradation of lactose.[207,208] The effects of heat treatment on pH can be quite significant in determining the subsequent pattern of change in salt distribution. Aoki and Imamura[209] found that, if skim milk was sterilized by a high-temperature short-time treatment (130–135°C for up to 75 s), [Ca] in centrifuged serum decreased further with storage over 15 months, whereas in milk heated at 120°C for 15 min the pH was lower (6·3 rather than 6·6) and, while the serum [Ca] had decreased more on sterilization, serum [Ca] increased on storage. In concentrated sterilized milk, serum [Ca] was observed to decrease with storage time, though here the effects of concentration and heating are confounded.[210] Whitney *et al.*[211] found that the decrease in serum [Ca] and [Mg] observed on storage of a concentrated skim milk was dependent on storage temperature, as was the direction and amount of change of serum [citrate] and [P_i].

There is some evidence of a possible change of state in the calcium phosphate of heated milks. Thus, the interaction of calcium phosphate with casein is weakened and serum casein content increases.[164,209,210,212,213] While Edmondson and Tarassuk[203] found that a dicalcium phosphate was formed on addition of 0·15% Na_2HPO_4 and heating milk to 88°C for 15 min, Doi and Niki[132] were able to distinguish between di- and tri-calcium phosphate by their solubility in sodium citrate solution and found that MCP in unheated milk resembled DCP in this respect. When the milk was heated at 85°C for up to 1 h, the colloidal salts remained DCP-like but at 95°C after 15 min and at 115°C the solubility decreased and colloidal [citrate] increased. Furthermore, their figures show that the transformation to the more insoluble form involved a conversion of the original MCP and not just precipitation of diffusible Ca and P_i. When milk is heated to 85°C for 30 min and then dialysed against unheated milk at 5°C the increased colloidal P_i content produced by heating is maintained and diffusible concentrations are restored to about their original level.[202] However, if the milk is more severely heated (120°C for 50 min) a further increase in colloidal P_i is observed after dialysis,[214] suggesting that a new equilibrium or quasi-equilibrium state is being approached. Mineral-rich deposits on heat exchanger surfaces have been found to contain β-TCP,[215–218] though it is not clear whether the phase is formed from a transformation of the MCP or directly from diffusible Ca and P_i. Lyster[131] has made calculations to show what phases may precipitate from milk as a function of temperature (0–150°C) and pH (2–10); HA should form at all temperatures above pH 6 and tricalcium citrate may form in the pH range 4·7–5·8 below about 40°C. If HA formation is inhibited, the most stable phase at a pH greater than about 6·8 is always β-TCP, and there are regions of the phase diagram in which dicalcium phosphate and tricalcium citrate phases should form. It is clearly necessary to identify the phases formed in milk systems so as to move from a qualitative to a quantitative description of the effects of milk processing variables. However, few of the mineral-rich deposits that cause problems in the manufacture of milk products are well defined crystalline substances, so the physicochemical techniques that have proved so useful in establishing the short-range structure in MCP may find further applications in this field.

ADDENDUM

A recent study[219] of variations in Ca^{2+} activity in milk merits inclusion in this review. In nine samples of bulk skim milk they found a mean Ca^{2+} activity of $0·85 \pm 0·02$ mM, corresponding to free $[Ca^{2+}] = 2·1$ mM. On adjusting milk pH to 6·0, the activity was about double and at pH 7·5 about half the activity at the natural pH. On heating at 115°C for various times, Ca^{2+} activity fell and recovered slowly on cooling, such that the activity was directly proportional to log(time). In a typical experiment the activities after 1 min, 30 min, 24 h and 50 h were 58%, 70%, 88% and 89% of the initial value. Likewise, on reconstituting skim milk powders, Ca^{2+} activity also increased linearly with log(time), and the authors present some evidence that the pattern of recovery reflects the thermal history of the powder. These results encourage the view that a general description of the kinetics of changes in salt equilibria might be possible; it would then be applicable to a wide range of milk samples and processing conditions.

ACKNOWLEDGEMENT

It is a pleasure to acknowledge the help of Dr D. T. Davies in preparing this chapter.

REFERENCES

1. PYNE, G. T., *J. Dairy Res.*, 1962, **29**, 101.
2. ROOK, J. A. F. and WHEELOCK, J. V., *J. Dairy Res.*, 1967, **34**, 273.
3. PEAKER, M., *IDF Bull.*, 1980, **125**, 159.
4. SCHMIDT, D. G., *Neth. Milk Dairy J.*, 1980, **34**, 42.
5. WHITE, J. C. D. and DAVIES, D. T., *J. Dairy Res.*, 1958, **25**, 236.
6. WRIGHT, N. C., *J. Agric. Sci. Camb.*, 1928, **18**, 478.
7. TAYLOR, J. E. and HUSBAND, A. D., *J. Agric. Sci. Camb.*, 1922, **12**, 111.
8. ROOK, J. A. F. and WOOD, M., *Nature*, 1959, **184**, 647.
9. DAVIES, W. L., *J. Dairy Res.*, 1933, **4**, 273.
10. PESKETT, G. L. and FOLEY, S. J., *J. Dairy Res.*, 1933, **4**, 279.
11. BARRY, J. M. and ROWLAND, S. J., *Biochem. J.*, 1953, **54**, 575.
12. OSHIMA, M. and FUSE, H., *J. Dairy Res.*, 1981, **48**, 387.
13. LINZELL, J. L. and PEAKER, M., *Physiol. Rev.*, 1971, **51**, 564.
14. LINZELL, J. L. and PEAKER, M., *J. Physiol.* (Lond.), 1971, **216**, 701.
15. LINZELL, J. L. and PEAKER, M., *J. Physiol.* (Lond.), 1971, **216**, 683.
16. KUHN, N. J. and WHITE, A., *Biochem. J.*, 1977, **168**, 423.
17. HOLT, C., *J. Theor. Biol.*, 1983, **101**, 247.
18. SASAKI, M. and KEENAN, T. W., *Exp. Cell Res.*, 1978, **111**, 413.
19. KUHN, N. J., CARRICK, D. T. and WILDE, C. J., *J. Dairy Sci.*, 1980, **63**, 328.
20. FOSTER, R. C., *Comp. Biochem. Physiol.*, 1978, **A61**, 253.
21. PATTON, S. and JENSEN, R. G., *Progr. Chem. Fats Lipids*, 1975, **14**, 163.
22. WOODING, F. B. P. and MORGAN, G., *J. Ultrastruct. Res.*, 1978, **63**, 323.
23. NEVILLE, M. C. and PEAKER, M., *J. Physiol.* (Lond.), 1979, **290**, 59.
24. LINZELL, J. L., MEPHAM, T. B. and PEAKER, M., *J. Physiol.* (Lond.), 1976, **260**, 739.
25. BAUMRUCKER, C. R., In: *Lactation*, Vol. 4, 1978, Academic Press, New York, p. 463.
26. WEST, D. W., *Biochim. Biophys. Acta*, 1981, **673**, 374.
27. ZULAK, I. M., EIGEL, W. N. and KEENAN, T. W., *J. Dairy Sci.*, 1978, **61**, 167.
28. POWELL, J. T. and BREW, K., *J. Biol. Chem.*, 1976, **251**, 3645.
29. NEVILLE, M. C. and WATTERS, C. D., *J. Dairy Sci.*, 1983, **66**, 371.
30. PEAKER, M., In: *The Biochemistry of Lactation*, T. B. Mepham (ed.), 1983, Elsevier, Amsterdam, p. 285.
31. LINZELL, J. L., *J. Physiol.* (Lond.), 1960, **153**, 510.
32. TALLAMY, P. T. and RANDOLPH, H. E., *J. Dairy Sci.*, 1970, **53**, 1386.
33. ROOK, J. A. F. and CAMPLING, R. C., *J. Dairy Res.*, 1965, **32**, 45.
34. KEOGH, M. K., KELLY, P. M., O'KEEFFE, A. M. and PHELAN, J. A., *Irish J. Food Sci. Technol.*, 1982, **6**, 13.
35. WHITE, J. C. D. and DAVIES, D. T., *J. Dairy Res.*, 1963, **30**, 171.
36. DAVIES, D. T. and WHITE, J. C. D., *J. Dairy Res.*, 1960, **27**, 171.
37. ANAGAMA, Y. and KAMI, T., *J. Fac. Fish Anim. Husb. Hiroshima Univ.*, 1964, **5**, 449.
38. D'AGOSTINO, A., BARBARO, A. and ZANNELLI, C., *Arch. Vet. Ital.*, 1966, **17**, 429.
39. PEAKER, M. and LINZELL, J. L., *Nature*, 1975, **253**, 464.

40. PEAKER, M., FAULKNER, A. and BLATCHFORD, D. R., *J. Dairy Res.*, 1981, **48**, 357.
41. ERHARDT, G., and SENFT, B., *Milchwissenschaft*, 1982, **37**, 20.
42. KITCHEN, B. J., *J. Dairy Res.*, 1981, **48**, 167.
43. BOGIN, E. and ZIV, G., *Cornell Veterin.*, 1973, **63**, 666.
44. HORVÁTH, G., MOHAMED, A. I. H., VARGA, J., SZEMERÉDI, G. and QUARINI, L., *Dairy Sci. Abstr.*, 1981, **43**, 370.
45. OSHIMA, M. and FUSE, H., *Jap. J. Dairy Food Sci.*, 1982, **31**, 43.
46. BLAXTER, K. L., REINEKE, E. P., CRAMPTON, E. W. and PETERSEN, W. E., *J. Anim. Sci.*, 1949, **8**, 307.
47. CHANDA, R., MCNAUGHT, M. L. and OWEN, E. C., *Biochem. J.*, 1952, **51**, 543.
48. TUCKER, H. A. and REECE, R. P., *J. Dairy Sci.*, 1961, **44**, 1751.
49. OSHIMA, M., FUSE, H., ISHII, T. and NAKAGAKI, K., *Jap. J. Zootech. Sci.*, 1978, **49**, 828.
50. FISHER, L. J., MACINTOSH, A. L. and CARSON, R. B., *Can. J. Anim. Sci.*, 1970, **50**, 121.
51. LEONHARD, I. and ŻYWCZOK, H., *Roczn. Naukrol. (Ser. Zootech.)*, 1963, **83**, 495.
52. GOLOVNIN, A. M. and MOLCHANOV, I. K., *Moloch. Prom.*, 1969, **30**, 24.
53. SLOBODYANIK, K. F. and GIRNA, O. V., *Fiziol. Biokhim. Silsk. Tvarin*, 1972, **19**, 39.
54. SHIMIS, V. YU, *Dairy Sci. Abstr.*, 1976, **38**, 230.
55. GAVRISH, V. G., *Sbornik Rabot*, 1974, **36**, 15.
56. EVANS, J. L. and KIM, C. W.,'*J. Dairy Sci.*, 1977, **60**, 115.
57. MITAMURA, K., *J. Fac. Agric. Sapporo*, 1937, **41**, 97.
58. LANDOU, L. and GAZO, M., *Vet. Easopia*, 1956, **5**, 403.
59. ATRAMENTOVA, V. G. and IOPA, V. A., *Dairy Sci. Abstr.*, 1972, **34**, 598.
60. JACOBSON, D. R., HEMKEN, R. W., HATTON, R. H., BUTTON, F. S. and ENLOW, C. M., *J. Dairy Sci.*, 1975, **58**, 750.
61. BRÁZDA, M. and DEDEK, J., *Dairy Sci. Abstr.*, 1977, **39**, 254.
62. SOMMER, H. H. and BINNEY, T. H., *J. Dairy Sci.*, 1923, **6**, 176.
63. LABUSCHAGNE, J. H. A. and LANDREY, J. S. A., *S. Afr. J. Dairy Technol.*, 1969, **1**, 31.
64. ECHENIQUE, L. and SUAREZ, B., *Compt. Rend. Soc. Biol.*, 1935, **120**, 570.
65. ECHENIQUE, L., *Compt. Rend. Soc. Biol.*, 1937, **124**, 589.
66. DAVIES, D. T. and WHITE, J. C. D., *J. Dairy Res.*, 1958, **25**, 256.
67. SEEKLES, L. and SMEETS, W. T. G. M., *Neth. Milk Dairy J.*, 1947, **1**, 7.
68. BOOGAERDT, J., *Nature*, 1954, **174**, 884.
69. YOSHIDA, S., *J. Fac. Appl. Biol. Sci. Hiroshima Univ.*, 1981, **20**, 71.
70. HOLT, C. and MUIR, D. D., *J. Dairy Res.*, 1979, **46**, 433.
71. HOLT, C., MUIR, D. D. and SWEETSUR, A. W. M., *J. Dairy Res.*, 1978, **45**, 183.
72. HOLT, C., MUIR, D. D., THOMAS, P. C., ZAMMIT, V. A. and PEAKER, M., *Ann. Rep. Hannah Res. Inst. (1979)*, 1980, 63.
73. REINERT, A. and NESBITT, J. M., *J. Dairy Res.*, 1959, **26**, 128.
74. DAVIDOV, R. and KRUGLOVA, L., *Moloch. Prom.*, 1960, **21**, 42.
75. BATRA, S. C. and DEMAN, J. M., *Milchwissenschaft*, 1964, **19**, 531.

76. BATTISTOTTI, B., *Latte*, 1964, **37**, 754.
77. MITCHELL, G. E., *Aust. J. Dairy Technol.*, 1979, **34**, 158.
78. MURONOVA, G. V. and KLIMOVSKII, I. I., *Moloch. Prom.*, 1982, **6**, 32.
79. KAMAL, T. H., JOHNSON, H. D. and RAGSDALE, A. C., *J. Dairy Sci.*, 1961, **44**, 1655.
80. ORMROD, I. H. L., THOMAS, P. C. and WHEELOCK, J. V., *Proc. Nutr. Soc.*, 1979, **38**, 121A.
81. ORMROD, I. H. L., THOMAS, P. C. and WHEELOCK, J. V., *Proc. Nutr. Soc.*, 1980, **39**, 33A.
82. FAULKNER, A. and CLAPPERTON, J. L., *Comp. Biochem. Physiol.*, 1981, **68A**, 281.
83. ORMROD, I. H. L., PhD Thesis, University of Glasgow, 1981.
84. CHAIYABUTR, N., FAULKNER, A. and PEAKER, M., *Brit. J. Nutr.*, 1981, **45**, 149.
85. FAULKNER, A. and PEAKER, M., *J. Dairy Res.*, 1982, **49**, 159.
86. PEAKER, M., *Symp. Zool. Soc. Lond.*, 1977, **41**, 113.
87. ORMROD, I. H. L., HOLT, C. and THOMAS, P. C., *J. Dairy Res.*, 1982, **49**, 179.
88. PYNE, G. T. and McGANN, T. C. A., *J. Dairy Res.*, 1960, **27**, 9.
89. HOLT, C., *J. Dairy Res.*, 1982, **49**, 29.
90. HOLT, C., DALGLEISH, D. G. and JENNESS, R., *Anal. Biochem.*, 1981, **113**, 154.
91. LYSTER, R. L. J., *J. Dairy Res.*, 1981, **48**, 85.
92. JENNESS, R. and KOOPS, J., *Neth. Milk Dairy J.*, 1962, **16**, 153.
93. KRONMAN, M. J., SINHA, S. K. and BREW, K., *J. Biol. Chem.*, 1981, **256**, 8582.
94. MURAKAMI, K., ANDREE, P. J. and BERLINER, L. J., *Biochemistry*, 1982, **21**, 5488.
95. PERMYAKOV, E. A., YARMOLENKO, V. V., KALINICHENKO, L. P., MOROZOVA, L. A. and BURNSTEIN, E. A., *Biofizika*, 1982, **27**, 380.
96. ZITTLE, C. A., DELLAMONICA, E. S., RUDD, R. K. and CUSTER, J. H., *J. Am. Chem. Soc.*, 1957, **79**, 4661.
97. HOSOYA, N., TAMURA, M. and OKU, T., *Developm. Biochem.*, 1980, **14**, 489.
98. SMEETS, W. T. G. M., *Neth. Milk Dairy J.*, 1955, **9**, 249.
99. TESSIER, H. and ROSE, D., *J. Dairy Sci.*, 1958, **41**, 351.
100. HOLT, C., *J. Dairy Sci.*, 1981, **64**, 1958.
101. DEMOTT, B. J., *J. Dairy Sci.*, 1968, **51**, 1008.
102. MULDOON, P. J. and LISKA, B. J., *J. Dairy Sci.*, 1969, **52**, 460.
103. KRÄMER, R. and LAGONI, H., *Naturwissenschaft*, 1969, **56**, 36.
104. VAN KREVELD, A. and VAN MINNEN, G., *Neth. Milk Dairy J.*, 1955, **9**, 1.
105. CHRISTIANSON, G., JENNESS, R. and COULTER, S. T., *Anal. Chem.*, 1954, **26**, 1923.
106. BELEC, J. and JENNESS, R., *J. Dairy Sci.*, 1963, **46**, 598.
107. PEARCE, K. N. and CREAMER, L. K., *Anal. Chem.*, 1974, **46**, 457.
108. NORDBÖ, R., *J. Biol. Chem.*, 1939, **128**, 745.
109. BOULET, M. and MARIER, J. R., *J. Dairy Sci.*, 1960, **43**, 155.
110. WOOD, G. B., REID, D. and ELVIN, R., *J. Dairy Res.*, 1981, **48**, 77.

111. BROWN, W. E., In: *Environmental Phosphorus Handbook*, E. J. Griffith (ed.), 1973, John Wiley, New York.
112. NANCOLLAS, G. H. (ed.), *Biological Mineralization and Demineralization*, 1982, Springer–Verlag, Berlin.
113. ROUFOSSE, A. H., LANDIS, W. J., SABINE, W. K. and GLIMCHER, M. J., *J. Ultrastruct. Res.*, 1979, **68**, 235.
114. GLIMCHER, M. J., *J. Dent. Res.*, 1979, **58B**, 790.
115. BROWN, W. E., *J. Dent. Res.*, 1979, **58B**, 857.
116. EANES, E. D. and MEYER, J. L., *Calcif. Tiss. Res.*, 1977, **23**, 259.
117. HOLT, C., HASNAIN, S. S. and HUKINS, D. W. L., *Biochim. Biophys. Acta*, 1982, **719**, 299.
118. BELTON, P. S., COX, I. J., HARRIS, R. K., HOLT, C., LYSTER, R. L. J., MANN, S., PARKER, S. B. and WILLIAMS, R. J. P., unpublished observations.
119. MCGANN, T. C. A. and PYNE, G. T., *J. Dairy Res.*, 1960, **27**, 403.
120. MCGANN, T. C. A., KEARNEY, R. D. and DONNELLY, W. J., *J. Dairy Res.*, 1979, **46**, 307.
121. ROSE, D. and COLVIN, J. R., *J. Dairy Sci.*, 1966, **49**, 351.
122. KNOOP, A. M., KNOOP, E., FREDE, E., PRECHT, D. and WIENCHEN, A., *Milchwissenschaft*, 1979, **34**, 129.
123. MCGANN, T. C. A., BUCHHEIM, W., KEARNEY, R. D. and RICHARDSON, T., *Biochim. Biophys. Acta*, 1983, **760**, 415.
124. PYNE, G. T., *Biochem. J.*, 1934, **28**, 940.
125. PYNE, G. T. and RYAN, J. J., *J. Dairy Res.*, 1950, **12**, 200.
126. MORR, C. V., JOSEPHSON, R. V., JENNESS, R. and MANNING, P. B., *J. Dairy Sci.*, 1971, **54**, 1555.
127. DE KADT, G. S. and VAN MINNEN, G., *Rec. Trav. Chim. Pays-Bas*, 1943, **62**, 257.
128. SCHMIDT, D. G., In: *Developments in Dairy Chemistry*, Vol. 1, P. F. Fox (ed.), 1983, Applied Science Publishers, London, p. 61.
129. DICKSON, I. R. and PERKINS, D. J., *Biochem. J.*, 1971, **124**, 235.
130. ALEKSEEVA, N. YU. and D'YACHENKO, P. F., *Dairy Sci. Abstr.*, 1976, **38**, 553.
131. LYSTER, R. L. J., *J. Dairy Res.*, 1979, **46**, 343.
132. DOI, J. and NIKI, T., *Jap. J. Zootech. Sci.*, 1966, **37**, 52.
133. IRLAM, J. C. and HUKINS, D. W. L., private communication.
134. MCGANN, T. C. A., KEARNEY, R. D., BUCHHEIM, W., POSNER, A. S., BETTS, F. and BLUMENTHAL, N. C., *Calcif. Tiss. Int.*, 1983, **35**, 821.
135. HOLT, C., IRLAM, J. C., HASNAIN, S. S. and HUKINS, D. W. L., unpublished observation.
136. JENNESS, R., *Neth. Milk Dairy J.*, 1973, **27**, 251.
137. WAUGH, D. F., In: *Milk Proteins*, Vol. 2, H. A. MCKENZIE (ed.), 1977, Academic Press, London, p. 3.
138. YAMAUCHI, K., TAKEMOTO, S. and TSUGO, T., *Agric. Biol. Chem.*, 1967, **31**, 54.
139. HO, C. and WAUGH, D. F., *J. Am. Chem. Soc.*, 1965, **87**, 889.
140. ONO, T., KAMINOGAWA, S., ODAGIRI, S. and YAMAUCHI, K., *Agric. Biol. Chem.*, 1976, **40**, 1717.
141. ONO, T., YAHAGI, M. and ODAGIRI, S., *Agric. Biol. Chem.*, 1980, **44**, 1499.

142. WAUGH, D. F., SLATTERY, C. W. and CREAMER, L. K., *Biochemistry*, 1971, **10**, 817.
143. IMADE, T., SATO, Y. and NOGUCHI, H., *Agric. Biol. Chem.*, 1977, **41**, 2131.
144. DALGLEISH, D. G. and PARKER, T. G., *J. Dairy Res.*, 1980, **47**, 113.
145. PARKER, T. G. and DALGLEISH, D. G., *J. Dairy Res.*, 1981, **48**, 71.
146. UPTON, J. R. and HOLT, C., unpublished observations.
147. GREEN, M. L. and CRUTCHFIELD, G., *J. Dairy Res.*, 1971, **38**, 151.
148. PEARCE, K. N., *J. Dairy Res.*, 1976, **43**, 27.
149. HILL, R. J. and WAKE, R. G., *Nature*, 1969, **221**, 635.
150. WALSTRA, P., *J. Dairy Res.*, 1979, **46**, 317.
151. CARROLL, R. J. and FARRELL, H. M. Jr, *J. Dairy Sci.*, 1983, **66**, 679.
152. HORNE, D. S., private communication.
153. HOLT, C., In: *International Conference on Colloidal and Surface Science Proceedings*, E. Wolfram (ed.), 1975, Akademiai Kiado, Budapest, p. 641.
154. DONATH, E. and PASTUSHENKO, V., *J. Electroanal. Chem.*, 1979, **104**, 543.
155. DEWAN, R. K. and BLOOMFIELD, V. A., *J. Dairy Sci.*, 1973, **56**, 66.
156. HOLT, C., KIMBER, A. M., BROOKER, B. E. and PRENTICE, J. H., *J. Coll. Interf. Sci.*, 1978, **65**, 555.
157. SOOD, S. M., GAIND, D. K. and DEWAN, R. K., *N.Z. J. Dairy Sci. Technol.*, 1979, **14**, 32.
158. McCAMMON, J. A., DEUTCH, J. M. and FELDERHOF, B. U., *Biopolymers*, 1975, **14**, 2613.
159. WALSTRA, P., BLOOMFIELD, V. A., WEI, G. J. and JENNESS, R., *Biochim. Biophys. Acta*, 1981, **669**, 258.
160. HESSELINK, F. TH., *Electroanal. Chem.*, 1972, **37**, 317.
161. LAMPITT, L. H., BUSHILL, J. H. and FILMER, D. F., *Biochem. J.*, 1937, **31**, 1861.
162. VUJIČIĆ, I., DEMAN, J. M. and WOODROW, I. L., *Can. Inst. Fd. Technol. J.*, 1968, **1**, 17.
163. ODAGIRI, S. and NICKERSON, T. A., *J. Dairy Sci.*, 1965, **48**, 19.
164. MORR, C. V., *J. Dairy Sci.*, 1967, **50**, 1038.
165. ROSE, D., *Dairy Sci. Abstr.*, 1963, **25**, 45.
166. SWEETSUR, A. W. M. and MUIR, D. D., *J. Soc. Dairy Technol.*, 1980, **33**, 101.
167. LEVITON, A., ANDERSON, H. A., VETTEL, H. E. and VESTAL, J. H., *J. Dairy Sci.*, 1963, **46**, 310.
168. DEMAN, J. M., *Intern. Dairy Congr. 17th Proc.*, Munich B, 1966, p. 365.
169. JEN, J. J. and ASHWORTH, U. S., *J. Dairy Sci.*, 1970, **53**, 1201.
170. BURDETT, M., *J. Dairy Res.*, 1974, **41**, 123.
171. HARWALKER, V. R. and VREEMAN, H. J., *Neth. Milk Dairy J.*, 1978, **32**, 94.
172. HALLIDAY, D. A., *Process Biochem.*, 1978, **13**, 6.
173. JONES, S. B., KALAN, E. B., JONES, T. C. and HAZEL, J. F., *J. Agric. Fd. Chem.*, 1972, **20**, 229.
174. NAKAJIMA, I., KAWANISHI, G. and FURUICHI, E., *Agric. Biol. Chem.*, 1975, **39**, 979.
175. MELNYCHYN, P. and WOLCOTT, J. M., In: *Phosphates in Food Processing*, J. M. deMan and P. Melnychyn (eds), 1971, Avi, Westport, Conn., p. 56.
176. VUJIČIĆ, I., BATRA, S. C. and DEMAN, J. M., *J. Agric. Fd. Chem.*, 1967, **15**, 403.

177. FUKUSHIMA, M. and DEMAN, J. M., *Milchwissenschaft*, 1972, **27**, 473.

178. KRUK, A., *Acta Alim. Polon.*, 1979, **5**, 147.

179. WILSON, H. K., VETTER, J. L., SASAGO, K. and HERREID, E. O., *J. Dairy Sci.*, 1963, **46**, 1038.

180. LEVITON, A. and PALLANSCH, M. J., *J. Dairy Res.*, 1962, **45**, 1045.

181. SAMEL, R. and MUERS, M. M., *J. Dairy Res.*, 1962, **29**, 269.

182. ROSE, D. and TESSIER, H., *J. Dairy Sci.*, 1959, **42**, 989.

183. LEVITON, A., *J. Dairy Sci.*, 1964, **47**, 670.

184. SCHARPF, L. G. and KICHLINE, T. P., *J. Agric. Fd. Chem.*, 1967, **15**, 787.

185. KLEPACKA, M., *Dairy Sci. Abstr.*, 1971, **33**, 731.

186. GLANDORF, K., THOMASOW, J., *Z. Lebensm.–Untersuch. Forsch.*, 1977, **163**, 178.

187. JENNESS, R., SHIPE, W. F., Jr and SHERBON, J. W., In: *Fundamentals of Dairy Chemistry*, B. H. Webb, A. H. Johnson and J. A. Alford (eds), 1974, Avi, Westport, Conn., p. 402.

188. VUJIČIĆ, I. and DEMAN, J. M., *Milchwissenschaft*, 1966, **21**, 346.

189. DEYSHER, E. F. and WEBB, B. H., *J. Dairy Sci.*, 1948, **31**, 123.

190. GOULD, I. A. and LEININGER, E., *Mich. Agr. Exp. Sta. Quart. Bull.*, 1947, **30**, 54.

191. DEYSHER, E. F. and WEBB, B. H., *J. Dairy Sci.*, 1952, **35**, 106.

192. FOX, K. K., HOLSINGER, V. H., POSATI, L. P. and PALLANSCH, M. J., *J. Dairy Sci.*, 1967, **50**, 1032.

193. PARKASH, S. and DEMAN, J. M., *Milchwissenschaft*, 1970, **25**, 167.

194. BRULÉ, G., MAUBOIS, J. L. and FAUQUANT, J., *Lait*, 1974, **54**, 600.

195. HIDDINK, J., DE BOER, R. and ROMIJN, D. J., *Neth. Milk Dairy J.*, 1978, **32**, 80.

196. BRULÉ, G. and FAUQUANT, J., *J. Dairy Res.*, 1981, **48**, 91.

197. PIERRE, A. and BRULÉ, G., *J. Dairy Res.*, 1981, **48**, 417.

198. CHEN, C. and YAMAUCHI, K., *Agric. Biol. Chem.*, 1969, **33**, 1333.

199. CHEN, C. and YAMAUCHI, K., *Agric. Biol. Chem.*, 1971, **35**, 637.

200. HILGEMAN, M. and JENNESS, R., *J. Dairy Res.*, 1951, **34**, 483.

201. EVENHUIS, N. and DEVRIES, TH. R., *Neth. Milk Dairy J.*, 1957, **11**, 111.

202. KANNAN, A. and JENNESS, R., *J. Dairy Sci.*, 1961, **44**, 808.

203. EDMONDSON, L. F. and TARASSUK, N. P., *J. Dairy Sci.*, 1956, **39**, 123.

204. ROSE, D. and TESSIER, H., *J. Dairy Sci.*, 1959, **42**, 969.

205. MULDOON, P. J. and LISKA, B. J., *J. Dairy Sci.*, 1972, **55**, 35.

206. VAN DEN BERG, M. G., *Neth. Milk Dairy J.*, 1978, **33**, 91.

207. TUMERMAN, L. and WEBB, B. H., In: *Fundamentals of Dairy Chemistry*, B. H. Webb and A. H. Johnson (eds), 1965, Avi, Westport, Conn., p. 555.

208. SWEETSUR, A. W. M. and WHITE, J. C. D., *J. Dairy Res.*, 1974, **42**, 73.

209. AOKI, T. and IMAMURA, T., *Agric. Biol. Chem.*, 1974, **38**, 1929.

210. AOKI, T. and IMAMURA, T., *Agric. Biol. Chem.*, 1974, **38**, 309.

211. WHITNEY, R. M., REDDY, G. V., LIN, J. V. and TYNER, S. J., *Illinois Res.*, 1976, **18**, 5.

212. AOKI, T., SUZUKI, H. and IMAMURA, T., *Milchwissenschaft*, 1975, **30**, 30.

213. AOKI, T. and KAKO, Y., *J. Dairy Res.*, 1983, **50**, 207.

214. PYNE, G. T., *J. Dairy Res.*, 1959, **25**, 467.

215. LYSTER, R. L. J., *J. Dairy Res.*, 1965, **32**, 203.

216. ITO, R. and NAKANASHI, T., *Jap. J. Dairy Sci.*, 1966, **15**, A129.
217. ITO, R. and NAKANASHI, T., *Jap. J. Dairy Sci.*, 1966, **15**, A78.
218. LUND, D. B. and BIXBY, D., *Process Biochem.*, 1975, **10**, 52.
219. GEERTS, J. P., BEKHOF, J. J. and SCHERJON, J. W., *Neth. Milk Dairy J.*, 1983, **37**, 197.

Chapter 7

NUTRITIONAL ASPECTS OF MINERALS IN BOVINE AND HUMAN MILKS

ALBERT FLYNN and PAUL POWER

*Department of Nutrition,
University College, Cork, Republic of Ireland*

1. INTRODUCTION

Nutritionally speaking, the term 'minerals' refers to all chemical elements, excluding carbon, oxygen, hydrogen and nitrogen, that are nutritionally essential for humans. While it is not strictly accurate to refer to all of these elements as minerals, this term is used here in deference to common usage. The minerals occur in the body in a number of chemical forms such as inorganic ions in solution, insoluble salts, or constituents of organic molecules such as proteins, fats, carbohydrates and nucleic acids. They serve a wide variety of essential physiological functions, ranging from structural components of body tissues to essential components of many enzymes and other biologically important molecules.

This chapter discusses the nutritional roles, recommended intakes and hazards of deficiency or excess of the 22 minerals that are considered to be nutritionally essential for man, all of which occur in both human and cow's milks. The content, chemical form, bioavailability and nutritional significance of these minerals in both human and cow's milk are considered together with comparative aspects of human and cow's milks. In addition, in view of the widespread use of infant formulae based on cow's milk, some nutritional aspects of minerals in these formulae are discussed. Various nutritional aspects of essential minerals in cow's and/or human milk have been reviewed in recent years by a number of authors.[1-13]

The 22 minerals that are considered essential in the human diet are sodium, potassium, chloride, calcium, magnesium, phosphorus, sulphur, iron, copper, zinc, manganese, selenium, cobalt, chromium, molybdenum, iodine, fluoride, nickel, vanadium, silicon, tin and arsenic. While some of these, e.g. tin and arsenic, have not been shown to be essential for humans, they are essential for experimental animals and probably are also essential for man. A number of other chemical elements occur in milk, e.g. lithium, boron, bromine, aluminium, strontium, silver, lead, mercury, cadmium, rubidium and caesium. These are not nutritionally essential and are not reviewed here, but many of them are toxic. However, their concentrations in milk are normally well below toxic levels, and these have been reviewed by others.[2,6,14]

The essential minerals are sometimes classified into two groups, i.e. the macrominerals and trace elements. The macrominerals (sodium, potassium, chloride, calcium, magnesium, phosphorus and sulphur) are present in the body in amounts greater than about 0·01% by weight while the trace elements (the remaining 15 essential minerals) occur in the body at much lower levels and are required in the diet in amounts less than about 100 mg day^{-1}.[15] Many of the essential minerals, particularly the trace elements, are toxic when ingested in excess of requirements. However, the levels at which they exert their toxic effects are generally well in excess of their normal levels in milk and other foods, and for this reason their toxicological aspects will only be discussed where they are considered relevant.

There is much less information on the nutritional aspects of some minerals than others, and a considerable amount of current research is being carried out to clarify the roles of minerals in nutrition. This chapter includes some of the more recent findings in relation to the nutritional significance of minerals, particularly calcium and trace elements, in milk.

Although the minerals are treated separately, it is important to realize that interactions of minerals with each other, with other constituents of milk, and with other food constituents occur, and such interactions are assuming an increasing importance in nutrition.

2. MINERAL COMPOSITION OF HUMAN AND BOVINE MILKS

The mineral composition of human and cow's milks is not constant but is influenced by a number of factors such as stage of lactation, nutritional status of the mother, environmental and genetic factors. Data on the average mineral composition of mature human and cow's milks are presented in Table 1.

TABLE 1

MINERAL COMPOSITION OF MATURE HUMAN MILK AND COW'S MILK

	Content per litre			
	Mature human milk[a]		Cow's milk[b]	
Mineral	Mean	Range[c]	Mean	Range
Sodium (mg)	150	110–200	500	350–900
Potassium (mg)	600	570–620	1500	1100–1700
Chloride (mg)	430	350–550	950	900–1100
Calcium (mg)	350	320–360	1200	1100–1300
Magnesium (mg)	28	26–30	120	90–140
Phosphorus (mg)	145	140–150	950	900–1000
Iron (μg)	760	620–930	500	300–600
Zinc (μg)	2950	2600–3300	3500	2000–6000
Copper (μg)	390	370–430	200	100–600
Manganese (μg)	12[17]	7–15[18]	30[19]	20–50[10]
Iodine (μg)	70	20–120	260[20]	—
Fluoride (μg)	77	21–155	—[d]	30–220[21]
Selenium (μg)	14	8–19	—[d]	5–67[22]
Cobalt (μg)	12[12]	1–27[12]	1[19d]	0·5–1·3[19]
Chromium (μg)	40[12]	6–100[12]	10[19]	8–13[19]
Molybdenum (μg)	8[23]	4–16[2]	73[24]	18–120[24]
Nickel (μg)	25[12]	8–85[12]	25[12]	0–50[12]
Silicon (μg)	700[12]	150–1200[12]	2600[12]	750–7000[12]
Vanadium (μg)	7[12]	tr–15[12]	—	tr–310[12]
Tin (μg)	—	—	170[12]	40–500[12]
Arsenic (μg)	50[12]	—	45[12]	20–60[12]

[a] Values taken from Ref. 3 except where otherwise indicated by superscript.
[b] Values taken from Ref. 16 except where otherwise indicated by superscript.
[c] Values from Ref. 3 are ranges of means of five pooled samples.
[d] Content in milk varies with the diet of the cow.

The total mineral content of cow's milk ($\sim 7{\cdot}3$ g litre^{-1}) is considerably higher than that of mature human milk ($\sim 2{\cdot}0$ g litre^{-1}).[12] This is mainly due to the higher concentrations of sodium, potassium, chloride, calcium, phosphorus and magnesium in cow's milk (Table 1).

3. SODIUM, POTASSIUM AND CHLORIDE

Mature human milk contains considerably less sodium, potassium and chloride than cow's milk (Table 1). The concentration of sodium in human milk decreases considerably in the early stages of lactation; sodium concentration is high in colostrum on the first day (1400 mg litre^{-1}), falls to about 460 mg litre^{-1} on the second and third days, remains unchanged at this level until the eleventh day, and then falls below 115 mg litre^{-1} during the third and fourth weeks.[25] The potassium concentration in human milk declines from 720 mg litre^{-1} at the beginning of lactation to 585 mg litre^{-1} after 10 days.[26] The concentrations of sodium, potassium and chloride have been reported to decrease from 480, 740 and 850 mg litre^{-1}, respectively, in colostrum to 160, 530 and 400 mg litre^{-1}, respectively, in mature human milk.[8] It has been suggested that the high sodium concentration in colostrum might constitute a protective mechanism against dehydration and sodium deficiency in the newborn during a period of relative thirst and hunger.[26] No relationship has been demonstrated between maternal dietary sodium intake and milk sodium concentration.[7]

Particularly high levels of sodium occur in cow's colostrum,[27] but these decrease within a few days to the levels shown in Table 1. The sodium content of cow's milk is not influenced by dietary sodium intake within the normal range,[27] but milk sodium concentration decreases with a sodium-deficient diet.[11] Sodium concentration in milk tends to be higher at the end of lactation when milk yields are low.[27] The potassium concentration in cow's colostrum is lower than that of mature milk and increases to normal values within the first 2–3 days of lactation.[28] Milk potassium is independent of potassium intake although higher levels are occasionally observed when cows go out to grass or in summer.[27]

The chloride concentration in cow's milk decreases from colostrum to mature milk but increases sharply towards the end of lactation.[27] It is independent of dietary intake, and high air temperatures increase the chloride content while low air temperatures have the opposite effect.[27]

The concentrations of sodium, potassium and chloride in milk are determined by the permeability and transport properties of the Golgi apparatus, its vesicles and of the apical cell membrane. There are small variations in the concentrations of these ions in milk from day to day, and the direction of the shifts is such that, if the concentration of lactose decreases, those of the major monovalent ions increase.[29] However, in mastitis, equilibrium of small molecules and ions between milk and extracellular fluid occurs and the concentrations of sodium, chloride and bicarbonate increase while those of potassium and lactose decrease.[30,31]

Sodium, potassium and chloride are believed to be present in milk almost entirely as free ions.[31] Practically all the sodium, potassium and chloride in milk is absorbed in the gastrointestinal tract although much of what is absorbed is not retained.[6]

The concentrations of sodium, potassium and chloride in milk are of physiological importance in the feeding of the young infant, and clinical problems may arise if there is an excessive intake of these nutrients, particularly of sodium. For about the first 10–14 days after birth the kidney of the young infant, compared to that of the adult, re-absorbs sodium ions from the glomerular filtrate more efficiently and also has a limited capacity to excrete water. An increased intake of sodium therefore leads to over-hydration and hypertonicity.[32,33] Although potassium and chloride ions are more readily excreted than sodium, an obligatory amount of water is needed for their excretion, and too great an intake of potassium and chloride may affect the volume and composition of body fluids.[6]

Renal solute load is determined mainly by these minerals and protein (which gives rise to urea). Cow's milk has a much higher renal solute load (228 mosm litre^{-1}) than human milk (79 mosm litre^{-1}).[34] The increase in renal solute load caused by cow's milk may be of relatively little significance in most circumstances because the kidney merely excretes a more concentrated urine. However, it does lead to a smaller margin of safety against dehydration which can occur in conditions of diarrhoea, fever and low water intake, and for this reason milk foods for young infants are now formulated with contents of sodium much closer to that of human milk.[6] Recommended concentrations of sodium, potassium and chloride in infant formulae are 15–35, 50–100 and 40–80 mg litre^{-1}, respectively.[6]

4. CALCIUM

The adult human body contains about 1200 g of calcium, which amounts to about 1·5–2% of body weight (3–6% dry weight basis). Of this, 99% is found in bones and teeth where it is present as calcium phosphate, providing strength and structure. The remaining ~1%, found in blood, extracellular fluid and various soft tissues, is responsible for a number of regulatory functions such as maintenance of normal heart beat, blood coagulation, hormone secretion, integrity of intracellular cement substances and membranes, nerve conduction, muscle contraction and activation of enzymes.[35,36]

Cow's milk is a very good source of dietary calcium, and dairy products such as cheese and yoghurt are also rich in this mineral.[16] Nutrient density, i.e. the contribution of nutrients relative to the contribution of energy by a specific food, is a useful way to express the nutritional quality of foods.[37] The nutrient density of calcium in whole cow's milk (1846 mg per 1000 kcal[16]) is 3·9–7·6 times the recommended ratio of calcium to energy in adult human diets (242–471 mg Ca per 1000 kcal, based on a Recommended Dietary Allowance of 800 mg day^{-1} for calcium and 1700–3300 kcal day^{-1} for energy[38]). Reduction of the fat content of milk further enhances its nutrient density for calcium by lowering the energy content while leaving the calcium content essentially unchanged.

In Western countries milk and dairy products provide up to 75% of the total calcium intake.[13] The contribution of dairy products to total calcium intake has been estimated as 75% in the Netherlands,[13] 72% in the USA,[39] 60% in the UK,[40] and 53% in Ireland.[41]

In the absence of milk and dairy products from the diet, calcium intakes in excess of 300 mg day^{-1} are difficult to achieve.[13] This is far below the Recommended Dietary Allowance (RDA) for calcium which has been set at 800 mg day^{-1} for adults and children aged 1–10 years and 1200 mg day^{-1} for adolescents and pregnant and lactating women.[38] On this basis, it might be considered that consumption of dairy products is very important in order to achieve an adequate calcium intake. However, there is still considerable disagreement on human calcium requirements, and this is reflected in the wide variation in adult RDAs for calcium (400–1200 mg day^{-1}) which have been set by different authorities.[42]

The calcium content of human milk is considerably lower than that of cow's milk (Table 1). In human milk, the calcium concentration varies considerably between individuals, and lower values have been reported in milk of malnourished women.[4] On the other hand, studies on the longitudinal changes in the mineral content of human milk have not shown any significant correlation between dietary intake of calcium and the calcium concentration in milk,[43,44] while Kirskey et al.[45] reported that increasing dietary calcium intake does not affect calcium concentration in milk. Vaughan et al.[43] reported that the calcium concentration in human milk increases between the first and second month of lactation and declines thereafter, while Kirskey *et al.*[45] claim that there is a peak in calcium concentration between the fifth and seventh months. In cow's milk, calcium concentration is higher in colostrum and at the end of lactation but varies little with feeding or season.[12]

In cow's milk, two-thirds of the total calcium occurs in colloidal form associated with the casein micelles; the remaining one-third is soluble.[31] Ionized calcium in the soluble phase accounts for about 10% of the total calcium.[46] In human milk, calcium is present mainly in a 'soluble' rather than a colloidal form,[47] and 35% of the total is present as ionized calcium.[48] Fransson and Lonnerdal[49] reported that the distribution of calcium in whole human milk is: fat, 20·2%; casein, 2·3%; whey proteins, 43·9%; low molecular weight compounds 33·6%. α-Lactalbumin has been identified as the major calcium-binding protein in human milk.[50]

Vitamin D plays an important role in controlling calcium metabolism and promoting calcium absorption. Vitamin D deficiency leads to poor calcium absorption, causing rickets in children and osteomalacia in adults. In the newborn infant, calcium deficiency resulting in hypocalcaemia may be caused by maternal vitamin D deficiency.[51]

Lactose promotes intestinal absorption and body retention of calcium both in rats[52,53] and in humans.[54,55] This effect of lactose is independent of vitamin D but the mechanism by which it occurs remains unresolved. Ambrecht and Wasserman[53] suggested that undigested lactose reaching the ileum interacts with the brush border membrane, increasing its permeability to calcium.

The bioavailability to rats of calcium from whole cow's milk is high with 88% of the calcium absorbed and 87% retained.[56] Wong and La Croix[57] reported that the bioavailability of calcium from non-fat dried milk is greater than from $CaCO_3$. This high bioavailability has been attributed to two factors: lactose (as already discussed), and phospho-peptides formed during the digestion of casein, which form soluble complexes with calcium in the gastrointestinal tract.[58] However, it is unclear if such phosphopeptides could survive sufficiently long in the small intestine to affect the absorption of calcium.[59]

Calcium is absorbed better by the human infant from human milk than from cow's milk.[60] Various explanations for this have been suggested including: the amount and nature of fatty acids present in milk fat;[61] the higher ratio of calcium to phosphorus in human milk;[62,63] the higher concentration of lactose in human milk.[64] It is also possible that calcium absorption is influenced by the distribution of calcium between different proteins, particularly the high proportion of calcium associated with casein in cow's milk.

Age-related osteoporosis, a common bone disease and a major cause of disability in Western countries,[65,66] is characterized by reduced bone density resulting in increased bone fragility and susceptibility to fracture. The condition is particularly common in elderly women; it has been estimated that $\sim 25\%$ of Caucasian women of 65 years or older have osteoporosis,[67] and that 15 million North Americans have some degree of osteoporosis.[68]

Osteoporosis is a multifactorial disorder but there is increasing evidence that inadequate calcium intake throughout life is a contributory factor.[66] Adequate calcium intake is required in early life in order to develop maximum bone mass at maturity (age 30–40) since there is evidence that the amount of bone mass present at maturity is an important factor influencing fracture susceptibility in the elderly.[66] Furthermore, there is evidence that the rate of age-related bone loss after maturity may be decreased by dietary calcium supplementation.[66] There is considerable evidence that calcium intake is inadequate in a large fraction of the populations of Western countries, particularly in women. For example, mean daily calcium intake for females over 12 years of age in the USA is less than 85% of the US RDA for adults (800 mg day^{-1}),[38] and calcium intake is less than the RDA in more than 66% of all US females aged 18–30 and in over 75% of females over age 35. More than 25% of US females of all ages ingest less than 300 mg of calcium per day.[66,69]

Recently attention has been focussed on the effect of dietary calcium intake on blood pressure. Ackley *et al.*[70] reported a negative correlation of blood pressure with calcium intake from liquid milk and also with total calcium intake from dairy products. They also found that hypertensive men consumed significantly less calcium from milk and dairy products than normotensive men, and suggested that the intake of calcium from milk and dairy products exerts a protective effect against hypertension. Association of low total dietary calcium intake (and by implication low intake of milk and dairy products) with raised blood pressure has also been observed in a number of other studies.[71–78] Such associations do not prove causality; however, when taken together with evidence of a reduction of blood pressure by calcium supplementation in clinical trials[79–81] and in experimental animal studies,[82–85] they suggest that inadequate calcium intake commonly observed in nutrition surveys[69,86] may be a risk factor for hypertension. The relationship of dietary calcium intake to blood pressure has been reviewed in detail.[87]

Given the large proportion of calcium intake obtained from dairy products, dietary restriction of these foods may significantly reduce calcium intake below levels that may be already inadequate. It has recently been shown in this laboratory that calcium intakes of diabetics who followed dietary recommendations that restricted milk and dairy products were significantly lower than those of non-diabetics; this was almost totally attributable to lower intake of milk, cheese and yoghurt among diabetics. Calcium intakes were inadequate in both groups, particularly among females; calcium intake was less than the RDA (800 mg) in 86% of diabetic females and in 70% of non-diabetic females.[88] These findings bear out the concern expressed by others[76,89] that clinical use of restricted diets may inadvertently reduce intake of essential nutrients below recommended levels.

5. MAGNESIUM

Magnesium has an essential role in a wide variety of physiological processes including protein and nucleic acid metabolism, neuromuscular transmission and muscle contraction, and it acts as a cofactor for many enzymes. Dietary deficiency of magnesium is uncommon except in conditions of severe malnutrition. The daily RDAs for magnesium are: 50 and 70 mg for infants 0–6 and 6–12 months of age, respectively; 150, 200 and 250 mg for children aged 1–3, 4–6 and 7–10 years, respectively; 350 mg for male adolescents and adults; 300 mg for female adolescents and adults; and 450 mg in pregnancy and lactation.[38]

Mature human milk in the UK contains about 28 mg of magnesium per litre, with a range of means of pooled samples of 26–30 mg (Table 1). This is similar to values reported by others: 31 ± 1.7 mg litre^{-1} in the USA;[43] 30 mg litre^{-1} in India.[90] Higher and more variable values (41.4 ± 15.2 mg litre^{-1}) have been reported in Sweden by Fransson and Lonnerdal.[49] Magnesium concentration has been reported to be higher in colostrum (40 mg litre^{-1}) than in mature milk (30 mg litre^{-1}),[90] but it appears to be unaffected by stage of lactation in mature milk.[43,49] There is no relationship between milk magnesium concentration and dietary magnesium intake within the normal range of dietary intake.[43]

The mean concentration of magnesium in cow's milk is 120 mg litre^{-1}, with a range of 90–140 mg (Table 1). Magnesium concentrations in colostrum are 2–3 times those in mature milk and decrease to the level in mature milk within the first 1–3 days of lactation,[27] remaining relatively constant thereafter. Magnesium concentration in cow's milk is unaffected by dietary magnesium intake.[27]

The distribution of magnesium in human milk has been reported as $\sim 2\%$ in the fat fraction, 0.8% in casein, 44% associated with the whey proteins, and 53.6% in a low molecular weight form that has not been characterized.[49] In cow's milk, about two-thirds of the magnesium is in a soluble form with about 15% of total magnesium present as free ions,[27] while the remaining one-third is colloidal and is associated with the casein micelles.[31,91]

Although cow's milk has a greater concentration of magnesium than human milk, hypomagnesaemia may result from feeding unmodified cow's milk to infants owing to its high phosphate content.[92,93]

6. PHOSPHORUS

Phosphorus is an essential nutrient for humans and serves a number of important biological functions. As calcium phosphate it is a major structural component of bones and teeth. Phosphorus occurs as organic and inorganic phosphates in all body tissues and fluids, is an essential component of many biological molecules and plays a central role in metabolism.

The daily RDAs for phosphorus are: 240 and 360 mg for infants aged 0–6 and 6–12 months, respectively; 800 mg for children aged 1–10 years and adults; and 1200 mg for adolescents and pregnant and lactating women.[38]

Cow's milk and milk products such as cheese and yoghurt are good dietary sources of phosphorus;[16] the contribution of milk and dairy products to total phosphorus intake has been reported as $30–45\%$ in Western countries.[12]

The phosphorus content of mature human milk is 145 mg litre^{-1}, with a range of 140–150 mg for five pooled samples (Table 1). This is considerably lower than the concentration in cow's milk (950, range 900–1000 mg litre^{-1}) (Table 1). In cow's milk, 20% of the phosphorus is esterified to casein, a further 40% is present as colloidal inorganic calcium phosphate, 30% occurs as phosphate ions in solution, and about 10% is associated with the lipid fraction.[12,94] In human milk about 15% occurs in an inorganic form and the remainder is associated with the lipids.[12]

In the first few weeks of life, the infant's ability to regulate plasma calcium concentration has not developed fully and hypocalcaemia may result in neonatal tetany which used to occur more frequently in artificially fed than in breast fed infants.[95,96] Excessive phosphorus intake may contribute to this condition, and feeding unmodified cow's milk, which is high in phosphorus, lowers plasma calcium in the newborn infant.[6] For this reason it is recommended that the calcium/phosphorus ratio in artificial infant formulae should be greater than that in cow's milk ($\sim 1.2/1$) and closer to that in human milk ($\sim 2.2/1$);[6] this practice is now followed in the manufacture of infant formulae. A calcium/phosphorus ratio of $1.5/1$ is recommended for infants under 1 year of age.[38]

A calcium/phosphorus ratio close to $1/1$ has been suggested as optimum for adult human diets[97] but there is no experimental evidence to support this.[13] In experimental animals, a high intake of phosphorus or a low calcium/phosphorus ratio in the diet can lead to bone loss.[98] However, it is generally concluded that broad variations in phosphorus intake or in the calcium/phosphorus ratio in the diet do not adversely affect bone in adult man.[66]

7. SULPHUR

Sulphur is an essential mineral and must be supplied in the diet in an organic form as the essential amino acids cysteine (or cystine) and methionine which are required for synthesis of body proteins. Human adults can readily synthesize cysteine from methionine supplied in the diet, so cysteine is not an essential amino acid for adults. However, the human infant is unable to carry out this conversion and thus both cysteine and methionine are essential for infants.[6]

The mean sulphur concentrations in cow's and human milks are 0.3 g litre^{-1} [16] and 0.14 g litre^{-1},[99] respectively, most of which is present in the amino acids cysteine and methionine which are constituents of milk proteins. Thus, sulphur concentrations mainly reflect the concentration and type of protein in the milk and are therefore higher in colostrum.[6] It can be estimated that, in mature human milk, cysteine and methionine together contribute about 108 mg of sulphur per litre, about 77% of the total sulphur (based on cysteine and methionine concentrations of 252 and 190 mg litre^{-1}, respectively[3]), while in cow's milk these two amino acids contribute about 284 mg litre^{-1}, about 95% of the total sulphur (based on cysteine and methionine concentrations of 312 and 936 mg litre^{-1}, respectively[16]).

Mature human milk also contains taurine, a sulphur-containing amino-sulphonic acid, at a concentration of 66–95 mg litre^{-1},[100] and this contributes about 17–24 mg litre^{-1} of sulphur, about 12–17% of the total sulphur. Taurine, which participates in the development of the central nervous system,[101] can be synthesized from cysteine and methionine and is not an essential amino acid for adults. However, there is evidence that the human infant is unable to synthesize adequate amounts of taurine from these precursors,[102,103] suggesting that taurine may be essential in the infant's diet. Cow's milk contains much less taurine ($\sim$28 mg litre^{-1} [104]) than human milk, and this amino acid contributes only about 2% of the total sulphur in cow's milk. The remaining sulphur in human and cow's milks is present as the sulphate ion which is not utilized to any significant extent by the human body.[15]

The cysteine and methionine contents of human milk, which are considered to be suited to the needs of the newborn infant,[9] are 120 and 91 mg per gram of N, respectively,[3] while for cow's milk the respective values are 60 and 180 mg.[16] Thus the ratio of cysteine to methionine is considerably lower in cow's milk ($\sim$0·3) than in human milk ($\sim$1·3). It has been estimated that a cow's milk based infant formula containing 1·5% (w/v) of protein would supply only 67% of the cysteine provided by human milk.[6] It is mainly for this reason that the casein/whey protein ratio in cow's milk based infant formulae is sometimes modified from that of cow's milk (80/20) to that of human milk (40/60).[6]

8. IRON

Iron, as a component of haem in haemoglobin, myoglobin, cytochromes and other proteins, plays an essential role in the transport, storage and utilization of oxygen. It is also a cofactor for a number of enzymes. Deficiency of iron is relatively common in both Western and underdeveloped countries and results in anaemia, increased tendency to fatigue and increased susceptibility to infection. Daily RDAs for iron are: 10 mg for infants under 6 months of age; 15 mg for age 6 months to 3 years; 10 mg for children aged 4–10, male adults and post-menopausal females; 18 mg for male and female adolescents and pre-menopausal female adults.[38] Iron supplements of 30–60 mg day^{-1} are usually recommended during pregnancy and lactation.[38]

Milk and milk products are very poor sources of iron, and cow's milk contributes little (only about 3% in Canada) to the total iron intake of adults.[105] Mean iron concentration in mature human milk (0·76 mg litre^{-1}, range 0·62–0·93) is slightly higher than that of cow's milk (0·5, range 0·3–0·6) (Table 1). A much greater variation in the iron content of human milk (0·1–1·6 mg litre^{-1}) has been reported by Picciano and Guthrie[106] who found that it varied with time of day, fat content of the milk, and age and parity of the mother. The iron content of human milk decreases with advancing lactation: 0·6–1·0 mg litre^{-1} in colostrum decreasing to 0·2–0·4 mg in mature milk.[43,107] This decrease of approximately 60% occurs predominantly in the first 3 weeks of lactation with little change thereafter.[10]

No correlation has been observed between the dietary intake of iron and its concentration in human milk,[43,107] and dietary supplementation with up to 30 mg of iron per day does not affect its concentration in milk.[106] Milk iron concentration is not affected by maternal iron status.[106]

Although several investigations have reported lack of a developmental pattern for iron in cow's milk, de Maria[108] showed that iron concentration decreased by 35–50% during the first 3 days of lactation and remained relatively constant thereafter. The iron content of cow's milk varies with the breed of cow, geographical location and season, but it is resistant to changes in dietary iron intake.[2,19,109] Contact with metal containers can increase the iron concentration in cow's milk.[5]

In cow's milk, substantial amounts (40–50%) of iron occur in milk fat[110,111] where it is probably associated with the fat globule membrane. Casein is the major iron-binding protein, binding 24% of the iron,[112] mainly to the phosphate groups of α-casein.[113]

The distribution of iron added to cow's milk has been widely studied and found to depend on the form of the iron supplement used. Ferrous sulphate donates less iron to casein micelles than other supplements tested.[113] Supplementation of unhomogenized milk with ferrous salts results in a milk with a high iron content in the fat which can result in lipid peroxidation and development of 'off flavour'.[114,115] Homogenization, by causing adsorption of casein onto the fat globule membrane, increases the affinity of milk fat for added iron. Ferric nitrilotriacetate donates iron specifically to the casein fraction, and iron added in this form causes relatively little lipid peroxidation.[114,115] Iron added as the chloride to skim milk was 85% bound to caseins, mainly α-caseins.[116]

In human milk, 30% (16–46%) of the iron is associated with the lipid fraction,[117] and there is evidence that it is bound to xanthine oxidase, an enzyme that is a component of the fat globule membrane.[118] A significant proportion ($\sim 30\%$, range 18–56%) of the iron is associated with a low molecular weight ($<15\,000$) fraction which has not yet been characterized, and only about 9% is bound to casein.[117] Lactoferrin, which is present in mature human milk at a concentration of 1·6–2 g litre^{-1},[119,120] binds about 10–30% of the total iron in human milk[117] and is only 1–4% saturated with iron. Previously it had been reported that lactoferrin is saturated with iron to the extent of 10–30%.[121] Considering the very high affinity of lactoferrin for iron,[122] it is difficult to understand why its degree of saturation with iron in milk is so low.

Lactoferrin is a glycoprotein (molecular wt $\sim 86\,000$) which can bind 2 molecules of iron (as ferric ion) per molecule of protein. Its concentration in human milk is ~ 15 g litre^{-1} in colostrum, decreasing to ~ 5 g on day 5 and to 1·6–2 g in mature milk.[117,119,120] The concentration of lactoferrin in bovine colostrum is ~ 6 g litre^{-1}, decreasing to $\sim 0·9$ g in mature milk.[123,124]

It has been suggested that lactoferrin in human milk, which is largely unsaturated with iron, has a bacteriostatic function, i.e. the strong affinity of lactoferrin for iron would enable this protein to sequester iron from its surroundings and thus prevent the growth of bacteria that require iron.[121] While there is evidence that this does occur *in vitro*[121,125] definitive evidence that it occurs *in vivo* is still lacking.[10] The incidence of gastrointestinal infections is much lower in breast fed than in bottle fed infants,[126,127] but whether this is influenced by lactoferrin is unknown since other constituents of breast milk, such as immunoglobulins and enzymes, are probably involved also[128] (cf. Chapter 10).

There is conflicting evidence regarding the role of lactoferrin in iron absorption. Fransson[129] reported that the lactoferrin concentrations were 3–4 times higher in anaemic women compared to well-nourished women, and suggested that the protein may play a role in iron absorption, at least in infants of undernourished women. Supplementation of weanling mice with lactoferrin-bound iron resulted in high iron uptake.[130] The mucosal surfaces of the human gastrointestinal tract are coated with a thin layer of lactoferrin,[131] and it has been suggested that lactoferrin may act by donating iron to the small intestine and thus play a role in iron absorption.[132]

Other authors have reported that lactoferrin actually decreases iron absorption.[133,134] Brock[135] suggested that lactoferrin may suppress the absorption of iron in the young infant and prevent iron overload, and may thus regulate iron absorption.

As most infant formulae are based on cow's milk, the relative bioavailability of iron in human and in cow's milks has been studied. Bioavailability of iron to the infant from human milk is reported to be in the range 50–100%[136,137] or 49–70%.[138,139] Considerably lower absorption of iron by human infants from cow's milk has been reported, usually around 10%.[139,140] To compensate for the relatively low bioavailability of iron in cow's milk, infant formulae are supplemented with iron in various forms, and levels of 6–12 mg litre^{-1} of reconstituted formula have been found to be adequate and are often recommended,[141]. although other authorities[6] recommend lower levels (0·7–7 mg litre^{-1}).

The reason for the exceptionally high bioavailability of iron from human milk is not understood. It has been suggested that it may be related to the high concentration of lactoferrin in human milk but, as previously discussed, evidence for this is conflicting. Only a small portion of the iron in human milk is associated with lactoferrin.[117]

There has been considerable debate on whether the iron content of human milk is sufficient for the breast fed infant and whether these infants should be supplemented with iron. Iron deficiency is one of the most common nutritional deficiencies in infancy and childhood due to a rapid growth and marginal supply of iron in the diet.[142] The iron stores of a breast fed infant at 4–6 months of age may become compromised if they are not replenished from dietary sources, and dietary iron supplementation has been recommended at no later than 4 months of age for full-term infants and no later than 2 months of age for premature infants.[142] On the other hand, it has been suggested that the amount of iron in human milk, taken together with the considerable stores of iron present at birth, is sufficient to meet the needs of full-term infants for the first 6 months of life.[6,140,143] The iron status of infants after 6 months of breast feeding has been reported to be similar to that of infants receiving iron-supplemented infant formulae.[140,144,145]

9. COPPER

Copper is essential for iron utilization and is a cofactor for enzymes involved in the metabolism of glucose and synthesis of haemoglobin, connective tissue and phospholipids. Dietary deficiency of copper is uncommon except in conditions of severe malnutrition. Dietary requirements for copper are not known precisely but estimated safe and adequate daily intakes are: 0·5–0·7 and 0·7–1·0 mg for infants aged 0–6 months and 6–12 months, respectively; 1·0–1·5, 1·5–2·0 and 2·0–2·5 mg for children aged 1–3, 4–6 and 7–10 years, respectively; and 2·0–3·0 mg for adolescents and adults.[38]

Milk and milk products are considered poor sources of copper, and cow's milk contributes little (only about 4% in Canada) to the total copper intake of adults.[105] The mean copper concentration in mature human milk (0·39 mg litre^{-1}, range 0·37–0·43) is higher than in cow's milk (0·2, range 0·1–0·6) (Table 1).

The copper content of cow's milk can be increased by contact with metal containers and processing equipment,[146] and concentrations as low as 0·02–0·04 mg litre^{-1} have been reported in uncontaminated raw cow's milk.[147] Copper concentration in cow's milk decreases by up to 50% during the first 3 days of lactation,[108] and Lonnerdal et al.[10] have reported that typical concentrations in colostrum and mature milk are 0·2–0·3 and 0·1–0·2 mg litre^{-1}, respectively. The copper concentration in cow's milk varies widely depending on individuality, season and milk yield, and can be increased by dietary copper supplementation.[2]

In human milk, copper concentration decreases during lactation; mean values decrease from 0·6 mg litre^{-1} at the second week of lactation to 0·25 mg at 20 weeks,[148] or from 0·36 mg litre^{-1} at 6–8 weeks to 0·21 mg at 17–22 weeks lactation.[107] There is no significant correlation between dietary copper intake and milk copper concentration;[43,45,107] there are no reports on the effect of inadequate dietary copper on milk copper concentration.

In cow's milk about 10–35% of copper is bound to the fat globule membrane.[110] Martin et al.[149] have reported that a major fraction (not quantified) of the copper in the skim milk phase is associated with citrate, while King et al.[110] have reported that 57–70% of total milk copper is bound to casein.

The distribution of copper in human milk has been reported by Lonnerdal et al.[150] to be: 9% (1–21%) in the fat fraction; 39% bound to serum albumin; 28% (7–48%) bound to casein; and 24% (15–47%) in a low molecular weight form. Fransson and Lonnerdal[117] reported the distribution as: 16% in fat (probably associated with the fat globule membrane); 56% bound to whey proteins; 4% bound to casein; and 26% in a low molecular weight form. On the other hand, Martin et al.[149] have shown that the major fraction (not quantified) of copper is associated with low molecular weight ligands (mainly citrate, but also including glutamate and other amino acids), with a significant but smaller amount associated with protein. The discrepancies between these results may be due, at least in part, to the use of different methodologies.

Infant formulae are normally supplemented with copper to 0·4–0·6 mg litre^{-1}, and even higher concentrations are present in some formulae for premature infants.[151] Nevertheless, a recent study found a wide variation in copper concentrations in infant formulae.[152] Information regarding the bioavailability of copper is limited,[10] and since copper deficiency has been reported in formula fed infants in some countries[153] there is a need for studies on the bioavailability of this mineral from human and cow's milks.

10. ZINC

Zinc is essential for growth and development, sexual maturation and wound healing, and it may also be involved in the normal functioning of the immune system and other physiological processes. It is a component of the hormone insulin and aids in the action of a number of hormones involved in reproduction, as well as being required for the synthesis of DNA, RNA and protein and as a cofactor for many enzymes. Zinc deficiency in humans was first reported in the Middle-East,[154] giving rise to dwarfism, impaired sexual development and anaemia. Mild deficiencies of zinc, although difficult to detect, have been shown to occur in Western countries, particularly in infants and young children.[155] Daily RDAs for zinc are: 3 and 5 mg for infants aged 0–6 and 6–12 months, respectively; 10 mg for children aged 1–10 years; 15 mg for adolescents and adults; 20 mg in pregnancy; and 25 mg during lactation.[38]

Dairy products such as milk, cheese and yoghurt are moderately good sources of zinc, and it has been estimated that dairy products contribute between 13%[105] and 25%[156] of the total zinc intake in Western countries.

Mean zinc concentration in mature human milk (2·95 mg litre^{-1}) is slightly lower than that in cow's milk (3·5 mg) (Table 1) but large variations in the zinc content of human (0·65–5·3 mg litre^{-1} [10]) and cow's milk (2·0–6·0 mg litre^{-1} [16]) have been reported. There is a pronounced decrease in the zinc concentration in human milk during lactation. Vuori and Kuitunen[148] reported that zinc concentration decreases from 4·0 mg litre^{-1} during the second week of lactation to 0·5 mg litre^{-1} by the sixth month, while a decrease from 5·3 to 1·1 from colostrum to mature milk at seven months lactation was reported by Rajalakshmi et al.[90]

No significant correlation was observed between dietary zinc intake and zinc concentration in human milk,[107] and zinc supplementation of a zinc-adequate diet does not appreciably affect the zinc concentration in milk.[45,106] There are no reports on the effect of dietary zinc deficiency on the zinc concentration in human milk.

There is a dramatic decrease (50%) in the zinc concentration in cow's colostrum during the first 72 hours of lactation, with little change thereafter.[108] Dietary zinc supplementation increases the zinc concentration of cow's milk only slightly.[157,158]

There are significant differences in the distribution of zinc in human and cow's milks, and these may influence the bioavailability of zinc from the milk. In cow's milk most of the zinc is in the skim milk fraction with only 3–14% in the lipid fraction.[159] Of the zinc in the skim milk fraction, over 95% is associated with the casein micelles,[160] with a small proportion associated with a low molecular weight compound which has recently been identified as citrate.[149,160]

Lonnerdal et al.[150] reported that zinc in human milk is distributed as follows: 29% (20–45%) in the lipid fraction; 14% (5–21%) associated with casein, 28% (7–48%) with serum albumin and 29% (24–36%) with a low molecular weight compound. The association of a significant proportion of zinc in human milk with a low molecular weight compound has been observed in a number of studies.[161–164] The identity of this low molecular weight zinc-binding ligand in human milk has been the subject of much controversy. Evans and Johnson[165] suggested that picolinic acid is the ligand, but this has been discounted by Rebello et al.[166] who found that the picolinic acid concentration in human milk is insufficient to bind a significant proportion of the zinc. The ligand has been identified by other workers as citrate.[149,167,168]

On the other hand, Blakeborough et al.[160] found that over 85% of the zinc in skim milk is associated with a high molecular weight (>150 000) protein complex whose major protein is lactoferrin, and about 13% is bound to a fraction with an average molecular weight of 30 000. Ainscough et al.[169] have also reported that lactoferrin is a zinc-binding protein. Thus, while there is still disagreement on the distribution of zinc in human milk, it is generally accepted that much more of the zinc is associated with casein in cow's milk than in human milk.

The bioavailability of zinc from human milk is greater than from cow's milk. Johnson and Evans[170] reported that the bioavailability to rats of zinc from human and cow's milk is 59% and 42%, respectively. The plasma zinc concentration of breast fed infants has been reported to be significantly higher than that in infants fed a cow's milk based formula, even with added zinc,[171] suggesting that zinc is more available from breast milk than from cow's milk. Human milk (but not cow's milk) has a therapeutic value in the treatment of acrodermatitis enteropathica, a hereditary zinc malabsorption syndrome.[162,163]

A number of theories have been proposed to explain the higher bioavailability of zinc from human milk compared with cow's milk. (1) Binding of a significant fraction of the zinc in human milk (but not in cow's milk) to a low molecular weight zinc-binding ligand (e.g. citrate) may enhance zinc absorption either by facilitating zinc transport across the gastrointestinal tract wall or by preventing the sequestration of zinc by other substances in the gastrointestinal tract that lower its availability for absorption.[161,163] (2) The binding of a large fraction of the zinc in cow's milk to casein (present at about 10 times its concentration in human milk)[128] may result in the entrapment of zinc in casein curds formed in the stomach which may be incompletely digested in the small intestine,[172] thus rendering a significant proportion of zinc in cow's milk unavailable for absorption.[160,162,173] (3) Lactoferrin, which is present in human milk in large quantities compared with cow's milk, may act to donate zinc to the small intestine in a manner similar to that proposed by Cox *et al.*[132] for its role in the absorption of iron.[160] On the basis of the evidence available, it is not possible to differentiate between these various hypotheses and it is possible that the effect may be related to one or more of the proposed mechanisms.

11. MANGANESE

Manganese is an essential cofactor for enzymes involved in the metabolism of glucose, lipids and cholesterol, and is also required for development of the pancreas. It is widely distributed in foods, and dietary deficiency is not known to occur in humans. The exact dietary requirements for manganese are unknown but recommended safe and adequate daily intakes are: 0·5–0·7 and 0·7–1·0 mg for infants aged 0–6 and 6–12 months, respectively; 1·0–1·5, 1·5–2·0 and 2·0–3·0 mg for children aged 1–3, 4–6 and 6–10 years, respectively; and 2·5–5·0 mg for adolescents and adults.[38]

Cow's milk is a poor source of manganese and contributes little (2–3% in Western countries) to the total dietary intake of this mineral.[105,156] The mean manganese concentration in cow's milk is 30 µg litre^{-1} (Table 1). The manganese concentration is higher in colostrum (100–160 µg litre^{-1}) than in mature milk (20–50 µg),[108,174] and a decrease of over 50% has been reported to occur during the first 3 days of lactation.[108] Oral supplementation of manganese to cows can increase the manganese content of milk provided that large doses are administered over a long period of time. However, the cow reacts slowly to fluctuations in dietary intake of manganese because of her capacity for manganese storage.[175]

The mean concentration of manganese in mature human milk is 12 µg litre^{-1}, with a range of 7–15 µg (Table 1). However, there are considerable differences in published values, and higher mean values than this (20 µg litre^{-1} [7] and 41 µg litre^{-1} [8]) have been reported. Manganese concentration varies with stage of lactation, with values averaging 6 µg litre^{-1} in early lactation, 4 µg in mid-lactation (2–6 months) and 6–8 µg in late-lactation.[43,176] A correlation between dietary manganese intake and milk manganese has been reported.[43,107]

There is little information on the localization of manganese in milk, and none at all on its bioavailability. In cow's milk, manganese is reported to be present as an 'organic compound', some of it in the fat globule membrane.[27] Lactose synthetase, present in milk, is reported to be a manganese-dependent enzyme.[177]

A wide variation in the manganese content of infant formulae has been reported, with concentrations in several formulae 100–1000 times higher than in human milk, while some had levels less than those recommended and others had undetectable concentrations.[152] These authors suggested that long-term intake of such formulae could lead to manganese toxicity or deficiency although no such effects have been reported.

12. SELENIUM

Selenium is an essential component of the enzyme glutathione peroxidase which occurs in many human tissues where, together with vitamin E and the enzymes catalase and superoxidedismutase, it functions as an antioxidant, protecting cells against oxidative damage. The recommended ranges of safe and adequate daily intakes of selenium are: 20–50 µg for age 0·5–1 year; 20–80 µg for age 1–3 years; 30–120 µg for age 4–7 years; and 50–200 µg for children over age 7, adolescents and adults.[38]

In areas of China where the soil content of selenium is low, selenium deficiency causes a disorder of heart muscle (Keshan disease) in young children,[178] and low selenium status has been reported in New Zealand[179] and Finland,[180] countries where the soil content of selenium is also low.

The contribution of dairy products (with which eggs were included) to daily dietary intake of selenium has been estimated as: 5 µg (8% of the total intake) in the UK;[181] 13 µg (10%) in the USA;[182] 13 µg (21–26%) in Finland;[183] and 11 µg (39%) in New Zealand.[179]

Selenium concentration in cow's milk is 5–67 µg litre^{-1} in the USA[22] and 7–11 µg litre^{-1} in Germany.[184] The concentration in milk depends on dietary intake, and in New Zealand, where the selenium content of soil and plants is low, concentrations as low as 3–5 µg litre^{-1} have been reported.[185,186]

In the UK, mature human milk contains an average of 14 µg of selenium per litre, with a range of 8–19 µg for five pooled samples (Table 1). Similar values have been reported in other countries: 16·3 ± 4·9 (range 8–34) µg litre^{-1} [187] and 18 (7–33) µg litre^{-1} [188] in the USA, and 10–15 µg litre^{-1} in Germany.[184] However, higher values have been reported in other studies: 23·3 in Germany,[189] 13–50 in the USA,[190,191] and 12–145 in New Zealand.[186] In Finland, lower milk selenium concentrations have been reported and these decrease with advancing lactation from 10·7 µg litre^{-1} at 4 weeks to 6 µg at 12 and 25 weeks.[192] Selenium concentration is higher in human colostrum (41·2 ± 17·3 µg litre^{-1}) than in mature milk (16·3 ± 4·9).[187] A geographical variation in selenium concentration in mature human milk has been reported in the USA.[188]

The selenium requirement of the human infant is not known but daily intakes of 10–40 µg have been recommended as safe and adequate for infants under 6 months of age.[38] The mean selenium intake of breast fed infants coincides with the lower limit of this recommendation.[187] A mean selenium concentration of 8·6 ± 0·9 µg litre^{-1} has been reported for infant formulae in the USA,[187] which is significantly lower than that in mature human milk, and Lombeck et al.[189] found that the mean selenium concentration in cow's milk based infant formulae in Germany was less than 33% of that in mature human milk. Daily selenium intakes of formula fed infants at 3 months of age have been reported to be significantly lower (~7·2 µg) than those of breast fed infants (10·1 µg), and selenium status, as reflected by serum selenium concentrations, was lower in the formula fed infants.[187]

The bioavailability of selenium from human and cow's milk is unknown.

13. IODINE

In the human body, iodine is an essential component of the thyroid hormones, thyroxine and triiodothyronine, which are important in controlling the rate of basal metabolism and in reproduction. Dietary deficiency of iodine causes enlargement of the thyroid gland and goitre, while a large excess of iodine in the diet reduces the uptake of iodine by this gland and also produces signs of thyroid deficiency (thyrotoxicosis).

Recommended daily dietary intakes of iodine are: 40 µg for infants under 6 months of age; 50 µg for infants aged 6–12 months; 70, 90 and 120 µg for children aged 1–3, 4–6 and 7–10 years, respectively; 150 µg for adolescents and adults; and 175 and 200 µg during pregnancy and lactation, respectively.[38] Adult daily intakes of up to 1000 µg are considered safe.[193]

The mean iodine concentration in mature human milk in the UK is 70 µg litre^{-1}, with a range for five pooled samples of 20–120 µg (Table 1). Higher concentrations than this (178 µg litre^{-1}, range 29–490) have been reported in the USA.[194] Milk iodine concentration decreases from 50–240 µg litre^{-1} in colostrum to 40–80 µg litre^{-1} in mature milk.[195] There is a correlation between the iodine content of milk and dietary iodine intake,[194] and the use of iodized salt by the mother can lead to milk iodine concentrations of up to 200 µg litre^{-1}.[7] Recommended levels in infant formulae are 20–80 µg litre^{-1} of reconstituted formula.[6]

The iodine content of cow's milk has been reviewed recently.[196,197] Iodine concentration in milk is influenced by season and is closely related to dietary intake.[198] The UK National Institute for Research in Dairying reported milk iodine concentrations ranging from 10 µg litre^{-1} in a herd in summer without mineral supplementation to 800 µg litre^{-1} in a herd fed winter rations containing mineral supplements, and a mean concentration of 260 µg litre^{-1} in winter milk.[20] Broadhead et al.[199] reported a maximum iodine concentration in cow's milk in the UK of 200 µg litre^{-1} in March/April and a minimum concentration of 10 µg litre^{-1} in August/September. Iodine concentration decreases from 200–300 µg litre^{-1} in colostrum to 70–136 µg litre^{-1} in mature milk.[200]

Dietary supplementation with inorganic iodine as potassium iodide does not increase the iodine content of milk as much as organic iodine fed on ethylenediamine dihydriodide (EDDI). High concentrations of iodine in cow's milk have been related to the addition of excessive amounts of EDDI to dairy cow rations.[201,202]

The use of iodophors for teat disinfection increases the iodine content in cow's milk by 120 to 400 µg litre^{-1},[20] and by 300 to 700 µg litre^{-1} with improper use.[197] Iodine contamination of milk occurs as a result of absorption through the skin rather than by surface contamination of the teat.[203] Use of iodophors for the disinfection of containers, milking machines and processing equipment can also cause iodine contamination of milk.[197]

There is evidence that the average iodine content of cow's milk is increasing in some countries.[197] Prior to 1970, few values for the iodine content of milk over 180 µg litre^{-1} had been reported.[204] More recent reviews from Australia[205] and the USA[206] have reported that the average iodine contents in milk are probably between 300 and 700 µg litre^{-1}. The International Dairy Federation[197] has recommended the standardization of mineral feed supplements and supervised and re-stricted use of iodophor disinfectants as measures to reduce the iodine content of milk and milk products.

Estimates of iodine consumption in the USA suggest that in Western countries iodine excess is more likely to occur than iodine deficiency.[196] Average daily iodine consumption of young adult males in the USA has increased from 100–150 µg in 1960 to 600 µg in 1974, with up to 1050 µg where iodized salt was used.[207] Since dairy products contribute 38–50% of the total iodine intake of adults in the USA,[208] the increase in iodine content of cow's milk has caused some concern. However, the use of iodized salt is still recommended in non-coastal regions in the USA.[38] Talbot *et al.*[209] reviewed the effects of dietary iodine on thyroid disorders in the United States and concluded that there was no relationship between excess dietary intake of iodine and such disorders.

Most (80–90%) of the iodine in cow's milk is in the inorganic form, mainly as iodide,[27] and up to 13% is bound to proteins and less than 0·1% to fat.[210]

14. FLUORIDE

Fluoride accumulates in the hard tissues of the body (bones and teeth), and this occurs more rapidly in young growing animals than in older ones.[6] Excessive intake of fluoride causes fluorosis but a deficiency is associated with dental caries.

The recommended safe and adequate daily intakes of fluoride are: 100–500 µg for infants up to 6 months of age; 200–1000 µg for infants aged 6–12 months; 500–1500 µg for children aged 1–3 years; 1000–

2500 µg for children aged 4–6 years; 1500–2500 µg for children aged 7–10 years and adolescents; and 1500–4000 µg for adults.[38]

Human milk contains an average of 77 µg of fluoride per litre, with a range of means for pooled samples of 21–155 µg (Table 1), but values as high as 200 µg litre^{-1} have been reported.[211] The concentration in milk does not seem to be increased in areas where water is fluoridated.[21]

The fluoride concentration in cow's milk ranges from 30 to 220 µg litre^{-1} (Table 1). About 60% of the fluoride in cow's milk occurs as fluoride ions.[212] Since the fluoride content of cow's milk is low relative to that required for the prevention of dental caries, it has been suggested that fluoridation of milk to concentrations of 1·5–2·5 mg litre^{-1} might help in the prevention of caries. There is some evidence that this is beneficial in children.[213]

Since the fluoride contents of breast milk and infant formulae are lower than recommended intakes for infants, fluoride supplementation has been recommended for infants who are breast fed or fed on artificial infant formula in areas where the fluoride concentration in drinking water is less than 0·3 µg litre^{-1}.[214] Reconstitution of infant formula with water fluoridated to a concentration of 1 mg litre^{-1} (as recommended for water supplies in the UK[215]) provides more than adequate fluoride to the infant. There is no evidence that such a practice is harmful to the infant,[6] and the work of Ericsson and Ribelius[21] suggests that there is a considerable margin of safety in the amount of fluoride that may be ingested by the infant.

15. COBALT

The mean cobalt concentration in cow's milk has been reported as 1 (range 0·5–1·3) µg litre^{-1},[24] 0·5 (range 0·4–1·1) µg litre^{-1},[19] or 0·7 (range 0·1–3·4) µg litre^{-1}.[2] The cobalt concentration in mature human milk has been reported as 4–16 µg litre^{-1}[2] or 12 (range 1–27) µg litre^{-1}.[12]

The only known function of cobalt in humans is its presence as an essential component of vitamin B_{12}. The amount of cobalt combined with vitamin B_{12} in human and cow's milk has been estimated as 0·004 and 0·012 µg litre^{-1}, respectively,[6] which is considerably less than the total cobalt concentrations in these milks, and thus most of the cobalt must be present in a form other than in vitamin B_{12}. It has been estimated that dairy products contribute 26·5% of total cobalt intake in Canada.[105]

16. CHROMIUM

Chromium is regarded as an essential nutrient for humans, and the earliest detectable effect of deficiency is an impairment of glucose tolerance.[216] The exact requirements for chromium are unknown but recommended safe and adequate daily intakes are: 10–40 and 20–60 µg for infants aged 0–6 and 6–12 months, respectively; 20–80 and 30–120 µg for children aged 1–3 and 4–6 years, respectively; and 50–200 µg for children aged 7–10 years, adolescents and adults. In Western countries dairy products contribute a significant amount of chromium (19·5% of total intake in Canada[105]).

The mean of chromium concentration in cow's milk has been reported as 10 (range 8–13) µg litre^{-1},[19] 20 µg litre^{-1} [8] or 17 (range 5–50) µg litre^{-1}.[12] The chromium content of cow's milk can be increased by contamination with chromium-containing detergents.[5] In human milk, chromium concentration is 6–100 (mean 40) µg litre^{-1},[12] and a concentration of 29 ± 21 µg litre^{-1} has been reported by Murthy.[2]

17. MOLYBDENUM

Molybdenum is an essential component of several enzymes including xanthine oxidase. Recommended safe and adequate daily intakes of molybdenum are: 30–60 µg for infants under 6 months of age; 40–80 µg for infants 6–12 months old; 50–100 µg for children aged 1–3 years; 60–150 µg for children aged 7–10 years; and 150–500 µg for adolescents and adults.

The molybdenum content of cow's milk is very variable, and values in the range 18–120 µg litre^{-1}, with a mean of 73 µg, have been reported.[24] Milk molybdenum concentration has been shown to be dependent on dietary intake; Archibald[24] reported that the molybdenum concentration increased from 73 to 371 µg litre^{-1} when cows were supplemented with ammonium molybdate. All of the molybdenum present in cow's milk is considered to be associated with xanthine oxidase.[217] However, when the molybdenum concentration in milk was increased by dietary supplementation there was no corresponding increase in milk xanthine oxidase activity.[24,217]

The molybdenum content of mature human milk has been reported as 8·4 µg litre^{-1},[23] and the xanthine oxidase activity of human colostrum is about one-tenth that of cow's milk.[218]

18. OTHER TRACE ELEMENTS

There is evidence from experimental animal studies that nickel, silicon, tin, vanadium and arsenic are nutritionally essential minerals which are required in trace amounts in the diet, and deficiency of any of these results in poor growth and development. They are probably also essential nutrients for humans although their physiological functions are unknown. Deficiencies of these minerals have not been demonstrated in humans, and human requirements for these elements are unknown. Some of these elements are toxic in excessive amounts but there is no evidence that toxic levels occur in milk or dairy products. Reported concentrations of these elements in human and cow's milk are presented in Table 1.

REFERENCES

1. FOMON, S. J., *Infant Nutrition*, 2nd edn, 1974, W. B. Saunders Company, Philadelphia.
2. MURTHY, G. K., *CRC Crit. Rev. Environ. Control*, 1974, **4**, 1.
3. DEPARTMENT OF HEALTH AND SOCIAL SECURITY, *The Composition of Mature Human Milk*, 1977, HMSO, London.
4. JELLIFFE, D. B. and JELLIFFE, E. F. P., *Human Milk in the Modern World*, 1978, Oxford University Press, Oxford.
5. JARRETT, J. D., *Aust. J. Dairy Technol.*, 1979, **34**, 28.
6. DEPARTMENT OF HEALTH AND SOCIAL SECURITY, *Artificial Feeds for the Young Infant*, 1980, HMSO, London.
7. AMERICAN ACADEMY OF PEDIATRICS, Committee on Nutrition, *Pediatrics*, 1981, **68**, 435.
8. BLANC, B., *Wld. Rev. Nutr. Diet.*, 1981, **36**, 1.
9. GURR, M. I., *J. Dairy Res.*, 1981, **48**, 519.
10. LONNERDAL, B., KEEN, C. L. and HURLEY, L. S., *Ann. Rev. Nutr.*, 1981, **1**, 149.
11. PACKARD, V., *Human Milk and Infant Formula*, 1982, Academic Press, London.
12. RENNER, E., *Milk and Dairy Products in Human Nutrition*, 1983, Volkswirtschaftlicher Verlag, München.
13. SCHAAFSMA, G., *Internat. Dairy Federation Bull.*, Document 166, 1984.
14. LEE, V. A. and LORENZ, K., *CRC Crit. Rev. Food Sci. Nutr.*, 1979, **11**, 41.
15. BRIGGS, G. M. and CALLOWAY, D. H., *Nutrition and Physical Fitness*, 10th edn, 1979, W. B. Saunders Company, London.
16. PAUL, A. A. and SOUTHGATE, D. A. T., McCance and Widdowson's *The Composition of Foods*, 1978, HMSO, London.
17. CAVELL, P. A. and WIDDOWSON, E. M., *Arch. Dis. Childh.*, 1964, **39**, 496.
18. McLEOD, B. E. and ROBINSON, M. F., *Brit. J. Nutr.*, 1972, **27**, 229.

19. UNDERWOOD, E. J., *Trace Elements in Human Nutrition*, 3rd edn, 1971, Academic Press, New York.
20. DODD, F. H., KINGWILL, R. G., SHEARN, M. F. H., MORANT, S. V. and LEWIS, G., *National Institute for Research in Dairying Report 1973–4*, 1975, p. 57.
21. ERICSSON, Y. and RIBELIUS, U., *Acta Pediatr. Scand.*, 1970, **59**, 424.
22. HADJIMARKOS, D. M. and BONHORST, C. W., *J. Pediatr.*, 1961, **59**, 256.
23. DANG, H. S., JAINSAL, D. D., WADHWANI, C. N. and SOMASUNDERAM, S., *Sci. Total Environ.*, 1983, **27**, 43.
24. ARCHIBALD, J. C., *Dairy Sci. Abstr.*, 1958, **20**, 712, 798.
25. APERIA, A., BROBERGER, O., HERIN, P., and ZETTERSTROM, R., *Acta Pediatr. Scand.*, 1979, **68**, 441.
26. ANSELL, C., MOORE, A. and BARRIE, H., *Pediatr. Res.*, 1979, **11**, 1177.
27. KIRCHGESSNER, M., FRIESECKE, H. and KOCH, G., *Nutrition and the Composition of Milk*, 1967, Crosby Lockwood, London.
28. GARRETT, O. F. and OVERMAN, O. R., *J. Dairy Sci.*, 1940, **23**, 13.
29. PEAKER, M. and FAULKNER, A., *Proc. Nutr. Soc.*, 1983, **42**, 419.
30. PEAKER, M., In: *Lactation*, Vol. 4, B. L. Larson (ed.), 1978, Academic Press, New York, p. 437.
31. HOLT, C., Chapter 6 (this volume).
32. MCCANCE, R. A. and WIDDOWSON, E. M., *Acta Pediatr. Scand.*, 1957, **46**, 337.
33. THODENIUS, K., *Acta Pediatr. Scand.*, Suppl. 253, 1974, 1.
34. TOMARELLI, R. M., *J. Pediatr.*, 1976, **88**, 454.
35. JAGERSTAD, M., In: *The Role of Calcium in Biological Systems*, Vol. III, L. J. Anghileri and A. M. Tuffett-Anghileri (eds), 1982, CRC Press, Boca Raton, Florida, p. 45.
36. *Dairy Council Digest*, 1984, **55**, 1.
37. LEVEILLE, G. A., *J. Dairy Sci.*, 1979, **62**, 1665.
38. FOOD AND NUTRITION BOARD, *Recommended Dietary Allowances*, 9th rev. edn, 1980, National Academy of Sciences, National Research Council, Washington, DC.
39. MARSTON, R. M. and WELSH, S. O., *Nat. Food Rev.*, 1983, **21**, 17.
40. DONOVAN, S., *Proc. Nutr. Soc.*, 1983, **42**, 375.
41. UPTON, P. K. and GIBNEY, M. J., *Ir. J. Food Sci. Technol.*, 1976, **1**, 79.
42. INTERNATIONAL UNION OF NUTRITIONAL SCIENCES, *Nutr. Abstr. Rev.*, 1983, **53**, 1075.
43. VAUGHAN, L. A., WEBER, C. W. and KEMBERLING, S. R., *Am. J. Clin. Nutr.*, 1979, **32**, 2301.
44. KARMARKAR, M. G. and RAMAKRISHNAN, C. V., *Acta Pediatr. Scand.*, 1960, **49**, 599.
45. KIRSKEY, A., ERNST, J. A., ROEPKE, J. F. and TSAI, T. L., *Am. J. Clin. Nutr.*, 1979, **32**, 30.
46. JENNESS, R., In: *Lactation: A Comprehensive Treatise*, Vol. 3, B. L. Larson and V. R. Smith (eds), 1974, Academic Press, London, p. 3.
47. KONDO, T., KIYOSAWA, I., RYOKI, T. and MUENO, M., *Nippon Nogeikagaku Kaishi*, 1964, **38**, 71.
48. ALLEN, J. C. and NELVILLE, M. C., *Clin. Chem.*, 1983, **29**, 858.

49. FRANSSON, G. B. and LONNERDAL, B., *J. Pediatr.*, 1982, **101**, 504.
50. AZUMA, N., KAMINOGAWA, S., KOBAYASHI, H. and YAMAUCHI, K., *J. Dairy Sci.*, 1982, **65**, 1401.
51. PURVIS, R. J., BARRIE, W. J., MacKAY, G. S., WILKINSON, E. M., COCKBURN, F., BELTAN, N. R. and FORFAR, J. O., *Lancet*, 1973, **ii**, 811.
52. ALI, R. and EVANS, J. L., *J. Agr. Univ. Puerto Rico*, 1973, **57**, 149.
53. ARMBRECHT, H. J. and WASSERMAN, R. H., *J. Nutr.*, 1976, **106**, 1265.
54. MILLS, R., BREITER, H., KEMPSTER, E., McKEY, B., PICKENS, M. and OUTHOUSE, J., *J. Nutr.*, 1940, **20**, 467.
55. ZIEGLER, E. E. and FOMON, S. J., *Pediatr. Res.*, 1980, **14**, 513.
56. KANSAL, V. K. and CHAUDHARY, S., *Milchwissenschaft*, 1982, **37**, 261.
57. WONG, N. P. and LA CROIX, D. E., *Nutr. Rep. Internat.*, 1980, **21**, 673.
58. MYKKANEN, H. M. and WASSERMAN, R. H., *J. Nutr.*, 1980, **110**, 2140.
59. SRINIVASAN, M. R. and RAO, M. V. L., *J. Fd. Sci. Technol.*, 1979, **16**, 95.
60. FOMON, S. J., OWEN, G. M., JENSEN, R. L. and THOMAS, L. N., *Am. J. Clin. Nutr.*, 1963, **12**, 346.
61. SOUTHGATE, D. A. T., WIDDOWSON, E. M., SMITS, B. J., COOKE, W. T., WALKER, C. H. M. and MATHERS, N. P., *Lancet*, 1969, **i**, 487.
62. WIDDOWSON, E. M., McCANCE, R. A., HARRISON, G. E. and SUTTON, A., *Lancet*, 1963, **ii**, 1250.
63. BARLTROP, D., MOLE, R. and SUTTON, A., *Arch. Dis. Childh.*, 1977, **52**, 41.
64. SCHAAFSMA, G. and DE WAARD, H., *Voeding*, 1982, **43**, 398.
65. THE AMERICAN SOCIETY FOR BONE AND MINERAL RESEARCH, *Osteoporosis*, 1982, Kelseyville, California.
66. HEANEY, R. P., GALLAGHER, J. C., JOHNSTON, C. C., NEER, R., PARFITT, A. M. and WHEDON, G. D., *Am. J. Clin. Nutr.*, 1982, **36**, 986.
67. SEEMAN, E. and RIGGS, B. L., *Geriatrics*, 1981, **36**, 71.
68. NATIONAL INSTITUTE OF ARTHRITIS, DIABETES AND DIGESTIVE AND KIDNEY DISEASES, *Osteoporosis: Cause, Treatment, Prevention*, NIH Publication No. 83–2226.
69. CARROLL, M. D., ABRAHAM, S. and DRESSER, C. M., DHSS Publication No. (PHS) 83–1681, 1983, National Centre for Health Statistics, Public Health Service, US Government Printing Office, Washington, DC.
70. ACKLEY, S., BARRETTE-CONNOR, E. and SUAREZ, L., *Am. J. Clin. Nutr.*, 1983, **38**, 457.
71. LANGFORD, H. C. and WATSON, R. L., *Clin. Sci. Mol. Med.*, 1973, **45**, 111.
72. McCARRON, D. A. and MORRIS, C. D., *Clin. Res.*, 1982, **30**, 338A (Abstr.).
73. McCARRON, D. A., MORRIS, C. D. and COLE, C., *Science*, 1982, **217**, 267.
74. McCARRON, D. A., *Ann. Intern. Med.*, 1983, **98**, 800.
75. McCARRON, D. A., STANTON, J., HENRY, H. and MORRIS, C., *Ann. Intern. Med.*, 1983, **98**, 715.
76. McCARRON, D. A., MORRIS, C. D., HENRY, H. J. and STANTON, J. L., *Science*, 1984, **224**, 1392.
77. MORRIS, C. D., HENRY, H. J. and McCARRON, D. A., *Clin. Res.*, 1984, **32**, 57A (Abstr.).
78. HARLAN, W. R., HULL, A. L., SCHMOUDER, R. L., LANDIS, J. R., THOMPSON, F. F. and LARKIN, F. A., *Am. J. Epidemiol.*, 1984, **120**, 17.

79. BELIZAN, J. M., VILLAR, J., PINEDA, O., GONZALEZ, A. E., SAINZ, E., GARRERA, G. and SIBRIAN, R., *J. Am. Med. Assoc.*, 1983, **249**, 1161.
80. BELIZAN, J. M., VILLAR, J., ZALAZAR, A., ROJAS, L., CHAN, D. and BRYCE, G. F., *Am. J. Obstet. Gynaecol.*, 1983, **146**, 175.
81. MCCARRON, D. A., HENRY, H. J. and MORRIS, C. D., *Clin. Res.*, 1984, **32**, 37A (Abstr.).
82. BARRY, G. D., *Fed. Proc.*, 1977, **36**, 492 (Abstr.).
83. AYACHI, S., *Metabolism*, 1979, **28**, 1234.
84. MCCARRON, D. A., YUNG, N. N., UGORETZ, B. A. and KRUTZIK, S., *Hypertension* (Suppl.), 1981, **1**, 162.
85. ANDERSON, S., GRADY, J. R., ELLISON, D. E. and MCCARRON, D. A., *Hypertension* (Suppl.), 1983, **1**, 59.
86. ABRAHAM, S., CARROLL, M. D., DRESSER, C. M. and JOHNSON, C. L., *Dietary Intake Findings, United States 1971–74*, National Center for Health Statistics.
87. PARROTT-GARCIA, M. and MCCARRON, D. A., *Nutr. Rev.*, 1984, **42**, 205.
88. MADDEN, A., FLYNN, A. and CREMIN, F. M., *Ir. J. Fd. Sci. Technol.*, 1984, **8**, 152.
89. ENGSTROM, A. M. and TOBLEMAN, R. C., *Ann. Intern. Med.*, 1983, **98**, 870.
90. RAJALAKSHMI, K. and SRIKANTIA, S. G., *Am. J. Clin. Nutr.*, 1980, **33**, 664.
91. DAVIES, D. T. and WHITE, J. C. D., *J. Dairy Res.*, 1960, **27**, 171.
92. ANAST, C. S., *Pediatrics*, 1964, **33**, 969.
93. COUSSONS, H., *Postgrad. Med.*, 1969, **46**, 135.
94. JENNESS, R. and PATTON, S., *Principles of Dairy Chemistry*, 1959, John Wiley and Sons, New York.
95. AMERICAN ACADEMY OF PEDIATRICS, COMMITTEE ON NUTRITION, *Pediatrics*, 1978, **62**, 826.
96. GITTLEMAN, I. F. and PINCUS, J. B., *Pediatrics*, 1951, **8**, 778.
97. JOWSEY, J., *Metabolic Diseases of Bone*, 1977, W. B. Saunders Company, London.
98. ALLEN, L. H., *Am. J. Clin. Nutr.*, 1982, **35**, 783.
99. MACY, I. G., KELLY, H. J. and SLOAN, R. C., *The Composition of Milks*, Nat. Res. Council Publ. No. 254, 1953, National Academy of Sciences, Washington, DC.
100. SVANBERG, U., GEBER-MEDHIN, M., LJUNGGQVIST, B. and OLSON, M., *Am. J. Clin. Nutr.*, 1977, **30**, 499.
101. STURMAN, J. A. and GAULL, G. E., *J. Neurochem.*, 1975, **25**, 831.
102. GAULL, G. E., RASSIN, D. K., RAIHA, N. C. R. and HEINONEN, K., *J. Pediatr.*, 1977, **90**, 348.
103. RIGO, J. and SENTERRE, J., *Biol. Neonate*, 1977, **32**, 73.
104. LAVANCHY, P. and BUHLMANN, C., unpublished results, 1977, cited in Ref. 8.
105. MERANGER, J. C. and SMITH, D. C., *Can. J. Publ. Hlth.*, 1972, **63**, 53.
106. PICCIANO, M. F. and GUTHRIE, H. A., *Am. J. Clin. Nutr.*, 1976, **29**, 242.
107. VUORI, E., MAKINEN, S. M., KARA, R. and KUITUNEN, P., *Am. J. Clin. Nutr.*, 1980, **33**, 227.
108. DE MARIA, C. G., *Ann. Rech. Vet.*, 1978, **9**, 277.

109. MURTHY, G. K., RHEA, U. and PEELER, J. T., *J. Dairy Sci.*, 1972, **55**, 1666.
110. KING, R. L., LUICK, J. R., LITMAN, I. I., JENNINGS, W. G. and DUNKLEY, W. L., *J. Dairy Sci.*, 1959, **42**, 780.
111. RICHARDSON, T. and GUSS, P. L., *J. Dairy Sci.*, 1965, **48**, 523.
112. FRANSSON, G. B. and LONNERDAL, B., *Pediatr. Res.*, 1983, **17**, 912.
113. HEGENAUER, J., SALTMAN, P., LUDWIG, D., RIPLEY, L. and LEY, A., *J. Agric. Food Chem.*, 1979, **27**, 1294.
114. HEGENAUER, J., SALTMAN, P., LUDWIG, D., RIPLEY, L. and BAJO, P., *J. Agric. Food Chem.*, 1979, **27**, 860.
115. HEGENAUER, J., SALTMAN, P. and LUDWIG, D., *J. Agric. Food Chem.*, 1979, **27**, 868.
116. DEMOTT, B. J. and DINCER, B., *J. Dairy Sci.*, 1976, **59**, 1557.
117. FRANSSON, G. B. and LONNERDAL, B., *J. Pediatr.*, 1980, **96**, 380.
118. FRANSSON, G. B. and LONNERDAL, B., *Naringsforskning*, 1980, **24**, 66 (Abstr.).
119. MCLELLAND, D. B. L., MCGRATH, J. and SAMSON, R. R., *Acta Pediatr. Scand.*, 1978, Suppl. 271, 3.
120. LONNERDAL, B., FORSUM, E. and HAMBREUS, L., *Am. J. Clin. Nutr.*, 1976, **29**, 1127.
121. BULLEN, J. J., ROGERS, H. J. and LEIGH, L., *Brit. Med. J.*, 1972, **1**, 69.
122. GROVES, M. L., In: *Milk Proteins*, Vol. 2, H. A. McKenzie (ed.), 1971, Academy Press, New York, p. 367.
123. REITER, B., *J. Dairy Res.*, 1978, **45**, 131.
124. MEYER, F. and SENFT, B., *Milchwissenschaft*, 1979, **34**, 74.
125. ARNOLD, R. R., COLE, M. F. and MCGHEE, J. R., *Science*, 1977, **197**, 263.
126. CUNNINGHAM, A. S., *J. Pediatr.*, 1979, **95**, 685.
127. FALLOTT, M. E., BOYD, J. L. and OSKI, F. A., *Pediatrics*, 1980, **65**, 1121.
128. HAMBREUS, L., *Pediatr. Clin. N. Am.*, 1977, **24**, 17.
129. FRANSSON, G-B., In: *Role of Milk Proteins in Human Nutrition*, W. Kaufman (ed.), 1983, Verlag Th. Mann, Germany, p. 441.
130. FRANNSON, G. B., KEEN, C. L. and LONNERDAL, B., *J. Pediatr. Gastroenterol. Nutr.*, 1983, **2**, 693.
131. MASSON, P. L., HEREMANS, J. F., SCHONNE, E. and CRABBE, P. A., *Protides in Biol. Fluids Proc. Colloq.*, 1969, **16**, 633.
132. COX, T. M., MAZURIER, J., SPIK, F., MONTREUIL, J. and PETERS, T. J., *Biochim. Biophys. Acta*, 1979, **588**, 120.
133. MCMILLAN, J., OSKI, F. A., LOURIE, G., TOMARELLI, R. M. and LANDOW, S. A., *Pediatrics*, 1977, **60**, 896.
134. DE VET, B. J. C. M. and VAN GOOL, J., *Acta Med. Scand.*, 1974, **196**, 393
135. BROCK, J. H., *Arch. Dis. Childh.*, 1980, **55**, 417.
136. GARBY, L. and SJOLIN, S., *Acta Pediatr. Scand.*, 1959, Suppl. 117, 24.
137. GORTEN, M. H., HEPPNER, R. and WORKMAN, J. B., *J. Pediatr.*, 1963, **63**, 1063.
138. SAARINEN, U. M., SIIMES, M. A. and DALLMAN, P., *J. Pediatr.*, 1977, **91**, 36.
139. SAARINEN, U. M. and SIIMES, M. A., *Pediatr. Res.*, 1979, **13**, 143.
140. MCMILLAN, J. A., LANDAW, S. A. and OSKI, E. A., *Pediatrics*, 1976, **58**, 686.
141. AMERICAN ACADEMY OF PEDIATRICS, COMMITTEE ON NUTRITION, *Pediatrics*, 1978, **62**, 246.

142. AMERICAN ACADEMY OF PEDIATRICS, COMMITTEE ON NUTRITION, *Pediatrics*, 1980, **66**, 1015.
143. DALLMAN, P. R., SIIMES, M. A. and STEKEL, A., *Am. J. Clin. Nutr.*, 1980, **33**, 86.
144. SAARINEN, U. M., *J. Pediatr.*, 1978, **93**, 177.
145. WOODRUFF, C. W., LATHAM, C. and McDAVID, S., *J. Pediatr.*, 1977, **90**, 36.
146. ROH, J. K., BRADLEY, R. L., RICHARDSON, T. and WECKEL, K. G., *J. Dairy Sci.*, 1976, **59**, 382.
147. MULDER, H., MENGER, J. W. and MEIJERS, P., *Neth. Milk Dairy J.*, 1964, **18**, 52.
148. VUORI, E. and KUITUNEN, P., *Acta Pediatr. Scand.*, 1979, **68**, 33.
149. MARTIN, M. T., JACOBS, F. A. and BRUSHMILLER, J. G., *J. Nutr.*, 1984, **114**, 869.
150. LONNERDAL, B., HOFFMAN, B. and HURLEY, L. S., *Am. J. Clin. Nutr.*, 1982, **36**, 1170.
151. WALREAVENS, P. A., *Clin. Chem.*, 1980, **26**, 185.
152. LONNERDAL, B., KEEN, C. L., OHTAKE, M. and TSUNENOBU, T., *Am. J. Dis. Childh.*, 1983, **137**, 433.
153. TANAKA, Y., HATANO, S., NISHI, Y. and USUI, T., *J. Pediatr.*, 1980, **96**, 255.
154. PRASAD, A. S. D., MIALE, A., FARID, Z., STANSTEAD, H. H. and SHUBERT, A. R., *J. Lab. Clin. Med.*, 1963, **61**, 537.
155. HAMBIDGE, K. M., HAMBIDGE, C., JACOBS, M. and BARUM, J. D., *Pediatr. Res.*, 1972, **6**, 868.
156. MURTHY, G. K., RHEA, U. and PEELER, J. T., *Environ. Sci. Technol.*, 1971, **5**, 436.
157. ARCHIBALD, J. G., *J. Dairy Sci.*, 1944, **27**, 257.
158. MILLER, W. J., CLIFTON, C. M., FLOWER, P. R. and PERKINS, H. F., *J. Dairy Sci.*, 1965, **48**, 450.
159. SANDSTORM, B., KEEN, C. L. and LONNERDAL, B., *Am. J. Clin. Nutr.*, 1983, **38**, 420.
160. BLAKEBOROUGH, P., SALTER, D. N. and GURR, M. I., *Biochem. J.*, 1983, **209**, 505.
161. HURLEY, L. S., DUNCAN, J. R., SLOAN, M. V. and ECKHERT, D., *Proc. Nat. Acad. Sci. USA*, 1977, **34**, 3547.
162. HURLEY, L. S., LONNERDAL, B. and STANISLOWSKI, A. G., *Lancet*, 1979, **i**, 677.
163. ECKHERT, C. D., SLOAN, M. W., DUNCAN, J. R. and HURLEY, L. S., *Science*, 1977, **195**, 789.
164. COUSINS, R. J. and SMITH, K. T., *Am. J. Clin. Nutr.*, 1980, **33**, 1083.
165. EVANS, G. W. and JOHNSON, P. E., *Pediatr. Res.*, 1980, **14**, 876.
166. REBELLO, T., LONNERDAL, B. and HURLEY, L. S., *Am. J. Clin. Nutr.*, 1982, **35**, 1.
167. LONNERDAL, B., STANISLOWSKI, A. G. and HURLEY, L. S., *J. Inorg. Biochem.*, 1980, **12**, 71.
168. MARTIN, M. T., LICKLIDER, K. and BRUSHMILLER, J. G., *J. Inorg. Biochem.*, 1981, **15**, 55.
169. AINSCOUGH, E. W., BRODIE, A. M. and FLOWMAN, J. E., *Am. J. Clin. Nutr.*, 1980, **33**, 1314.

170. JOHNSON, P. E. and EVANS, G. W., *Am. J. Clin. Nutr.*, 1975, **31**, 416.
171. HAMBIDGE, K. M., WALREAVENS, P. A., CASEY, C. E., BROWN, R. M. and BENDER, C., *J. Pediatr.*, 1979, **94**, 607.
172. FOMON, S. J. and FILER, L. J., In: *Infant Nutrition*, S. J. Fomon (ed.), 1974, W. B. Saunders Company, Philadelphia, p. 359.
173. HARZER, G. and KAUER, H., *Am. J. Clin. Nutr.*, 1982, **35**, 981.
174. ARCHIBALD, J. G. and LINDQVIST, H. G., *J. Dairy Sci.*, 1943, **26**, 325.
175. ARCHIBALD, J. G., *Milk Plant Monthly*, 1941, **9**, 36.
176. VUORI, E., *Acta Pediatr. Scand.*, 1979, **68**, 571.
177. RICHARDSON, R. H. and BREW, K., *J. Biol. Chem.*, 1980, **255**, 3377.
178. CHEN, X., YANG, G., CHEN, J., CHEN, X., WEN, Z. and GE, K., *Biol. Trace Elem. Res.*, 1980, **2**, 91.
179. THOMSON, C. D. and ROBINSON, M. F., *Am. J. Clin. Nutr.*, 1980, **33**, 303.
180. *Nordic Symposium on Mineral Elements '80*, 1981, Academy of Finland, Helsinki.
181. THORN, J., ROBERTSON, T. and BUSS, B. H., *Brit. J. Nutr.*, 1978, **39**, 391.
182. LEVANDER, O. A., In: *Proceedings of the Symposium on Selenium-Tellurium in the Environment*, 1976, Industrial Health Foundation, Pittsburgh, p. 26.
183. VARO, P. and KOIVISTOINEN, P., *Int. J. Vitam. Nutr. Res.*, 1981, **51**, 79.
184. LOMBECK, I., KASPEREK, K., HARBISCH, H. D., FEINENDEGEN, L. E. and BREMER, H. J., *Eur. J. Pediatr.*, 1977, **125**, 81.
185. GRANT, A. B. and WILSON, G. F., *N. Z. J. Agric. Res.*, 1968, **11**, 733.
186. MILLAR, K. R. and SHEPPARD, A. D., *N. Z. J. Sci.*, 1972, **15**, 3.
187. SMITH, A. M., PICCIANO, M. F. and MILNER, J. A., *Am. J. Clin. Nutr.*, 1982, **35**, 521.
188. SHEARER, T. R. and HARJIMARKOS, D. M., *Arch. Environ. Health*, 1975, **30**, 230.
189. LOMBECK, I., KASPEREK, K., BONNERMANN, B., FEINENDEGEN, L. E. and BREMER, H. J., *Eur. J. Pediatr.*, 1978, **129**, 139.
190. HADJIMARKOS, D. M., *J. Pediatr.*, 1963, **63**, 273.
191. HADJIMARKOS, D. M. and SHEARER, T. R., *Am. J. Clin. Nutr.*, 1973, **26**, 583.
192. KUMPULAINEN, J., VUORI, E., KUITUNEN, P., MAKINEN, S. and KARA, R., *Int. J. Vitam. Nutr. Res.*, 1984, **53**, 420.
193. FOOD AND NUTRITION BOARD, NATIONAL RESEARCH COUNCIL, *Iodine Nutrition in the United States*, 1970, National Academy of Sciences, Washington, DC.
194. GUSHURST, C. A., MUELLER, J. A., GREEN, J. A. and SEDOR, F., *Pediatrics*, 1984, **73**, 354.
195. SALTER, W. T., In: *The Hormones: Physiology, Chemistry and Applications*, Vol. 2, G. Pincus and K. V. Thimann (eds), 1950, Academy Press, New York, p. 181.
196. HEMKEN, R. W., *J. Am. Vet. Med. Assoc.*, 1980, **176**, 1119.
197. INTERNATIONAL DAIRY FEDERATION, *Iodide in Milk and Milk Products*, Document 152, 1982.
198. MILLER, J. K., SWANSON, E. W. and SPALDING, G. E., *J. Dairy Sci.*, 1975, **58**, 1578.
199. BROADHEAD, G. D., PEARSON, I. B. and WILSON, G. M., *Brit. Med. J.*, 1965, **1**, 343.

200. LEWIS, R. C. and RALSTON, N. P., *J. Dairy Sci.*, 1951, **36**, 33.
201. CONVEY, E. M., CHAPIN, L., KESNER, J. S., HILLMAN, D. and CURTIS, A. R., *J. Dairy Sci.*, 1977, **60**, 975.
202. BRUHN, J. C. and FRANKE, A. A., *J. Dairy Sci.*, 1978, **61**, 1557.
203. CONRAD, L. M. and HEMKEN, R. W., *J. Dairy Sci.*, 1978, **61**, 770.
204. DUNSMORE, D. G. and LUCKHURST, A. M. L., *N. S. Wales Dep. Agric. Ann. Rep.*, 1975, 142.
205. DUNSMORE, D. G., *Aust. J. Dairy Technol.*, 1976, **31**, 125.
206. HEMKEN, R. W., *J. Anim. Sci.*, 1979, **48**, 981.
207. TALBOT, J. M., FISHER, K. D. and CARR, C. J., *A Review of Dietary Iodine in Certain Thyroid Disorders*, 1976, Life Sciences Research Office, Federation of Associated Societies of Experimental Biologists, Bethesda, Maryland.
208. VANDERVEEN, J. E., *Proceedings of Workshop on Iodine in American Foods*, 1979, American Medical Association, cited from Ref. 196.
209. TALBOT, J. M., FISHER, K. D. and CARR, C. J., *A Review of the Significance of Untoward Reactions to Iodine in Foods*, 1974, Life Sciences Research Office, Federated Societies of Experimental Biologists, Bethesda, Maryland.
210. MURTHY, G. K. and CAMPBELL, J. E., *J. Dairy Sci.*, 1969, **43**, 1042.
211. HODGE, H. C., SMITH, F. A. and GEDALIA, I., In: *Fluorides and Human Health*, 1970, WHO Monograph, Series No. 59, WHO, Geneva, p. 59.
212. ESALA, S. E. and VUORI, E., *Brit. J. Nutr.*, 1982, **48**, 201.
213. STEPHEN, K. W., BOYLE, I. T., CAMPBELL, D., MCNEE, S., FYFFE, J. A., JENKINS, A. S. and BOYLE, P., *Brit. Dent. J.*, 1981, **151**, 287.
214. HOWAT, A. P. and NUNN, J. H., *Brit. Dent. J.*, 1981, **150**, 276.
215. ROYAL COLLEGE OF PHYSICIANS OF LONDON, *Fluoride, Teeth and Health*, 1976, Pitman Medical, London.
216. NATIONAL ACADEMY OF SCIENCES, *Recommended Dietary Allowances*, 1974, National Academy of Sciences, Washington, DC.
217. HART, L. I., OWEN, E. C. and PROUDFOOT, R., *Brit. J. Nutr.*, 1967, **21**, 617.
218. OLIVER, I., SPERLING, O., LIEBERMAN, U. A., FRANK, M. and DE BRES, A., *Biochem. Med.*, 1971, **5**, 279.

Chapter 8

FLAVOUR OF MILK AND MILK PRODUCTS

D. J. Manning and H. E. Nursten

*Food Research Institute and
Department of Food Science, University of Reading, UK*

1. INTRODUCTION

Flavour is composed principally of the sensations of smell and taste, though the senses of sight, touch, and hearing interact with those of smell and taste. Whereas the sense of taste is relatively simple in having only four modalities—bitter, salt, sour, sweet (with perhaps astringent, soapy, and MSG-like)—the modalities of smell are still very much under discussion; the 44 qualities described by Harper *et al.*[1] are a good summary of the current state of knowledge.

For a substance to evoke a taste response, it has to come into contact with the taste buds, located mainly on the tongue. This presupposes solubility in aqueous media. For a substance to evoke a smell, it must come into contact with the receptors located in the olfactory epithelium in the higher regions of the nasal cavity. This presupposes volatility, but volatility alone is not sufficient; volatile substances range in odour potency from zero to detectability in the vapour above an aqueous solution at 2 parts per 10^{14}.[2]

To separate and identify sapid substances, traditional methods of chemistry have been employed. However, in general, the volatile constituents of foods are much more complex mixtures of components, present at widely different concentrations, with significant contributions being possible even below parts per million million. The methods of investigating the volatile components of foods have, therefore, become

TABLE 1

TYPES AND NUMBERS OF VOLATILE COMPONENTS IDENTIFIED IN MILK AND DAIRY PRODUCTS[9]

	Milk and milk products	Blue cheese	Cheddar cheese	Swiss cheese	Other cheeses
Hydrocarbons	43	3	33	16	54
Alcohols	16	13	12	8	29
Aldehydes	33	7	12	10	28
Ketones	28	12	14	12	30
Acids	43	30	27	32	46
Esters	21	17	20	15	52
Lactones	28	7	9	5	4
Bases	16	10	14	14	48
Sulphur compounds	14	2	6	2	25
Acetals	4				
Ether	1		1	1	1
Halogen compounds	5		5	3	12
Nitriles	2				1
Amides		6	6	6	6
Phenols	6	1	1		13
Furans	13	1			3
Pyrone	1				
Dilactide			1		
Total	274	109	161	124	352

relatively sophisticated and rely heavily on gas chromatography combined with mass spectrometry.[3,4]

To avoid artefact formation, the simplest technique is direct injection of a sample of the 'headspace' above a food into the gas chromatograph by means of an air-tight syringe, but the sensitivity of this method is not usually sufficient. Concentration of the headspace is therefore carried out, mainly by adsorption onto synthetic polymers, such as Tenax GC, or onto charcoal, followed by desorption by heat or solvents. Alternatively, isolation by steam-distillation (perhaps under reduced pressure) can be combined with concurrent solvent extraction of the distillate, as in the technique of Likens and Nickerson.[5] The relative advantages and disadvantages of these and other methods need to be considered in depth for each particular foodstuff to be investigated.[3,4]

Methods such as these have been applied to milk and milk products for many years, the most useful reviews of the results obtained being those of Forss,[6] Dumont and Adda,[7] and Badings and Neeter.[8]

The types and numbers of volatile components of milk and milk products identified are listed in Table 1.[9] Such data are impressive but by themselves convey nothing in terms of odour (or flavour) impact. In order for them to do so, the odour contributions, both quantitative and qualitative, of each substance at the specific concentration in the specific medium, and their interactions, would have to be known. This is hardly the case for any foodstuff, and certainly not for any food as complex as milk and dairy products.

As regards odour intensity, the odour number, i.e. the concentration of an odorant divided by its detection threshold in the same medium, has proved a useful concept,[10,11] even if not a theoretically justifiable one.[12] Unfortunately, the number of detection thresholds of odorants in milk and milk products that has been determined is limited; useful data are available (see e.g. Refs 13–16).

The origin of the significant flavorants is of considerable interest.[6,7,16,17] The main sources are the milk itself and degradation of milk constituents during processing. The normal volatiles of milk probably originate via the cow's metabolism, since cows fed with an odourless synthetic diet produce milk of normal flavour.[7] However, there is potential for manipulating flavour by means of feed.[18] Unusual components may appear in the milk connected with the feed (Table 2) or due to contaminants (Table 3). The milk may be affected by endogenous enzymes or by microbial activity before or after undergoing processing. Non-enzymic changes may also take place. Processing may involve deliberate use of chemical, enzymic, and microbial manipulation, and

TABLE 2
EFFECTS OF FEED ON MILK FLAVOUR

Effect	Compounds involved	Plant involved	Ref.
Cowy flavour	Dimethyl sulphide[a]	Lucerne: *Medicago sativa* L.	19,20
Faecal odour	Skatole[b]	Peppercress: *Lepidium* spp.	21
Scorched flavour	Benzylthiol, benzyl methyl sulphide	Landcress: *Coronopus didymus* (L.) SM	22
Fishy flavour	Trimethylamine	Wheat pasture	23

[a] Low concentrations impart a desirable flavour.
[b] Lower levels contribute to the desirable flavour of butter.[14]

TABLE 3
TAINTS IN DAIRY PRODUCTS

Effect	Compounds involved	Source of odorants	Ref.
Catty/blackcurrant odour in cheese	2-Methylpentane-2-thiol-4-one	Mesityl oxide + H_2S	24, 25
Spray-dried milk powders	E-6-Nonenal	O_3 (8,15- and 9,15-octadienoic acids)	26
Raw potato off-flavour in Gruyère de Comté	3-Methoxy-2-propylpyridine		27
Fishy, cod liver oil-like odour in caramel containing butterfat	Deca-E-2,Z-4,Z-7-trienal and related ones	Oxidation of polyunsaturated fats induced by Cu and α-tocopherol	28
Metallic odour in caramel containing butterfat	Octa-1,Z-5-dien-3-one and 1-octen-3-one	Oxidation of polyunsaturated fats induced by Cu and α-tocopherol	29

changes brought about in these ways are part of maturation and storage.

Physicochemical aspects, such as the effects of homogenization and the adsorption of flavorants on proteins, are of importance too, but have so far been subjected only to limited study (see e.g. Ref. 30).

In this chapter, the salient features of the flavour of milk and milk products will be pointed out in the sequence: milk, UHT milk, butter, cream, fermented milks, cheese. The last section is the longest and is subdivided into soft, semi-soft, semi-hard, and hard cheeses.

2. MILK

Thomas[31] has stated that consumer acceptance and preference for milk as a beverage is influenced by its flavour more than any other attribute. Good quality fresh milk has a bland yet characteristic flavour. It has a pleasant mouth-feel, determined by its physical nature, i.e. an emulsion of fat globules in a colloidal aqueous solution, and a slightly salty and sweet taste due to the presence of salts and lactose. That there is a real fresh milk aroma is belief rather than fact. Because of its blandness, milk certainly is an effective vehicle for off-odours; owing to the lability of some of its components, odorous compounds are readily generated in milk by hydrolysis, oxidation, and enzymic and microbial activity.

Exposure to light (sunlight or fluorescent light in cold cabinets) can lead to off-flavours in milk, due to methional at first and to methanethiol subsequently.[32,33]

3. UHT MILK

UHT milk is described as 'cooked' and 'cabbagy' initially, but these notes decrease, giving maximum flavour acceptability after a few days. Acceptability declines slowly after about six days, the milk being described as more 'stale';[34] 'heated' or 'sterile' milk flavour persists for several months.[31] Shelf-life of directly heated UHT milk is eventually limited by gelation and the development of bitterness. Proteolysis, not too surprisingly, correlates well with bitterness and could be used as a predictor of shelf-life.[35]

Badings and Neeter[8] have detected at least 400 volatiles in milk processed in different ways; they pinpointed 47 gas chromatographic peaks, involving 57 substances (see Table 4), as making a strong contribution to the flavour of indirectly heated UHT milk (UHT-i; 4·6 s, 142°C, Alfa Laval VTS), a further 46 peaks making a moderate contribution. Table 4 also indicates the degree to which the different substances, and some additional ones, contribute to the flavour differences between UHT-i milk and low temperature pasteurized milk (LP; 13 s, 75°C, Stork), and between UHT milk indirectly or directly (UHT-d; 4·6 s, 145°C, Alfa Laval VTIS) heated.[36]

Badings et al.[36] found that a synthetic UHT flavour (also given in Table 4), when added to LP, causes the flavour of that milk to change into a typical UHT milk flavour. Considerable understanding of UHT milk flavour has thus been achieved.

From a study of the volatiles of milk treated for 3 s or 90 s at 140°C and stored at ambient temperatures, Jaddou et al.[37] concluded that cabbagy defects are correlated with total volatile sulphur, and that the compounds hydrogen sulphide, carbon oxysulphide, methanethiol, carbon disulphide, and dimethyl sulphide could be responsible.

Important attempts to control such cooked flavour in dairy products by chemical manipulation have been made both by Ferretti[38] and by Badings et al.[39] They used additions of thiosulphonates and thiosulphates to react with mercaptans, and of cystine to react with hydrogen sulphide. These experiments are very promising, even though they raise the questions of safety and of the legality of such additions.

Biochemical manipulation of volatiles is also possible. Skudder et al.,[40] in collaboration with Swaisgood,[41] are working with an immobilized sulphydryl oxidase to remove cooked flavour from heat-treated milk.

TABLE 4

SUBSTANCES MAKING A STRONG CONTRIBUTION TO THE FLAVOUR OF INDIRECTLY HEATED UHT MILK, THOSE CONTRIBUTING TO DIFFERENCES IN FLAVOUR OF MILK HEAT-TREATED IN DIFFERENT WAYS, AND THOSE USED IN A SYNTHETIC UHT FLAVOUR

	UHT-i[a]	UHT-i −LP[b]	UHT-i −UHT-d[c]	Synthetic UHT flavour[d] (mg/kg LP)
Dimethyl sulphide	+	0	1	
3-Methylbutanal	+	1	1	
2-Methylbutanal	+	0	1	
2-Methyl-1-propanethiol	+	1	1	0·008
Pentanal	+	1	1	
3-Hexanone	+			
Hexanal	+	1	1	
2-Heptanone	+	4	2	0·40
Styrene	+			
Z-4-Heptenal[e]	+	1	0	
Heptanal	+			
2-Acetylfuran	+			
Dimethyl trisulphide	+	2	0	
Cyanobenzene	+			
1-Heptanol	+			
1-Octen-3-one[e]	+			
Octanal	+	1	1	
p-Cymene	+			
Phenol	+			
Indene	+			
2-Ethyl-1-hexanol	+			
Benzyl alcohol	+			
Unknown	+			
Acetophenone	+	1	0	
1-Octanol	+			
2-Nonanone	+	4	2	0·21
Nonanal	+			
p-Cresol	+			
m-Cresol	+			
E-2,Z-6-Nonadienal	+			
E-2-Nonenal	+			
3-Methylindene	+			
Methylindene	+			
Ethyldimethylbenzene	+			
Decanal	+			
Tetraethylthiourea	+			
Benzothiazole	+	1	0	0·005
γ-Octalactone	+	1	0	0·025
2,3,5-Trimethylanisole	+			
δ-Octalactone	+	1	0	

TABLE 4—*contd.*

	UHT-i^a	UHT-i −LP[b]	UHT-i −UHT-d[c]	Synthetic UHT flavour[d] (mg/kg LP)
1-Decanol	+	1	1	
2-Undecanone	+	2	1	0.18
2-Methylnaphthalene	+			
Indole	+			
δ-Decalactone	+	1	0	0·650
Hydrogen sulphide		2	1	0·03
Diacetyl		2	1	0·005
Dimethyl disulphide		2	1	0·002
2-Hexanone		2	1	
γ-Dodecalactone		2	1	0·025
δ-Dodecalactone		2	1	0·1
Methanethiol		1	1	0·002
2-Pentanone		1	1	0·29
Methyl isothiocyanate		1	1	0·01
Ethyl isothiocyanate		1	1	0·01
Furfural		1	1	
Benzaldehyde		1	0	
2-Octanone		1	0	
Naphthalene		1	0	
γ-Decalactone		1	0	
2-Tridecanone		1	0	
Acetaldehyde		−1	0	
1-Cyano-4-pentene		−1	0	
2-Methyl-1-butanol		−1	1	
Ethyl butyrate		−1	0	
3-Buten-1-yl isothiocyanate		−1	0	
E-2,E-4-nonadienal		−1	0	
2,4-Dithiapentane			1	
Maltol				10·00

[a] Indirectly heated UHT milk; +indicates a component that makes a strong contribution to the flavour. In addition to the components listed, a further 12 unknowns made strong contributions.[8]

[b] Components contributing to a difference in flavour between indirectly heated UHT milk and low temperature pasteurized milk. Scale for difference: 1 = slight, 2 = moderate, 3 = strong, 4 = very strong.[36]

[c] Components contributing to a difference in flavour between indirectly and directly heated UHT milks. Scale for difference: as UHT-i–LP.[36]

[d] Composition of synthetic UHT flavour.[36]

[e] Tentative identification.

4. BUTTER

Urbach *et al.*[14] found that decanoic acid, δ-octanolide, δ-decanolide, *p*-cresol, indole, and skatole are present in sweet-cream butter in sufficient concentrations to contribute sensorily, whereas the concentrations of octanoic and dodecanoic acids, γ- and δ-dodecanolides, phenol, *m*-cresol, and guaiacol are borderline. However, a simulated butter flavour based on these findings was described as 'coconut' and 'chemical'. Addition of skim-milk solids altered the description to 'condensed milk plus cream flavour'. In subsequent work, the quantitative estimation of these compounds was improved,[42,43] which allowed affirmation of the importance of indole and skatole and of the borderline contribution of the phenols.

The reproduction of cultured-butter flavour has proved closer to the mark.[44,45] Table 5 gives a composition that, when added to sweet-cream butter, produces a flavour similar to that of ripened-cream butter.

TABLE 5

FORMULATION FOR CULTURED-BUTTER FLAVOUR (mg kg^{-1} ADDED TO SWEET-CREAM BUTTER)[45]

Diacetyl	4	γ-Decalactone	3
3-Methylbutanal	0·01	δ-Decalactone	3·6
Z-4-Heptenal	0·006	δ-Dodecalactone	6·9
2-Phenylethanal	0·002	Hydrogen sulphide	0·01
Acetic acid	57	Methanethiol	0·01
Valeric acid	0·15	Dimethyl sulphide	0·06
Phenol	0·01	Indole	0·006
p-Cresol	0·005	Skatole	0·067
Guaiacol	0·002	Glutamine, sodium salt	1
Ethyl butyrate	0·002	Lactic acid	to pH 4·6

Continental cooking often uses butter, the flavour of browned butter being highly prized. Lactones, which play the dominant role here,[14] are derived from appropriate hydroxy acids present in dairy products, but not in margarine. Additives that give similar flavours on heating are:

Such compounds ($x = 1$ or 2, $R = C_4H_9$ to C_9H_{19}) are stable at room temperature but, being derivatives of β-carbonyl acids, decarboxylate to odorous lactones at 80–120°C.[46] Surprisingly, lactones are also associated with the development of stale flavour in some butters stored at $-10°C$.[47]

Butterfat, when present in caramels, together with copper (from the boiling pans) and α-tocopherol, has been shown to give rise to fishy (cod liver oil-like)[28] and metallic taints[29] (see Table 3).

5. CREAM

The flavour of cream is expected to differ from that of sweet-cream butter mainly owing to greater contributions from the aqueous phase of milk and the fat-globule membrane.[8] Whipping may slightly increase the concentrations of oxidation products and, provided it is not overdone, the flavour is improved. In accordance with this, Haverkamp Begemann and Koster[48] found that Z-4-heptenal in particular, if present at about 1 part per 10^9, contributes to a full cream flavour. This compound is apparently formed from Z-10- and 11,Z-15-octadienoic acids, which are themselves present in milk fat only to the extent of 0·02%.

6. FERMENTED MILKS

Flavour development in fermented milks has been reviewed recently by Marshall.[49] Since fermentation, the oldest method for preserving milk, developed independently in different parts of the world, the organisms responsible differ, and the differences are reflected in the types of cultured milk that they produce. Marshall[49] classified fermented milks into four types (Table 6).

The predominant reaction is fermentation of lactose to lactic acid, which is responsible for the sharp, refreshing taste. Minor components determine the aroma, and formation of proteins or polysaccharides may be required for the appropriate consistency, in which carbon dioxide formation may also be of significance.

There are several pathways of lactose metabolism; their interrelationships are complex and by no means fully understood. Nevertheless, whatever the route of catabolism, the important intermediates are pyruvate, acetyl-coenzyme A, and acetyl phosphate. Each starter has its

TABLE 6

RELATIONSHIP BETWEEN TYPE OF CULTURED MILK AND GEOGRAPHICAL AREA[49]

Type	Culture(s) used	Region
I	Streptococcal and leuconostoc species	Norway, Sweden, Finland, Iceland
II	Lactobacillus species	Bulgaria, Japan
III	Streptococcal and lactobacillus species	Egypt, Iraq, Lebanon, Syria, Turkey, India
IV	Streptococcal and lactobacillus species and yeasts	USSR, Lebanon, Finland

own way of metabolizing them, incidentally producing flavour compounds, and this metabolism may be affected by factors such as the growth phase of the organism, increased formation of intermediates from other sources (e.g. citrate), and differences between mutants.

In fermented milks of all four types, formation of both acetaldehyde and diacetyl is important. In type IV, much if not all the acetaldehyde is reduced to ethanol. The flavour of yoghurt (type III) has been examined in greater detail than that of any other fermented milk. Optimum flavour and aroma are obtained with acetaldehyde, the principal aroma component, at 23–41 ppm.[50] Other volatiles of significance are diacetyl, acetoin, acetone, and 2-butanone. Typical values found in German yoghurts were: acetaldehyde, 5–7; ethanol and acetone, 0·6–2·4; diacetyl, 4 mg kg^{-1}.[51] No sensory differences were detected between yoghurts made from UHT or batch-treated milks.[52]

TABLE 7

MAJOR PATHWAYS TO CHEESE FLAVOUR COMPOUNDS[53]

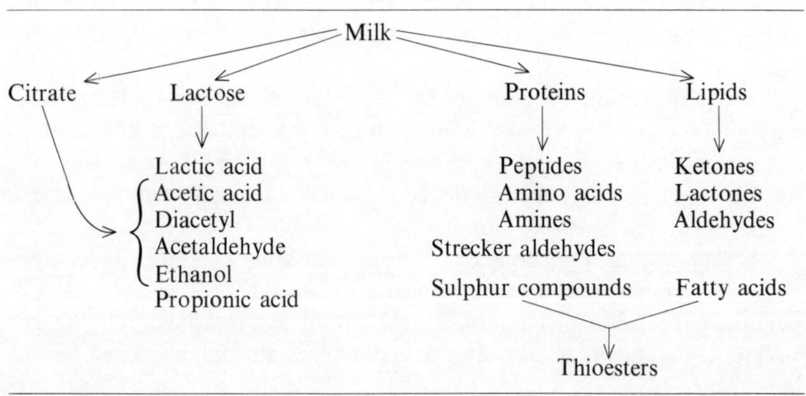

7. CHEESE

The different types of cheese span the whole range of volatiles obtained from dairy products (Table 1). Here attention will be paid to the compounds of known sensory significance. Law[53] has indicated the major pathways to cheese flavour compounds (Table 7). Following Law,[54] the different types of cheese will be considered in turn, starting with those of highest moisture content.

7.1. Soft Cheeses

7.1.1. Unripened Cheese

Cottage cheese is a typical product in this category. Here diacetyl is more important than acetaldehyde, ratios of 5:1 to 3:1 being acceptable.[55] Too high a proportion of acetaldehyde gives a green, yoghurt-like flavour; diacetyl in excess is unacceptably harsh. Lindsay *et al.*[44] found that diacetyl ($1·0 \, \mathrm{mg \, kg^{-1}}$), acetaldehyde ($0·2$), dimethyl sulphide ($0·02$), and acetic acid ($30$) simulate the naturally produced flavour very closely.

7.1.2. Ripened Cheese

Here the development of distinctive flavours depends primarily on the secondary microflora, the moisture level, the method and extent of salting, and the acidity, the emphasis on diacetyl formation being much less.

Most work has been done on Camembert. Moinas *et al.*[56,57] claimed that its flavour can be reproduced in a bland cheese base with 2-heptanone, 2-nonenone, 2-heptanol, 2-nonanol, phenol, butyric acid, methyl cinnamate, and 1-octen-3-ol, the role of the last compound being to mask partially the harsh blue cheese flavour of the ketones. 1-Octen-3-ol (threshold concentration $0·01$ ppm) has an aroma of mushrooms, and Dumont *et al.*[58] consider it to play a positive, specific role only at low concentrations; at higher concentrations it gives rise to a defect. 3-Methyl-1-butanol, phenethanol, and phenol make a general but important contribution to surface-ripened cheeses.[59]

7.1.3. Brined Cheese

Feta cheeses originating from Bulgaria, Romania, or Greece gave head-space volatiles that were qualitatively similar, with relatively large amounts of ethanol, 1-propanol, 2-butanol, and 2-butanone, as well as smaller amounts of pentane, 2-propanol, ethyl acetate, 2-methyl-1-propanol, toluene, and ethyl butyrate.[60] Significant lipolysis in the Greek cheese resulted in much higher levels of ethyl butyrate. Australian Feta cheeses gave a different, less complex profile; fat composition indicated that they were made entirely from cow's milk, whereas the European cheeses were based on ewe's milk, at least in part. A 'plastic paint, kerosene' off-flavour in Feta was caused by $\sim$200 ppm of E-1,3-pentadiene (threshold in Feta slurry, 4 ppm), associated with use of sorbic acid.[61]

7.2. Semi-soft Cheeses

These cover a wide range of flavours, from the mainly acid-flavoured English cheeses, such as Caerphilly and Cheshire, to the very strongly flavoured surface-smear cheeses, such as Limburg. The latter are of more interest here. The surface-smear introduces lactate-utilizing yeasts, which raise the pH and provide growth factors for *Brevibacterium linens*. These microorganisms are very proteolytic[62] and able to convert methionine into methanethiol,[63] which contributes to the characteristic putrid odour. Compounds such as acetyl methyl disulphide have been claimed to have aromas characteristic of Limburg cheese.[64]

7.3. Semi-hard Cheeses

Both Gouda and the blue cheeses fall into this group. The aroma of the latter is much more characteristic and very much better understood. Methyl ketones are the key compounds, with C_5, C_7, C_9, and C_{11} most prominent. The mould *Penicillium roqueforti*, with which these cheeses are spiked, produces fatty acids, which are partially converted into the methyl ketones via the ketoacyl-coenzyme A and the β-keto acid. The thiohydrolase prefers β-keto-octanoyl-coenzyme A and thus ensures a preponderance of 2-heptanone in the cheese, even though the decarboxylase prefers β-ketododecanoic acid. The rate of fatty acid release governs the rate of methyl ketone formation.[65] Addition of fatty acids to *P. roqueforti* cheese slurry systems greatly increases the amount of methyl ketones formed.[66] Roquefort is traditionally made with ewe's milk, and Hall and Kosikowski[67] attributed its superiority over other blue varieties to relatively high levels of hexanoic acid and 2-nonanone. δ-Lactones are thought to contribute to the flavour also,[68] and the quality of blue cheese is improved by increasing the concentrations of the C_{14} and the C_{12} homologues.[69] Two routes to such δ-lactones are possible: (1) spontaneous or enzymic ring-closure of 4-hydroxy acids liberated from milk glycerides by lipases; (2) via reduction of 4-oxo acids liberated similarly from milk glycerides.[70]

Blue cheese flavouring can be based on alkanoic acids (C_4 to C_{10}), lactones (C_4 to C_{22}), methyl ketones (C_2 to C_{13}), and phenol,[71] or on the rapid fermentation of lipolysed fats (milk fat, maize oil, or coconut oil) with *P. roqueforti*.[72–74] It has also been claimed that 1-octen-3-ol improves blue cheese flavourings.[75] Diacetyl, acetoin, and acetaldehyde levels of up to 0·44, 2, and 13 mg kg^{-1}, respectively, have been recorded in blue cheese.[76]

Anethole and bismethylthiomethane are considered to be important contributors to the odour of Gouda cheese.[77] The odour of the latter compound has been described as that of garlic.[15]

7.4. Hard Cheeses

The three principal types of hard cheese are represented by Cheddar, Emmental, and Italian. Whereas Cheddar is made with the same starters as the softer cheeses above, thermophiles such as *Streptococcus thermophilus* and *Lactobacillus bulgaricus, helveticus,* or *casei* must be used for the high-cook Emmental and Italian cheeses.

7.4.1. Cheddar Cheese

The flavour of Cheddar cheese, despite being the most intensely investigated cheese flavour, still remains something of an enigma.

The flavour of very young Cheddar is similar to that of other, internally salted cheeses made with mesophilic starters, being acid, slightly buttery, and salty. At this stage, the flavour compounds are largely derived by lactose fermentation, resulting mainly in lactic acid, but alternative pathways of pyruvate metabolism allow formation of acetic acid, ethanol, and acetaldehyde. These alternative pathways are significant since, when pyruvate dehydrogenase, a key enzyme in diverting pyruvate to more flavorous compounds, in starter streptococci was inhibited in cheese made from milk containing high levels of polyunsaturated fat, the cheese turned out to be rather bland and was low in acetate, acetaldehyde, and diacetyl.[78] It is also of interest to note that cheeses made under conditions that exclude microorganisms other than the starter are invariably mild in flavour and not typical of commercial cheeses, implying that adventitious bacteria play a significant role.[79]

The importance of diacetyl is supported by Manning and Robinson,[80] who identified it as one of ten compounds (the others are hydrogen sulphide, methanol, acetaldehyde, methanethiol, ethanol, acetone, dimethyl sulphide, 2-butanone, and 2-pentanone) that contribute to the typical aroma in low-boiling Cheddar distillates. However, the role of acetic acid is questionable, since its concentration can differ considerably between cheeses without a noticeable effect on quality or intensity of the typical flavour.[81] Indeed, it can lead to 'vinegary' off-flavours.

Amounts of volatile fatty acids, other than acetic, increase during maturation. Although such acids are included in synthetic cheese flavourings,[71,82] evidence for their contribution to typical Cheddar flavour is equivocal.

On the one hand, studies with enzyme-modified cheese suggest that increased free volatile fatty acids increase the flavour intensity of American Cheddar, provided that rancidity-inducing levels of lipase are avoided.[83,84] Since the normal levels of these acids (~500 ppm) are well above their thresholds (0·3–100 ppm),[85] they must contribute to Cheddar flavour.

On the other hand, Manning and Price[86] showed that removal of volatile fatty acids from Cheddar headspace left the aroma unaltered and concluded that these acids were only important to the background taste. Additional evidence playing down the role of fatty acids comes from the analysis of New Zealand Cheddars, where 23 out of 41 contained no homologues higher than butyric.[87]

The methyl ketones do not appear to be vital to Cheddar flavour, but 2-pentanone in normal cheese is a good index of cheese age,[88] without necessarily contributing to flavour.

There is no definite evidence linking lactones with Cheddar flavour, but they are included in synthetic flavourings.[71,89]

Protein degradation is a further source of flavour compounds. Some amino acids have distinctive tastes, but their role in Cheddar is one of conferring a savoury background.[79,90] The evidence for contributions from peptides is not definite although addition of exogenous proteinases can lead to intensification of flavour and/or accelerated flavour development.[84,91] The correlation between increased proteolysis and increased intensity of typical flavour only holds over a limited range and the type of proteinase in critical, neutral proteinases being ideal, whereas acid proteinases produce excessive amounts of bitter peptides.[54]

There are two ways in which proteolysis can favour flavour formation. The first is by increasing the free amino acid concentration, thus fortifying the savoury taste. The second is by providing substrates for the release of volatile sulphur compounds, the unique contribution of which was demonstrated by McGugan et al.[92] These workers compared the neutral volatile compounds from mature, normal Cheddar with those from a flavourless cheese produced without starter. The profiles of volatiles were very similar, both qualitatively and quantitatively, except for sulphur compounds.

Manning and Price[86] subsequently accumulated evidence that methanethiol was the key compound, hydrogen sulphide also contributing. The latter is an essential component of good quality Cheddar, which contains levels well above threshold. It is only at very high levels that 'sulphide' defects become evident. In fact, the quality of Cheddar can be related to the ratio of hydrogen sulphide to methanethiol. Synthetic Cheddar flavours based on these compounds have not proved successful, their great volatility and strength of aroma raising problems of judiciously balancing flavour binding and release. McGugan *et al.*[93] place considerable importance on these factors in relation to the overall flavour intensity of cheese. Proteolysis may facilitate flavour formation, not only enzymically and chemically, but also physically through reduced flavour-binding capacity.

In assessing the extent to which the identity of the essential flavour compounds has been established, it has been a common practice to imitate the flavour by mixing the known components. One cannot deny that optimism underlies this practice, since almost always essential components remain to be identified, the mixture will not necessarily have the same sensory effect without incorporation into the food, and, even if added to a base, it is unlikely that the correct distribution of components will be achieved.

Two other procedures are often more practical:

1. Removal of essential compounds must result in the flavour's being changed or removed altogether, as with hydrogen sulphide or methanethiol.[86]
2. Addition of precursors or manipulating ripening conditions to accelerate the production of essential flavour compounds, e.g. lowering of redox potential of non-starter cheeses containing no sulphur volatiles induced production of hydrogen sulphide and methanethiol, and of flavour.[94] Similarly, the addition of reducing agents stimulated production of these two volatiles in young cheeses, simultaneously increasing flavour.[95]

Lamparsky and Klimes[96] have added 41 compounds to the list given in Table 1: 11 alcohols, 1 aldehyde, 3 ketones, 12 acids, 10 esters, 3 lactones, and a phenol. They state that phenylacetaldehyde, phenylacetic acid and phenethanol contribute notes of importance to the mild Cheddar flavour they investigated. Two other interesting compounds were included among those newly identified: δ-7-decenolactone (jasmine lactone) and 1-(*p*-hydroxyphenyl)but-2-en-1-one.

There are particular problems associated with attempting to pinpoint the components responsible for the characteristic flavour of long ripening cheeses such as Cheddar, simply because it is impossible to specify what is a typical Cheddar. Unlike those many food products that are assumed to deteriorate with age, Cheddar merely changes its characteristic flavour and, provided it is stored under suitable conditions, can be kept in excess of 2 years. Cheddar is marketed any time between 2 months and 18 months, and it would be quite wrong to specify at which stage a cheese has its optimum flavour. In fact, cheeses of different ages can almost be regarded as different products. In addition to the age problem, there is the difference in adventitious bacteria and milk composition, which can influence the flavour of the cheese made in different regions. It is these differences in flavour encountered within a single variety of cheese that contribute to the problems of the cheese grader and explain why different graders have their own views on the constituents of good flavour. The relationship between a commercial grader's opinion of good flavour often conflicts with that of the consumer. McBride and Hall[97] in Australia have shown that among older people there is a significant preference for second-grade cheese. A consequence seems to be that, in order to minimize differences of opinion between graders, standards tend to be based upon absence of off-flavour. Off-flavours in most foods are usually obvious and are invariably associated with deterioration due to age or contamination. While it is true that one does encounter off-flavours in cheese, which are almost certainly unacceptable to virtually all consumers, i.e. phenolic or rancid notes, most flavours considered to be defects are not only acceptable but actually desirable to a large number of the population. Commercial creameries usually steer clear of cheeses of this type in order to maximize the appeal among consumers; in practice they probably offend nobody, but also possibly please nobody.

Recent work in the UK, Australia and New Zealand on the non-sensory assessment of cheese flavour should enable graders to predict the characteristic flavour of mature cheeses, possibly control their production and enable them to select cheeses for particular retail outlets where there is a demand for a particular type of Cheddar.[98]

Formation of bitterness has already been mentioned. The cause lies in peptides, containing a high proportion of hydrophobic side-chains (proline, leucine, and valine), as well as of glutamic acid, with a ratio of aliphatic to acidic amino acids of 0·8–1·3,[99] but the situation is complex. Residual chymosin can produce them, as also can starter proteinases.

Most significant seems to be the ability of starter proteinases to produce small bitter peptides from non-bitter, high molecular weight peptides, themselves derived from casein,[100,101] particularly α_{s1}-casein. (There is relatively little α_{s1}-casein in ewe's and goat's milk, cheeses from which are less susceptible to bitterness.[102]) Proteinase-deficient variants of starters are an interesting development.[103] Bitter-peptide degrading enzymes are also of significance, as are the different abilities of starter cultures to produce them.[104,105]

Fruity defect is relatively common in Cheddar,[106] occurring when ethanol levels are high (147–1527 ppm). Ethyl butyrate and hexanoate are the main cause.[106,107] The association of fruity flavour in Cheddar with high ethanol has been confirmed by Manning,[95] but it is not a reliable method of assessing fruitiness in commercial cheeses made in the UK. The formation of the esters responsible for fruitiness is dependent on ethanol concentration, provided sufficient acidity is present for esterification. Recent studies at NIRD have shown that cheeses of high ethanol content can be devoid of fruity flavours; in such cases, acid defects were always absent. Not only is the amount of ethanol in the cheese important, but also the stage at which it is produced. The ethanol must be present during the first few months of ripening for esterification to take place sufficiently for fruitiness to manifest itself. It should be noted that sweetness, a characteristic of some commercial Cheddar, has been strongly correlated with concentration of ethyl acetate.[108]

7.4.2. Emmental and Gruyère

Biede and Hammond[109,110] identified three major flavour fractions in mature Emmental cheese:

1. Water-soluble volatiles (acetic, propionic and butyric acids and diacetyl), which give the basic sharpness and general cheesy note.
2. Water-soluble non-volatiles (amino acids, especially proline, peptides, lactic acid, salts), which provide a mainly sweet note.
3. Oil-soluble fraction (short-chain fatty acids other than the water-soluble volatile ones).

Large peptides are important for the brothy flavour.

Fraction 3 contained nutty flavour notes, thought to be due to alkylpyrazines[111,112] (cf. American processed cheese[113]). Sweet Emmental-like flavour can be produced in Cheddar by including a proline-producing strain of *Lactobacillus bulgaricus* in the starter.[114] An Emmental-like flavour, not of highest quality, was produced in processed cheese spread[115] using acetic acid, propionic acid, and proline (500, 5000, and 1200–1500 ppm, respectively).

The flavour of Gruyère is augmented by a surface flora of yeasts, micrococci, and *Brevibacterium linens*. Their concerted action produces methanethiol and thioesters of acetic and propionic acids, which together can account for the mildly putrid aroma.[116,117] Beaufort, a Gruyère-type cheese from a limited area of the French Alps, has a delicate aroma which is highly valued. Dumont and Adda[117] identified 140 components, including 9 sesquiterpenes, a class of compound not previously found in dairy products. Their presence is associated with cows grazing at altitudes above 1500 m in summer; winter cheese is characterized by their absence or much lower concentrations. A Gruyère aroma defect, likened to musty potatoes, has been identified as due to 3-methoxy-2-propylpyridine.

7.4.3. Italian Hard Cheeses

Although there are differences between cheeses such as Provolone, Romano, Grana, and Fontina, they all contain volatile fatty acids in relatively high concentrations. This is traditionally achieved by use of lipase-containing rennet paste or pregastric lipase from lambs.

It is claimed that Italian cheese flavour can be imparted to fresh curds or processed cheese by addition of 1-phenylpropionic acid and isovaleric acid (100 and 20–300 ppm, respectively) or a mixture of butyric, hexanoic, and octanoic acids (600–10 000 ppm) to give Fontina or Provolone flavour, respectively.[118,119] Steam-distillation or degassing, followed by extraction with ethyl chloride, gave α-keto acids (0·01–0·25 μmol kg^{-1}) in Fontina, about twice the amount obtained from Provolone.[120] It is likely that methyl ketones play a role in the flavour of Grana.[121]

8. CONCLUSION

A great deal is known about the flavour of milk and milk products, but much remains to be discovered, understood, and integrated, particularly with commercial practice. The main pathways to flavorous compounds are reasonably clear, but the explanations of the differences in flavour between different products, and even more so between different products of the same type, are not sufficiently precise yet to allow scientific specifications for flavour optimization to be drawn up.

REFERENCES

1. HARPER, R., LAND, D. G., GRIFFITHS, N. M. and BATE-SMITH, E. C., *Brit. J. Psychol.*, 1968, **59**, 231.
2. DEVOLE, E., ENGGIST, P. and OHLOFF, G., *Helv. Chim. Acta*, 1982, **65**, 1785.
3. TERANISHI, R., FLATH, R. A. and SUGISAWA, H. (eds), *Flavor Research: Recent Advances*, 1981, M. Dekker, New York.
4. MAARSE, H. and BELZ, R., *Isolation, Separation, and Identification of Volatile Compounds in Aroma Research*, 1982, D. Reidel, Dordrecht.
5. LIKENS, S. T. and NICKERSON, G. B., *Proc. Amer. Soc. Brew. Chem.*, 1964, 5.
6. FORSS, D. A., *J. Dairy Res.*, 1979, **4**, 691.
7. DUMONT, J. P. and ADDA, J., In: *Progress in Flavour Research*, Land, D. G. and Nursten, H. E. (eds), 1979, Applied Science Publishers, London, p. 245.
8. BADINGS, H. T. and NEETER, R., *Neth. Milk Dairy J.*, 1980, **34**, 9.
9. VAN STRATEN, S., MAARSE, H., DE BEAUVESER, J. C. and VISSCHER, C. A. (eds), *Volatile Compounds in Food: Qualitative Data.* 5th edn, 1983, Division for Nutrition and Food Research TNO, Zeist.
10. ROTHE, M. and THOMAS, B., *Z. Lebensm. Unters.-Forsch.*, 1963, **119**, 302.
11. GUADAGNI, D. G., BUTTERY, R. G. and HARRIS, J., *J. Sci. Fd. Agric.*, 1966, **17**, 142.
12. FRIJTERS, J. E. R., In: *Progress in Flavour Research*, Land, D. G. and Nursten, H. E. (eds), 1979, Applied Science Publishers, London, p. 47.
13. STAHL, W. H. (ed.), *Compilation of Odor and Taste Threshold Values Data*, 1973, American Society for Testing and Materials, Philadelphia.
14. URBACH, G., STARK, W. and FORSS, D. A., *J. Dairy Res.*, 1972, **39**, 35.
15. CUER, A., DAUPHIN, G., KERGOMARD, A., ROGER, S., DUMONT, J. P. and ADDA, J., *Lebensm. Wiss. Technol.*, 1979, **12**, 258.
16. SHIBAMOTO, T., MIHARA, S., NISHIMURA, O., KAMIYA, Y., AITOKU, A. and HAYASHI, J., In: *The Analysis and Control of Less Desirable Flavors in Foods and Beverages*, Charalambous, G. (ed.), 1980, Academic Press, New York, p. 241.
17. ADDA, J., GRIPON, J. C. and VASSAL, L., *Fd. Chem.*, 1982, **9**, 115.
18. URBACH, G. and STARK, W., *J. Dairy Res.*, 1978, **45**, 223.
19. REDDY, M. C., BASSETTE, R., WARD, G. and DUNHAM, J. R., *J. Dairy Sci.*, 1967, **50**, 147.
20. GORDON, D. T. and MORGAN, M. E., *J. Dairy Sci.*, 1972, **55**, 905.
21. PARK, R. J., *J. Dairy Res.*, 1969, **36**, 31.
22. WALKER, N. J. and GRAY, I. K., *J. Agric. Fd. Chem.*, 1970, **18**, 347.
23. MEHTA, R. S., BASSETTE, R. and WARD, G., *J. Dairy Sci.*, 1974, **57**, 285.
24. BADINGS, H. T., *J. Dairy Sci.*, 1967, **50**, 1347.
25. STEINSHOLT, K. and SVENSEN, A., *Milchwiss.*, 1979, **34**, 598.
26. PARKS, O. W., WONG, N. P., ALLEN, C. A. and SCHWARTZ, D. P., *J. Dairy Sci.*, 1969, **52**, 953.
27. DUMONT, J. P., ROGER, S. and ADDA, J., *Lait*, 1975, **55**, 479.
28. SWOBODA, P. A. T. and PEERS, K. E., *J. Sci. Fd. Agric.*, 1977, **28**, 1010.
29. SWOBODA, P. A. T. and PEERS, K. E., *J. Sci. Fd. Agric.*, 1977, **28**, 1019.

30. GUBLER, B. A., In: *Flavour '81*, Schreier, P. (ed.), 1981, de Gruyter, Berlin, p. 717.
31. THOMAS, E. L., *J. Dairy Sci.*, 1981, **64**, 1023.
32. ALLEN, C. and PARKS, O. W., *J. Dairy Sci.*, 1975, **58**, 1609.
33. BRADLEY, JR, R. L., *J. Fd. Protect.*, 1980, **43**, 314.
34. THOMAS, E. L., BURTON, H., FORD, J. E. and PERKIN, A. G., *J. Dairy Res.*, 1975, **42**, 285.
35. MCKELLAR, R. C., FROEHLICH, D. A., BUTLER, G., CHOLETTE, H. and CAMPBELL, C., *Can. Inst. Fd. Sci. J.*, 1984, **17**, 14.
36. BADINGS, H. T., VAN DER POL, J. J. G. and NEETER, R., In: *Flavour '81*, Schreier, P. (ed.), 1981, de Gruyter, Berlin, p. 683.
37. JADDOU, H. A., PAVEY, J. A. and MANNING, D. J., *J. Dairy Res.*, 1978, **45**, 391.
38. FERRETTI, A., *J. Agric. Fd. Chem.*, 1973, **21**, 939.
39. BADINGS, H. T., NEETER, R. and VAN DER POL, J. J. G., *Lebensm. Wiss. Technol.*, 1978, **11**, 237.
40. SKUDDER, P. J., YOUNG, P. and ANDREWS, G. R., Report, N.I.R.D., 1982, Shinfield, p. 113.
41. SWAISGOOD, H. E., *Enzyme and Microbial Technology*, 1980, **2**, 265.
42. STARK, W., URBACH, G. and HAMILTON, J. S., *J. Dairy Res.*, 1976, **43**, 469.
43. STARK, W., URBACH, G. and HAMILTON, J. S., *J. Dairy Res.*, 1976, **43**, 479.
44. LINDSAY, R. C., DAY, E. A. and SATHER, L. A., *J. Dairy Sci.*, 1967, **50**, 25.
45. BADINGS, H. T., Neth. Patent Appl. 11820, 1973.
46. STOLL, M., DEMOLE, E., FERRERO, C. and BECKER, J., *Nature* (London), 1964, **202**, 350.
47. STARK, W. and URBACH, G., *Aust. J. Dairy Tech.*, 1976, **31**, 80.
48. HAVERKAMP BEGEMANN, P. and KOSTER, J. C., *Nature* (London), 1964, **202**, 552.
49. MARSHALL, V. M. E., In: *Advances in the Microbiology and Biochemistry of Cheese and Fermented Milk*, Davies, F. L. and Law, B. A. (eds), 1984, Applied Science Publishers, London, p. 153.
50. GÖRNER, F., PALO, V. and BERTRAN, M., *Milchwiss.*, 1968, **23**, 94.
51. HILD, J., *Milchwiss.*, 1979, **34**, 281.
52. LABROPOULOS, A. E., PALMER, J. K. and TAO, P., *J. Dairy Sci.*, 1982, **65**, 191.
53. LAW, B. A., In: *Advances in the Microbiology and Biochemistry of Cheese and Fermented Milk*, Davies, F. L. and Law, B. A. (eds), 1984, Elsevier Applied Science Publishers, London, p. 187.
54. LAW, B. A., *Perfumer Flav.*, 1982, **7** (Oct/Nov), 9.
55. LINDSAY, R. C., DAY, E. A. and SANDINE, W. E., *J. Dairy Sci.*, 1965, **48**, 863.
56. MOINAS, M., GROUX, M. and HORMAN, I., *Lait*, 1973, **53**, 601.
57. MOINAS, M., GROUX, M. and HORMAN, I., *Lait*, 1975, **55**, 414.
58. DUMONT, J. P., ROGER, S., CERF, P. and ADDA, J., *Lait*, 1974, **54**, 501.
59. DUMONT, J. P., ROGER, S. and ADDA, J., *Lait*, 1976, **56**, 595.
60. HORWOOD, J. F., LLOYD, G. T. and STARK, W. *Aust. J. Dairy Technol.*, 1981, **36**, 34.
61. HORWOOD, J. F., LLOYD, G. T., RAMSHAW, E. H. and STARK, W., *Aust. J. Dairy Technol.*, 1981, **36**, 38.

62. FOISSY, H., *Milchwiss.*, 1978, **33**, 221.
63. SHARPE, M. E., LAW, B. A., PHILLIPS, B. A. and PILCHER, D. G., *J. Gen. Microbiol.*, 1977, **101**, 345.
64. KATZ, I., PITTET, A. O., WILSON, R. A. and EVERS, W. J., US Patent 4045587, 1977.
65. KINSELLA, J. E. and HWANG, D., *Biotechnol. Bioeng.*, 1976, **18**, 927.
66. KING, R. D. and CLEGG, G. H., *J. Sci. Fd. Agric.*, 1979, **30**, 197.
67. HALL, R. and KOSIKOWSKI, F. V., *XVIII Int. Dairy Congress*, 1970, **1E**, 385.
68. WONG, N. P., ELLIS, R., LA CROIX, D. E. and ALFORD, J. A., *J. Dairy Sci.*, 1973, **56**, 636.
69. JOLLY, R. C. and KOSIKOWSKI, F. V., *J. Agric. Fd. Chem.*, 1975, **23**, 1175.
70. BOLDINGH, J. and TAYLOR, R. J., *Nature* (London), 1962, **194**, 909.
71. HENNING, G. J. US Patent 3520699, 1970.
72. DWIVEDI, B. K. and KINSELLA, J. E., *J. Fd. Sci.*, 1974, **39**, 620.
73. KOSIKOWSKI, F. V. and JOLLY, R. C., US Patent 4133895, 1979.
74. AMU, B. J. and JARVIS, B., British Patent 1361817, 1974.
75. NEY, K. H., POETOE, I., WIROTAMA, G. and FREYTAG, G., US Patent 3865952, 1975.
76. ABOSHAMAA, K., WASHAM, P. C., VASAVADA, P. C. and TOLIBIA, J. R., *J. Dairy Sci.*, 1977, **60** (Suppl. 1), 57.
77. SLOOT, D. and HARKER, P. D., *J. Agric. Fd. Chem.*, 1975, **23**, 356.
78. CZULAK, J., HAMMOND, L. A. and HORWOOD, J. F., *Aust. J. Dairy Technol.*, 1974, **29**, 124.
79. FRYER, T. F., *Dairy Sci. Abstr.*, 1969, **31**, 471.
80. MANNING, D. J. and ROBINSON, H. M., *J. Dairy Res.*, 1973, **40**, 63.
81. LAW, B. A., CASTANON, M. J. and SHARPE, M. E., *J. Dairy Res.*, 1976, **43**, 117.
82. NEY, K. H., WIROTAMA, I. P. G. and FREYTAG, W. G., Neth. Patent 7204792, 1972.
83. KOSIKOWSKI, F. V. and IWASAKI, T., *J. Dairy Sci.*, 1979, **58**, 963.
84. SOOD, V. K. and KOSIKOWSKI, F. V., *J. Dairy Sci.*, 1979, **62**, 1865.
85. BALDWIN, R. E., CLONINGER, M. R. and LINDSAY, R. C., *J. Fd. Sci.*, 1973, **38**, 528.
86. MANNING, D. J. and PRICE, J. C., *J. Dairy Res.*, 1977, **44**, 357.
87. LAWRENCE, R. C., *New Zealand J. Dairy Sci. Technol.*, 1967, **2**, 55.
88. MANNING, D. J., *J. Dairy Res.*, 1979, **46**, 523.
89. SMITH, A. Y., DIETRICH, P. and PICKENHAGEN, W., British Patent 1495227, 1977.
90. LIEBICH, H. M., DOUGLAS, D. R., BAYER, E. and ZLATKIS, A., *J. Chromatogr. Sci.*, 1970, **8**, 355.
91. LAW, B. A. and WIGMORE, A. S., *J. Dairy Res.*, 1982, **49**, 137.
92. MCGUGAN, W. A., HOWSAM, S. G., ELLIOT, J. A., EMMONS, D. B., REITER, B. and SHARPE, M. E., *J. Dairy Res.*, 1968, **38**, 237.
93. MCGUGAN, W. A., EMMONS, D. B. and LARMOND, E., *J. Dairy Res.*, 1979, **62**, 398.
94. GREEN, M. L. and MANNING, D. J., *J. Dairy Res.*, 1982, **49**, 737.
95. MANNING, D. J., Thesis, Reading University, 1976.

96. LAMPARSKY, D. and KLIMES, I., In: *Flavour '81*, Schreier, P. (ed.), 1981, de Gruyter, Berlin, p. 557.

97. MCBRIDE, R. L. and HALL, C., *Aust. J. Dairy Technol.*, 1979, **34**, 66.

98. MANNING, D. J., RIDOUT, E. A. and PRICE, J. C., In: *Advances in the Microbiology and Biochemistry of Cheese and Fermented Milk*, Davies, F. L. and Law, B. A. (eds), 1984, Elsevier Applied Science Publishers, London, p. 229.

99. EDWARDS, J. and KOSIKOWSKI, F. V., *J. Dairy Sci.*, 1983, **66**, 727.

100. LOWRIE, R. J. and LAWRENCE, R. C., *New Zealand J. Dairy Sci. Technol.*, 1972, **7**, 51.

101. LOWRIE, R. J., LAWRENCE, R. C. and PERBERDY, M. F., *New Zealand J. Dairy Sci. Technol.*, 1974, **9**, 116.

102. PÉLISSIER, J. R. and MANCHON, P., *J. Fd. Sci.*, 1976, **41**, 231.

103. MILLS, O. E. and THOMAS, T. D., *New Zealand J. Dairy Sci. Technol.*, 1980, **15**, 131.

104. STADHOUDERS, J. and HUP, G., *Neth. Milk Dairy J.*, 1975, **29**, 335.

105. CHIBA, Y. and SATO, Y., *Jap. Dairy Fd. Sci.*, 1980, **29**, 161.

106. MCGUGAN, W. A., BLAIS, J. A., BOULET, M., GIROUX, R. N., ELLIOT, J. A. and EMMONS, D. B., *Can. Inst. Fd. Sci. Technol. J.*, 1975, **8**, 196.

107. BILLS, D. D., MORGAN, M. E., LIBBEY, L. M. and DAY, E. A., *J. Dairy Sci.*, 1965, **48**, 1168.

108. MANNING, D. J., unpublished.

109. BIEDE, J. L. and HAMMOND, E. G., *J. Dairy Sci.*, 1977, **60** (Suppl. 1), 41.

110. BIEDE, J. L. and HAMMOND, E. G., *J. Dairy Sci.*, 1979, **62**, 227.

111. DUMONT, J. P., PRADEL, G., ROGER, S. and ADDA, J., *Lait*, 1976, **56**, 18.

112. SLOOT, D. and HOFMAN, H. J., *J. Agric. Fd. Chem.*, 1975, **23**, 358.

113. LIN, S. S., *J. Agric. Fd. Chem.*, 1976, **24**, 1252.

114. LLOYD, G. T., HORWOOD, J. F. and BARLOW, I., *Aust. J. Dairy Technol.*, 1980, **35**, 137.

115. MITCHELL, G. E., *Aust. J. Dairy Technol.*, 1981, **36**, 21.

116. CUER, A., DAUPHIN, G., KERGOMARD, A., DUMONT, J. P. and ADDA, J., *Agric. Biol. Chem.*, 1979, **43**, 1782.

117. DUMONT, J. P. and ADDA, J., *J. Agric. Fd. Chem.*, 1978, **26**, 364.

118. FREYTAG, W. G., NEY, K. H. and WIROTAMA, I. P. G., British Patent 1470256, 1977.

119. HINDUSTAN LEVER LTD., Indian Patent 136717, 1976.

120. NEY, K. H. and WIROTAMA, I. P. G., *Fette, Seifen, Anstrichm.*, 1978, **80**, 249; *Fd. Sci. Technol. Abstr.*, 1979, **11**, 3 P 503.

121. PIERGIOVANNI, L. and VOLONTERIO, G., *Industria del Latte*, 1977, **13** (1), 31; *Fd. Sci. Technol. Abstr.*, 1977, **9**, 11 P 1784.

Chapter 9

INDIGENOUS MILK ENZYMES

Barry J. Kitchen*

*Gilbert Chandler Institute of Dairy Technology,
Werribee, Victoria, Australia*

1. INTRODUCTION

Bovine milk contains a large number of enzymes (about 60 have been identified) which originate from the mammary gland tissue cells, blood plasma and blood leucocytes. Although they represent a minor fraction of the total milk protein, they have received in-depth scientific investigation over the past 50 years. Several major reviews[1-6] have been published during this period, together with hundreds of papers on detailed scientific studies on many of these individual enzymes.

Milk enzymes have been studied for many reasons. For instance, milk has been a convenient source for the isolation of sufficient quantities of certain enzymes whose properties could be studied and then compared with those of similar enzymes from other sources, e.g. alkaline phosphatase,[7,8] lactoperoxidase.[9] Certain flavour and stability problems in dairy products have stimulated considerable research[2] on enzymes such as lipase,[10] xanthine oxidase,[11] protease[12] and phosphatases.[7,8,13] Also, many biosynthetic pathways in the mammary gland could be studied by examining enzymes shed into milk during the normal secretory process (e.g. lactose synthetase).[14] The use of some milk enzymes in the diagnosis of udder disorders has induced further studies on their properties, e.g.

*Present address: Cadbury Schweppes Pty Ltd, PO Box 200, Ringwood, Victoria 3134, Australia.

    BARRY J. KITCHEN

TABLE 1
TECHNOLOGICALLY IMPORTANT MILK ENZYMES

Enzymes	Use/Function/Defect	Ref.
Protease(s)	UHT milk gelling	18
Lipase(s)	Hydrolytic rancidity	19
Alkaline phosphatase	Pasteurization test	20
γ-Glutamyltranspeptidase	Amino acid uptake into mammary gland	21
Sulphydryl oxidase	Removal of 'burnt'/'cooked' flavours	22
Superoxide dismutase	Oxidation deterrent	23, 24
Xanthine oxidase	Oxidation catalyst	25
Lactoperoxidase	Bacterial destruction	26
Lactose synthetase	Lactose biosynthesis in mammary gland	14
N-Acetyl-β-D-glucosaminidase	Mastitis diagnosis	15

catalase, arylesterase and glycosidases.[15,16] Table 1 summarizes some of the more important bovine milk enzymes, together with their use and function in milk and dairy products. Milks from other mammalian species also contain most of these enzymes, and considerable interest in them has been shown, particularly in human milk. The benefit of such enzymes to new-born infants, who have an incomplete digestive system, has been reviewed recently by Shahani *et al.*[17] and will not be discussed in the present chapter.

When examining a milk enzyme, the main question usually asked refers to its technological significance to the dairy industry. This cannot be answered with any certainty until its basic biochemical properties have been studied. Also, knowledge of its location in milk and the environment in which it resides is important. As enzymes in milk can be located in four different environments—soluble fraction, micellar fraction, fat globule membrane material, cellular material—it is important to know where most of the enzyme is located, and whether it is accessible to natural substrates. Their location in milk usually points to their origins within the animal (i.e. mammary gland secretory cells, leucocytes or blood), and this, in some cases, suggests possible physiological functions of such enzymes.

2. PROTEOLYTIC ENZYMES (EC 3.4.–.–)

Extensive reviews on the natural milk proteinase system have appeared recently[27,28] and, in an earlier volume of this series,[29] proteolysis in milk and the properties of some of the proteinases were further examined. Milk proteinases have been implicated in many technological aspects of the dairy industry; these include gelation of UHT milk,[30] formation of amino acids during Cheddar cheese ripening[31,32] and production of bitter flavours in milk and dairy products.[29] It is generally accepted that there are two main indigenous proteinase systems in milk, one operating optimally at neutral to slightly alkaline pH and the other at acid pH values. However, recent evidence suggests that there is more than one proteinase operating at neutral to slightly alkaline pH values.[29]

2.1. Neutral-Alkaline Proteinases

A summary of the properties of the main enzyme in this system is shown in Table 2. It is a typical serine protease with trypsin-like activity. A recent review by Humbert and Alais[27] considers the latest research on

TABLE 2

PROPERTIES OF MILK PROTEINASE (ALKALINE)[33,45,47,48]

Property	Comments
pH optimum	Range, 6·5–9·0 with optimum at 8
Heat stability	Survives pasteurization; 60% activity remains after 60°C/10 min and 20% after 70°C/10 min
Molecular weight	48 000 at pH 7
Inhibitors[a]	
DFP	Strong inhibition
pCMB	No effect
TPCK	No effect
TLCK	Strong inhibition
Trypsin inhibitor	Strong inhibition
Substrates	Hydrolysis of peptide bonds on C-terminal side of lysyl and arginyl residues; β-casein > α-casein > κ-casein
Origin	Similar to blood serum plasmin and therefore presumed to originate in blood

[a] DFP, diisopropyl fluorophosphate; pCMB, p-chloromercuribenzoate; TPCK, L-(1-tosylamido-2-phenyl)ethyl chloromethyl ketone; TLCK, N-α-p-tosyl-L-lysine chloromethyl ketone.

this enzyme, which has centred mainly around its similarity to blood serum plasmin, its role in the formation of minor caseins and proteose-peptones, and its significance in the gelation of UHT-treated milk.

The work of Kaminogawa *et al.*,[33] Halpaap *et al.*[34] and Eigel *et al.*[35] supports the view that this milk enzyme is actually blood plasmin. Milk normally contains both active plasmin and its inactive precursor, plasminogen, generally in the ratio of about 1:5.[36,37] Plasminogen can be converted into plasmin by specific activators (e.g. urokinase) which have not yet been identified in bovine milk but are present in human milk.[38] However, Eigel *et al.*[35] have observed that in bovine milk there is a slow conversion of inactive plasminogen into plasmin. Proteolysis in milk is also influenced by indigenous heat-labile inhibitors,[39,40] so overall the 'proteolytic potential' of a milk is controlled by a balance between a number of factors such as the initial levels of active plasmin and inactive plasminogen, the rate of conversion of plasminogen into plasmin, and the levels of indigenous inhibitors. Improved assay procedures such as that described by Richardson and Pearce,[36] Rollema *et al.*[41] and Korycka-Dahl *et al.*[37] should lead to a more rapid and sensitive method of determining plasminogen/plasmin ratios in milks, and therefore help in determining the significance of this system in proteolysis of dairy products. Already it has been shown in model systems that proteinases from some pseudomonads and coliforms are capable of transforming plasminogen into plasmin,[42,43] but this has not been examined with proteases for the more common types of bacteria found in refrigerated milk (e.g. *Pseudomonas fluorescens*).

The effect of heat on milk proteinase has been studied by a number of workers, and in general there has been little agreement concerning its heat stability. For instance, Kiermeier and Semper[39] observed that heat treatment at 70°C for 2 min inactivated the enzyme, while Whitney[1] stated that 80°C/10 min was required for its destruction. Noomen[44] found that pasteurization (72°C/15 s) of raw milk actually activated the proteinase, as did heating at 63°C for 30 min. Also, it has been reported that milk proteinase survives UHT processing (142°C/3 s) and it may be a factor in age gelation.[30] These apparent anomalies can probably be explained by the fact that the degree of purity of the enzymes studied by various workers has varied considerably. Also, it is possible that destruction of indigenous proteinase inhibitors occurs faster than the denaturation of the proteinase itself, thus causing an apparent activation effect. Further, a wide variation of heat stability can be obtained depending on the pH at which the experiments are conducted;[45] at acid and neutral pH values the enzyme is stable to pasteurization but marked inactivation occurs at alkaline pH values.

The reported molecular weight for the enzyme differs considerably: Dulley[45] reported a value of 65 000 with shoulder at 120 000, while Kaminogawa et al.[33] found the molecular weight to be 48 000. These values may be a function of the ionic strength and pH of the eluting buffers used during Sephadex chromatography or of the degree of purity of the preparations examined. However, the molecular weight of the most highly purified form of the milk enzyme (i.e. 48 000) does not agree with that reported for blood serum plasmin (i.e. 78 000).[46]

Table 3 shows the breakdown products of β-casein produced by plasmin or milk proteinase; the main products are the γ-caseins and

TABLE 3

SUMMARY OF PLASMIN-MEDIATED HYDROLYSIS OF β-CASEIN[49-54]

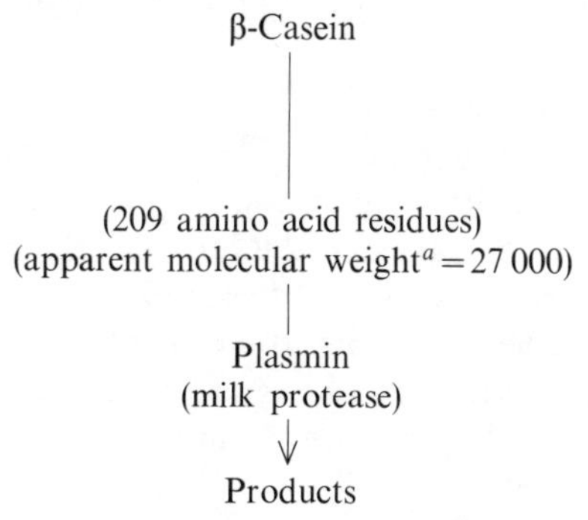

	Casein			PP-5^b	PP-8 (fast)	Residual fragment^c
	γ₁	γ₂	γ₃			
Residues	29–209	106–209	108–209	1–105 1–107	1–28	29–105 29–107
Apparent molecular weight^a	21 000	14 900	14 900	11 500 to 13 000	—	—

^{a} By sodium dodecyl sulphate electrophoresis.
^{b} PP, proteose-peptone.
^{c} This fragment was originally referred to as PP-8(slow),[51] but recent work by Andrews[53] has shown this to be incorrect.

proteose-peptones. Recent electrophoretic studies[53] have shown that PP-5 (residues 1–105, 1–107) is really an intermediate product of plasmin attack on β-casein and is subject to further proteolysis producing a fragment representing residues 29–105 (or 29–107), which Eigel and Keenan[51] referred to as protease-peptone-8-slow (PP-8$_s$). Andrews[53] considers this nomenclature to be incorrect.

Plasmin attacks α_{s1}- and κ-caseins at slower rates than β-casein[31] which is hydrolysed at approximately the same rate as α_{s2}-casein.[55] In general, the proteolysis products of the α-caseins are not as well defined as those from β-casein. However, Aimutis and Eigel[56] have shown that some of the peptides of the λ-casein fraction of milk may be plasmin-derived fragments of α_{s1}-casein. Milk proteinase (plasmin) also appears to be associated with the milk fat globule membrane (MFGM),[57] because autoproteolysis of purified MFGM results in the rapid degradation of a membrane protein of apparent molecular weight 77 000, and slower degradation of other MFGM polypeptides. Immunological methods show that there is material in the MFGM that cross-reacts with antisera to bovine plasminogen.[57]

Recent work by Reimerdes *et al.*[58] has shown that there is another blood serum-derived proteinase in milk, viz. thrombin. Andrews[54] confirmed the presence of other indigenous proteinases in normal milk by observing the effects of various inhibitors on the protein breakdown patterns as shown by polyacrylamide gel electrophoresis. This pattern was more complex when abnormal (mastitic) milk was examined, and it was suggested that proteinases originating from both blood serum (plasmin, thrombin) and proteinases from leucocytes contribute to overall proteolysis.

2.2. Acid Proteinase

Kaminogawa and Yamauchi[59,60] purified a proteinase from milk which has an optimum pH of 4; its properties are summarized in Table 4. It is more heat-labile than the alkaline protease and it has a lower apparent molecular weight. It hydrolyses α-casein in preference to β- and κ-casein. Kaminogawa *et al.*[61] showed that it produces breakdown products from all caseins which are similar to those produced by chymosin (rennin). The properties of this enzyme suggest that it may be cathepsin D, a protease usually found in the lysosome granules of PMN leucocytes. It has been suggested that this protease may have some significance in cheese ripening.[59,60]

TABLE 4
PROPERTIES OF MILK PROTEINASE (ACID)[59,60]

Property	Comments
pH optimum	4
Heat stability	70% activity remains after 60°C/10 min and less than 1% remains after 78°C/10 min
Molecular weight	36 000
Inhibitors[a]	
EDTA	No effect
DFP	No effect
pCMB	No effect
Casein substrates	$\alpha > \beta,\kappa$
Origin	Similar to cathepsin D from leucocytes

[a] EDTA, ethylene diamine tetra-acetic acid; DEP, diisopropyl fluorophosphate; pCMB, p-chloromercuribenzoate.

2.3. Other Peptidases

Mellors[62] purified an enzyme from bovine milk which hydrolysed amino acid β-naphthylamides. It was located in the 'microsomes' of the milk fat globule membrane and could be released from them by freeze-drying and sodium chloride extraction. Further purification was achieved by DEAE-cellulose chromatography which produced about a 13 000-fold purification of the enzyme. The optimum pH for the hydrolysis of lysyl-β-naphthylamide was 7·7. The enzyme was inhibited by EDTA, and calcium restored activity to the EDTA-inactivated enzyme. The enzyme was active only if preincubated with or assayed in the presence of dithiothreitol which indicates that its activity apparently depends on free sulphydryl groups. Thus, in its native state in milk, it is a 'latent' enzyme; it was suggested that it may be important in milk and milk products, presumably owing to its possible activation as a result of any processing or storage treatment of milk which reduces protein disulphide bridges.

Grundig and Hanson[63] purified a dipeptidase from human milk which hydrolyses L-L dipeptides at alkaline pH values. The enzyme shows greatest activity towards glycyl-L-methionine and L-alanyl-L-methionine; peptides containing D-amino acids were not hydrolysed. The temperature optimum was between 45 and 50°C and the pH optimum between 7·8 and 8·3. Heavy metals, EDTA and p-chloromercuribenzoate inhibited the enzyme. The molecular weight as determined by gel filtration on

Sephadex G-200 was approximately 77 000. The purified enzyme was devoid of arylamidase, leucine-aminopeptidase and γ-glutamyl-transpeptidase activities.

3. LIPASES AND ESTERASES

3.1. Lipases (EC 3.1.1.3)

Milk lipase has received considerable attention by research workers in the dairy industry owing to its potential to produce free fatty acids from milk triglycerides, thus causing unpalatable flavours. Many reviews have been written on this subject;[10,19,64-68] a chapter in a previous volume of this series examined lipolysis in general,[69] and this aspect will not be reviewed again here. Instead, biochemical aspects of the enzyme itself and more recent information on the mode of lipase activation by apo-lipoprotein (LP)-CII will be discussed.

From earlier studies on milk lipase it was concluded that the lipolytic activity observed in bovine milk was due to more than one enzyme.[70,71] Recent work[19,64] has shown that milk contains only one lipase which, depending on the assay system employed, can act either as a lipoprotein lipase (LPLase) or a true lipase. The main distinction between a lipoprotein lipase and a true lipase is that the former requires the presence of blood serum cofactors for maximal activity. The co-identity of the two activities in bovine milk has been established by comparing the susceptibility of each of the activities towards various inhibitors such as sodium chloride, protamine sulphate, ultraviolet light, and specific antisera.[19] Also, a comparison of more basic properties such as pH optima, temperature stability, molecular weight, and release from casein micelles by heparin and sodium chloride has supported the contention that these two activities can be attributed to a single enzyme.[65]

Egelrud and Olivecrona[72] purified lipoprotein lipase about 7000-fold from bovine skim milk. This was accomplished mainly by the use of an affinity chromatography step utilizing heparin bound to Sepharose 4B. The enzyme is a glycoprotein with a minimum molecular weight of 62 000–66 000. Other workers have shown that the native enzyme is composed of two equal sub-units of apparent molecular weight 50 000.[73] More recently a minimum molecular weight of 41 000 has been reported.[74] This polypeptide was cleaved by trypsin into a 19 000 and a 23 000 fragment; on the basis of the separation of these fragments by electrophoresis in the presence and absence of reducing agents, it was suggested that they were joined by a disulphide bond.[74] A lipase, with a molecular weight of 7000,[75] has been isolated from separator slime; it probably originates from somatic cells but more research is required to confirm this. A summary of the properties of bovine milk LPLase is shown in Table 5.

TABLE 5
PROPERTIES OF LIPOPROTEIN LIPASE[19,67,73,74]

Property	*Comments*
Molecular weight	Native, 100 000
	Monomer, 50 000, 41 700
Partial specific volume	$0.71 \pm 0.007 \, \text{ml g}^{-1}$
pH optimum	8.5–9.0
Carbohydrate content	8.3%
Activity in Milk	
Potential	$2 \, \mu\text{mol FFA ml}^{-1} \, \text{min}^{-1}$
Actual	$0.002 \, \mu\text{mol FFA ml}^{-1} \, \text{min}^{-1}$
Distribution in milk	80% in casein micelles
Heat stability in milk	Destroyed by HTST pasteurization
Substrates	Triglycerides, 1- and 3-monoglycerides, phosphatidylcholine, *p*-nitrophenyl esters
Inhibitors	1M NaCl, pyrophosphate, protamine, apo-lipoproteins CI and CIII, polyanions, free fatty acids
Activator	Apo-lipoprotein CII

The amount of lipoprotein lipase present in milk should be able, under optimum conditions, to release about $2 \, \mu\text{mol}$ of free fatty acid (FFA) per minute. However, in practical situations the actual activity that this enzyme can exert in milk is probably about 1000 times lower than this figure,[68] owing to factors such as storage temperature of milk (5°C), the pH of milk, access of the enzyme to the natural substrate (intact fat globules), and the presence of indigenous inhibitors and activators.[68]

Bovine milk LPLase is a rather non-specific enzyme capable of hydrolysing long-chain triglycerides, phospholipids, tributyrin, synthetic esters (*p*-nitrophenyl butyrate) and monoglycerides.[76,77] However, the properties of the enzyme vary considerably when examined against these various substrates. For instance, the rate of hydrolysis of tributyrin is about the same as that against long-chain triglycerides, but many of the agents (e.g. 1M NaCl, heparin, protamine, Ca^{2+}) that strongly affect LPLase activity against long-chain triglycerides in the presence of albumin and cofactor proteins have very little effect when tributyrin is used as substrate.[78] Further studies by Bengtsson and Olivecrona[79] demonstrate the complex mechanism of hydrolysis of monoglycerides by LPLase in the presence of activator protein apo-lipoprotein (LP)-CII; under appropriate conditions, in the presence of activator, the rate of monoglyceride hydrolysis is similar to that of triglyceride hydrolysis but substantial product (free fatty acid) inhibition occurs which abolishes the activating effects of the apo-LP CII.

Much of the recent research on milk LPLase has been aimed at understanding the mechanism of its interaction with fat emulsions, with particular emphasis on the role of blood serum lipoproteins. The main activator is apo-LP CII which can interact with, and stabilize, the enzyme to a much greater degree than serum albumin. Enzyme-activator interaction is still significant in the presence of 1M NaCl, indicating that forces other than electrostatic are involved. Actual binding of the enzyme to artificial fat globules is not influenced by the activator protein, and thus it appears that the role of the activator is to make the enzyme itself a more effective catalyst by orientating the enzyme and/or the lipid substrate to achieve maximal hydrolysis.[80] Similar conclusions were reached by Shirai et al.[81] who suggested that apo-LP CII increases the affinity of the active site of LPLase for the triacylglycerol substrate and this is accompanied by a conformational change in the enzyme molecule. Further work by Bengtsson and Olivecrona[82] demonstrates the mechanism of interaction of LPLase and apo-LP CII with natural milk fat globules; in contrast to studies on model systems, apo-LP CII causes a 10–20-fold increase in the amount of LPLase bound to the fat globules.

Other studies on the inhibition of LPLase by benzeneboronic acid, both in the absence and in the presence of apo-LP CII, have led to alternative hypotheses regarding the mechanism of LPLase activation by apo-LP CII. For example, it has been suggested that apo-LP CII acts as an esterase (rate-limiting step) which deacylates the acyl-LPLase intermediate.[83] In addition, the studies with benzeneboronic acid have indicated the presence of serine and histidine residues in the active site of the enzyme.[83]

The activation of LPLase by apo-LP CII with triacylglycerol as substrate is independent of the reaction pH (5·5–10·5), provided conditions are such that the binding of the enzyme to the lipid droplets and the stability of the enzyme are not limiting.[84] This finding implies that the enzyme–activator–lipid globule interaction is not related to groups that ionize in the pH range 5·5–10·5. The effect of pH on LPLase action in the presence of apo-LP CII is quite different on different substrates, e.g. natural milk fat globules and Intralipid (a commercially available lipase substrate). In addition, there is little inhibition of Intralipid hydrolysis in the presence of NaCl (1·1M) whereas almost complete inhibition occurs when milk fat globules are used as substrate. These findings lend further support to the hypothesis that the enzyme molecule itself is the main target for these physicochemical influences.[85]

3.2. Esterases (EC 3.1.–.–)

Esterases in general are distinguished from lipases by virtue of their preference for soluble rather than emulsified substrates. Esterases are further subdivided into distinct types on the basis of substrate specificity and inhibitor sensitivity. The main types identified in bovine milk include arylesterases (EC 3.1.1.2) and cholinesterases (EC 3.1.1.7, 3.1.1.8) but there are reports on the existence of carboxylesterases (EC 3.1.1.1) and acetylesterases (EC 3.1.1.6). Table 6 outlines the findings regarding the various types of esterases identified in bovine milk.

TABLE 6
TYPES OF ESTERASES FOUND IN BOVINE MILK

Esterase	Substrate	Ref.
Arylesterase	Phenyl acetate	86, 87
	β-Naphthyl acetate	
Carboxylesterase	β-Naphthyl acetate	88, 90
	4-Methylumbelliferyl heptanoate	
Cholinesterase	Phenyl acetate	86–88
	Phenyl propionate	
	β-Naphthyl acetate	
Acetylesterase	Indoxyl acetate	91, 92

[a]See Ref. 94 for classification of different types of esterases.

The relative levels (in terms of activity against a single substrate) of the various milk esterases are difficult to ascertain, but from the work of Forster et al.[86] it is possible to calculate that, of the esterolytic activity not attributable to milk lipase, arylesterase represents 70% and cholinesterase 30% of the total. Values of 84% and 10% for arylesterase and cholinesterase, respectively, have also been reported.[87] In contrast to these findings, Nakanishi and Tagata[88] reported that milk esterase distribution is 63% carboxylesterase, 13% cholinesterase and 12–15% arylesterase. Arbabi[89] could detect no cholinesterase in milk, cream or whey using a variety of sensitive titrimetric and histochemical procedures. Significant levels of carboxylesterase[88,90] and acetylesterase[91,92] activities have also been recorded. Thus, the whole issue of the levels and types of esterases present in milk is quite unclear, and more studies are required to clarify the situation. It is possible that much of the confusion has arisen because different types of substrates under varying assay conditions have been used. Standardization of assay procedures and care in interpreting the degree of inhibition obtained with selected inhibitors are needed before more progress in this area will be forthcoming. Interpretation of inhibition data is most important because it is the main basis for type classification, and in crude systems, such as milk or whey, some esterases are protected from inhibition by other agents present in milk. For instance, the degree of pCMB inhibition of arylesterase is influenced by the level of free Ca^{2+} present, so this fact has to be taken into account when attempting to determine the proportion of arylesterase in milk using selective inhibitors (Kitchen, unpublished). Also, non-enzymic hydrolysis of simple synthetic ester substrates by native and heat-denatured milk proteins[93] can complicate the issue even further, and care must be taken to ascertain this factor when examining esterase types in milk.

The best-characterized milk esterase is the arylesterase which has been partially purified by ammonium sulphate fractionation, ultrafiltration and gel permeation chromatography. This preparation contained both arylesterase and cholinesterase (90% and 10% of total esterase activity, respectively) which could be separated by polyacrylamide gel electrophoresis.[87] The arylesterase was sensitive to p-chloromercuribenzoate, EDTA and La^{3+}, which is characteristic of arylesterase studied from other sources.[87,94]

Arylesterase levels in milk are influenced by the stage of lactation, with high levels occurring in colostrum. Activity falls rapidly and reaches a constant level after about 1 month of lactation.[86,95] Udder infection also affects the level of arylesterase in milk.[96] From these and other studies it has been suggested that milk arylesterase originates from blood serum,[96,97] and there is some comparative biochemical evidence to support these claims. For instance, milk arylesterase has a molecular weight ($\sim 500\,000$)[87] similar to that found for the blood serum enzyme, which in turn appears to be strongly associated with the high density lipoprotein (HDL) fraction of blood serum.[98] Enzymes from both sources also have similar requirements for activators and are sensitive to the same inhibitors.[99]

As indicated above, arylesterase levels in milk can vary considerably, and since HDL's (which apparently contain this enzyme) are the major blood serum activators of milk lipolysis,[100] it would be of interest to determine the relationship between arylesterase activity in milk and the milk's ability to lipolyse. Since recent evidence suggests that the activation of LPLase is indeed a result of intrinsic esterase activity[83,101] of the activator polypeptide, determination of arylesterase levels in milk should be a practical means of identifying milk samples likely to undergo both spontaneous and agitation-induced lipolysis.

All the esterases found in bovine milk have pH optima in the range 7–8, and they are probably not significant technologically in milk or dairy products. All but the so-called acetylesterase[91] are extremely sensitive to heat; thus pasteurization results in considerable loss of activity. Since no information is available on their natural substrates in milk, no function in milk can be assigned to them. A suggestion that they may be involved in ester metabolism in Cheddar cheese maturation remains unsubstantiated.[87]

4. PHOSPHATASES

This group of enzymes hydrolyse esters of phosphoric acid and are generally classified by either the substrate acted upon or the pH at which optimum activity occurs. There are several enzymes in milk that split phosphate ester bonds causing the release of inorganic phosphate from the substrate. The acid and alkaline phosphatases have been studied in greatest detail as they have been implicated in stability problems in some dairy products.

4.1. Alkaline Phosphatase (EC 3.1.3.1)

Alkaline phosphatase has been studied extensively in milk because of its use in the milk pasteurization test. Much interest has also centred around the mechanism of post-heating reactivation[102,103] which could result in false positive results. Alkaline phosphatase is found mainly in membrane structures of both skim milk and cream but a small amount of activity is found free in serum prepared by ultracentrifugation of skim milk.[13] Recently the release of this enzyme from milk fat globule membrane into skim milk has been used to monitor the severity of agitation treatments of raw milk.[104] This enzyme splits most phosphomonoesters in the pH range 9–10·5. It does not hydrolyse ADP or ATP, but it does attack AMP, glycerophosphate, and the phosphates of glucose, serine and threonine. It splits phosphoserine optimally in the pH range 8·5–9·0, but its action towards casein is greatest at pH 6·5–7·0.[105–108] The action of this enzyme on native casein micelles in milk would be expected to be very low as it has been shown that inorganic phosphate, lactose and β-lactoglobulin (β-LG) all cause a certain amount of inhibition.[107] It has been suggested that alkaline phosphatase, together with β-LG, may play an important role in phosphorus metabolism in the mammary gland as the enzyme is inhibited to a greater degree by the A variant of β-LG than by the B or C variants.[109] However, Lorrient and Linden[107] have shown that the effect of β-LG is pH- and substrate-dependent, and thus its involvement in phosphorus metabolism remains unresolved.

Milk alkaline phosphatase has been purified by several workers.[7,110,111] It has a molecular weight in the range 160 000–190 000[112,113] and it appears to contain no phosphorus or carbohydrate.[7] However, Sharma and Ganguli[114] have claimed that it is a sialoglycoprotein. The possibility that it is a glycoprotein is further supported by the work of Peereboom[115] who showed that treatment of a high molecular weight form of the enzyme (called the β isozyme) with neuraminidase results in an increase in its electrophoretic mobility and a reduction in its molecular weight from 570 000 to 140 000 (called the α isozyme). Peereboom[115] has studied these different isoenzymes of alkaline phosphatases in cream and skim milk and has used the results to propose a mechanism for reactivation of the enzyme after pasteurization. Reactivation mechanisms for milk alkaline phosphatase have been reviewed recently by Fox and Morrissey.[5]

The mode of action of milk alkaline phosphatase is similar to that of phosphatases from other sources in that a phosphoryl-enzyme intermediate is formed which is rapidly hydrolysed at alkaline pH values.[116] A highly purified preparation of this enzyme has recently been reported by Linden et al.,[117] and electrophoretic and ultracentrifugation analysis confirmed the homogeneity of the preparation. The native molecular weight was in the range 160 000–190 000 and it was composed of two subunits of molecular weight 85 000.[108] The amino acid composition of the purified enzyme is characterized by high levels of threonine, serine, glutamic acid and lysine. There is no indication that the purified enzyme contains sialic acid or other carbohydrates,[117] a finding that is of interest in view of the claims of Sharma and Ganguli[114] and Peereboom.[115] Also, Linden et al.[117] indicated that their purified preparation probably represents only one of the reported isoenzymes of milk alkaline phosphatase, and they claim that it is probably the α isozyme described by Peereboom.[115] The term isozyme is probably misleading and it is most likely that the isoenzymes described by Peereboom[115] and Burviana and Dema[118] just represent different degrees of association of a single enzyme with various sized membrane fragments, which therefore have different electrophoretic mobilities and molecular weights. However, Peereboom[115] has shown that these 'isoenzymes' still exist after extraction of membrane preparations with n-butanol, a treatment that would be expected to produce 'soluble', unassociated forms of the enzyme. This was apparently the case during the preparation procedure described by Linden et al.[117] where the membrane preparation, after butanol extraction, showed only a single peak on both a Sephadex G-200 and a DEAE-cellulose column. Thus, the apparent anomalies in these reports concerning the electrophoretic and molecular weight heterogeneity have yet to be resolved.

Table 7 summarizes the properties of the purified bovine milk enzyme; it is similar in many ways to the enzyme obtained from the bovine

TABLE 7
PROPERTIES OF BOVINE MILK ALKALINE PHOSPHATASE

Property	Comments	Ref.
Native molecular weights		
Gel filtration	190 000	112, 113, 117, 120
Ultracentrifugation	160 000–170 000	117
Subunit molecular weight	85 000	108
Sedimentation constant (S)	6·0	117
Diffusion constant (D)	$3\cdot4 \times 10^{-7}\ cm^2\ s^{-1}$	117
Metal ion cofactor	Zn^{2+}, $4\cdot9 \pm 0\cdot6$ g-atoms mol^{-1}	108
Specific activity	96 units mg^{-1}	117
(pNPP[a] as substrate)	43 units mg^{-1}	121
Inhibitors		
L-Phenylalanine (20 mM)	39% (57%)[b]	122, 125
D-Phenylalanine (20 mM)	Nil	122, 125
Sucrose	—	123
Activators		
Zn^{2+}	Essential	108, 120
Mg^{2+}	Stimulatory	108, 120, 124, 125

[a] p-Nitrophenyl phosphate.
[b] Inhibition in absence of Mg^{2+}.

mammary gland.[119] For instance, the mammary gland enzyme is a dimer composed of two disulphide-linked monomers, each with a molecular weight of approximately 80 000. Mammary gland alkaline phosphatase, however, is definitely a glycoprotein, possessing up to 30% by weight of carbohydrate, and binds strongly to a concanavalin A–agarose column. The pH optimum of the enzyme is extremely dependent on the substrate concentration, decreasing from pH 10 at 2 mM p-nitrophenyl phosphate to pH 9 at 0·1 mM. The apparent Michaelis constant (K_m) for p-nitrophenyl phosphate is 0·7 mM, a value similar to that reported for the milk enzyme.[108] Despite these many physical and chemical similarities, the immunological cross-reactivity of these two enzymes has not been determined, and thus the exact origin of the milk enzyme and its homology with the mammary gland enzyme remain unresolved.

4.2. Acid Phosphatase (EC 3.1.3.2)

The presence of an enzyme in bovine milk capable of hydrolysing phosphomonoesters at acid pH (4·5) has been known for some time. It acts on aromatic phosphomonoesters (p-nitrophenyl phosphate), pyrophosphates (ADP, ATP, PP_i) and the phosphoserine residues of casein but shows virtually no activity against free serine phosphate or β-glycerophosphate.[126] Shahani *et al.*[3] have suggested that this enzyme is really a phosphoprotein phosphatase similar to that found in bovine spleen. Milk acid phosphatase is found in skim milk, both in a free form and associated with membrane material, and in cream associated with the milk fat globule membrane material.[13,127] No comparisons have been made between the properties of the free and bound forms of this enzyme. Also, the enzyme associated with the membrane material is strongly bound, and is not released in appreciable amounts by non-ionic detergents.[128]

The technological significance of this enzyme possibly lies in its action against casein from which phosphate groups can be readily removed.[126,129] This results in marked increases in the isoelectric point of casein and changes in micelle structure.[130] It has been suggested that such dephosphorylation reactions could affect milk gelling[129] or coagulation[2] following UHT and rennet treatment, respectively. Knoop and Peters[131] reported that acid phosphatase may play an important role in the disintegration of casein micelles into submicelles during acidification of milk, because addition of sodium fluoride, a potent phosphatase inhibitor, suppresses micelle disintegration.

Acid phosphatase is found in Cheddar cheese and, depending on the starter used, it may represent from 50% to 100% of the total acid phosphatase activity in this cheese.[132,133] The phosphatase activity in cheese is very stable during maturation, and it has been suggested that it may play an important role in dephosphorylation of protease-resistant phosphopeptides.[132,134] However, it appears that the production of free inorganic phosphate in cheese during maturation is extremely low[132] and therefore hydrolysis of phosphate esters in casein may not be important in cheese ripening.

Milk acid phosphatase is very heat-stable; it survives normal pasteurization conditions, and heating for 30 min at 88°C or for 1 min at 100°C is required to inactivate the enzyme fully.[135–137] It can, under certain conditions, survive UHT sterilization.[129] However, its heat stability is pH-dependent, with greater stability at lower pH values.[137] The enzyme is sensitive to light, with up to 60% loss of activity after 3 days at 17°C.[5,135]

Purification of the enzyme is usually achieved by batchwise extraction of skim milk with an Amberlite ion-exchange resin,[126] followed by gel filtration, cellulose phosphate column chromatography and affinity chromatography on immobilized α_{s1}-casein A.[138] Purification factors from 10 000 to 1 000 000 have been reported, with a wide range of specific activities (0·27–30 IU mg^{-1}). Affinity chromatography on a trypan blue–Sepharose column has also been used to purify the enzyme.[139]

Kinetic studies show that the overall reaction mechanism using p-nitrophenyl phosphate as substrate involves the rapid formation of an enzyme–substrate complex which then decomposes releasing initially p-nitrophenol and then free phosphate.[129] The effect of pH on the Michaelis constant (K_m) shows that a histidine residue is near to or involved in the active site.[129,137] Further studies, which showed that the enzyme is inhibited by iodoacetate, support this view.[129]

Table 8 summarizes the properties of the purified enzyme. Acid phosphatase is a basic protein, with an isoelectric point of 7·9 and a molecular weight of $42\,000 \pm 2000$.[129] Its optimum pH is 4·9; it is activated by ascorbic acid and sulphydryl compounds (β-mercaptoethanol, L-cysteine, thioglycollic acid) and inhibited by heavy metals (Hg^{2+}, Cu^{2+}, Al^{3+}, Fe^{3+}) and fluoride.[129] No effect was observed with EDTA, Pb^{2+}, or tartrate. Similar findings were reported by Kuzuya *et al.*,[139] except for the effect of Pb^{2+} which inhibited the enzyme by 95%. Amino acid composition of the purified enzyme shows that there is a high level of basic amino acids and no methionine present.[138]

TABLE 8

PROPERTIES OF BOVINE MILK ACID PHOSPHATASE

Property	Comment	Ref.
Molecular weight	$42\,000 \pm 2000$	129, 138
Carbohydrate	Glycoprotein containing galactose, mannose, and glucosamine	138
Extinction coefficient (280 nm)	9·4	138
Inhibitors	Most heavy metals, F^-, oxidizing agents, light, inorganic phosphate, iodoacetate	126, 129, 140
Activators	Ascorbate, mercaptoethanol, cysteine, thioglycollic acid	126, 129, 140

This enzyme has been studied also because of its apparent importance in pathological conditions in the udder.[15,141,142] Electrophoretic examination of milk from healthy animals shows the presence of only one band of acid phosphatase activity while mastitic milk shows the presence of two additional bands of activity which are of leucocyte origin.[141] The properties of the leucocyte enzymes have not been studied in detail and it would be of great interest to have more information on the enzymes from this source. It has been shown that the leucocyte acid phosphatase is more sensitive to heat than the enzyme present in normal milk,[143] and thus it would be of little technological significance in pasteurized or heated milk products.

4.3. 5′-Nucleotidase (EC 3.1.3.5)

This enzyme is located in membrane structures in cream and skim milk,[127] and Huang and Keenan[144] have partially purified it from milk fat globule membranes (MFGM) using detergent treatment, ammonium sulphate precipitation, sonication, and Sepharose 4B chromatography. Two peaks of activity were obtained which had different substrate specificities, kinetic properties and different phospholipid contents. The higher molecular weight form is enriched in sphingomyelin which may play some role in the biological activity of the enzyme.

The pH optima of the two forms of the enzyme are in the range 7–7·8 using a variety of substrates including AMP, GMP, and CMP. In the presence of Mg^{2+}, a double pH optimum profile occurs with maxima at pH 7 and 10. Temperature optimum is in the range 60–70°C, and only about 20% of activity is lost following heat treatment at 60°C for 30 min. Both forms of the enzyme are inhibited by Hg^{2+}, and the higher molecular weight form is activated by 1mM Mn^{2+} but inhibited by higher concentrations of this metal. However, the lower molecular weight form is inhibited only slightly by 1mM and considerably (by 60%) by 5mM Mn^{2+}. The K_m values for AMP are 0·94 mM and 5·0 mM for the high and low molecular weight forms, respectively.[144] Caulini *et al.*[145] reported K_m values for AMP of between 13 and 15 mM, but this was obtained by using a different assay procedure from that of Huang and Keenan.[144] ATP is a strong competitive inhibitor ($K_i = 5 \times 10^{-6}$ M) for the higher molecular weight form and a non-competitive inhibitor ($K_i = 4·7 \times 10^{-5}$ M) for the lower molecular weight form, using AMP as substrate.[144] Other workers[145] have also reported competitive inhibition using ATP, but apparent K_i values were considerably lower than the above figures. These kinetic discrepancies need resolving. Different forms of 5′-nucleotidase have been reported from other mammalian sources,[146] and the different kinetic properties have been explained on the basis of the association of a single type of enzyme molecule with different types of membranes. A similar situation may occur in bovine milk where this enzyme resides mainly in membrane structures.[127] 5′-Nucleotidase purified from buffalo MFGM was resolved[147] into two molecular weight forms and had essentially the same properties as the bovine enzyme, although there were some major differences in apparent K_i values for ATP inhibition.

5′-Nucleotidase can be released from MFGM material by deoxycholate, and the soluble form has different kinetic properties from the membrane-bound form, particularly with respect to concanavalin A inhibition.[148] It has been proposed that 5′-nucleotidase is located on the surface of the MFGM and is situated close to, and possibly interacts with, a major membrane polypeptide of apparent molecular weight 150 000 (i.e. xanthine oxidase).[148,149]

The main significance and use of 5′-nucleotidase in milk lies in its value as a 'marker' enzyme for cellular plasma membrane material, and it is useful in studying the biogenesis of MFGM material and the mechanism of fat globule secretion.

4.4. Adenosine Triphosphatase (ATPase) (EC 3.6.1.3)

Bovine milk contains an ATPase which hydrolyses CTP, GTP and UTP to about the same extent.[150] It is activated by Mg^{2+} but not by Na^+ or K^+, and it is not inhibited by ouabain. Its pH optimum is about 8·5 and it operates at relatively high temperatures (optimum, 55°C). The K_m for ATP in the presence of 5mM Mg^{2+} is 5.68×10^{-5} M. Other workers[151] have indicated that ATPase activity from bovine MFGM is made up of more than one enzyme because of biphasic plots obtained when enzyme activity is measured against increasing substrate concentration. The milk fat globule membrane ATPases lack any Na^+ ATPase activity, and this has led to the suggestion that Na^+ is not actively transported through the apical plasma membrane of mammary secretory cells during lactation.[150] Other studies support this view.[152]

4.5. Other Phosphatases

The presence of nucleotide pyrophosphatase,[127] inorganic pyrophosphatase[153] and glucose-6-phosphatase[127,153] and phosphodiesterase[154] has been reported in bovine milk but there is little information available concerning their properties and significance. Most likely, they would be useful enzymes to monitor and study in investigations relating to assembly of the components of milk within the secretory cell, the mode of secretion of such components and their fate in the final secreted milk.

5. γ-GLUTAMYL TRANSPEPTIDASE (EC 2.3.2.1)

γ-Glutamyl transpeptidase or transferase catalyses the transfer of γ-glutamyl residues from substrates such as glutathione (oxidized or reduced form) to an acceptor such as another glutamyl residue, hydroxylamine, glycylglycine or water. Baumrucker et al.[155] reported that it is located on the exterior surface of the mammary gland secretory cells and probably functions in the uptake of amino acids from the blood into the secretory cell for use in milk protein synthesis.

The properties of this enzyme in milk have been extensively studied.[21,156–160] It has a pH optimum in the range 8–9 depending on the type of acceptor present, and it is only partially destroyed by pasteurization. A summary of its properties is shown in Table 9. The enzyme is membrane-bound with nearly all the activity located either in skim milk membranes or in the milk fat globule membrane (MFGM).[127,161] Solubilization can be achieved by detergent and solvent treatments which have been routinely used in various purification procedures.[159,160] The pure enzyme is composed of two subunits with molecular weights of 57 000 and 25 000.[21] Both are glycoproteins and appear to be similar to polypeptides 14 and 18 of the MFGM.[162]

TABLE 9
PROPERTIES OF BOVINE MILK γ-GLUTAMYL TRANSPEPTIDASE

Property	Comments	Ref.
pH optimum (γ-glu-p-NA)[a]	9·0 (H_2O as acceptor), 8·5 (glycylglycine as acceptor)	159–161
Temperature optimum	45°C	159
Molecular weight		
Native	80 000 (after papain treatment)	
Subunits	55 000 and 25 000 (glycopeptides)	21, 160, 161
Isoelectric point (pH)	3·85	160
Inhibitors[b]		
Metals	Cu^{2+}, Fe^{3+}	159
Iodoacetamide	Strong effect	159
DFP	Strong effect	159
pCMB	No effect	159
EDTA	No effect	159
L-Serine/borate	Strong effect	159
2,3-Butanedione/borate	Strong effect	169

[a]Substrate γ-glutamyl-p-nitroanilide.
[b]DFP, diisopropyl fluorophosphate; pCMB, p-chloromercuribenzoate.

Recently, research has been directed towards determining the co-identity or otherwise of this enzyme with sulphydryl oxidase.[163–165] Results at present are conflicting; some workers support the view that these enzyme activities were due to separate active sites on a single protein,[164] while others have shown that sulphydryl oxidase activity is completely independent of γ-glutamyl transpeptidase.[165] The latter opinion has been further substantiated by the separation of γ-glutamyl transpeptidase activity from sulphydryl oxidase using covalent chromatography on cysteinylsuccinamidopropyl-glass columns.[166]

γ-Glutamyl transpeptidase levels are high in colostrum, and the activity in milk falls considerably during the first 2 weeks of lactation.[157,158] Earlier reports suggested that this enzyme is probably identical to the free secretory component of IgA. However, it was subsequently shown that the enzyme could be completely separated from the secretory component and other immunoglobulins by such procedures as ultracentrifugation and gel filtration.[167] High levels of this enzyme have also been found in human milk, and it has been suggested that it may be important for early metabolic requirements of the new-born.[168]

6. SULPHYDRYL OXIDASE

Kiermeier and Petz[170,171] first reported the presence of an enzyme in milk which oxidized sulphydryl groups to form disulphides. In model systems it was shown that the addition of a partially purified enzyme solution to heated skim milk considerably reduced its SH content, and its use in the elimination of cooked flavours in milk was suggested.

The purification and properties of this enzyme have been studied,[172–174] and a summary of its properties is shown in Table 10. The

TABLE 10

PROPERTIES OF BOVINE MILK SULPHYDRYL OXIDASE

Property	Comments	Ref.
pH optimum	7·0–8·0	172, 173
Temperature optimum	45–50°C	172
Heat stability	60% survives HTST pasteurization	22
Molecular weight		
Native	> 1 000 000 (large aggregates)	173
Subunit	89 000 (glycopeptide)	173
Inhibitors		
EDTA	Strong inhibition	172
Cyanide	Strong inhibition	172
Ascorbate	Strong inhibition	172
Metal ion requirement	Fe^{2+}, Cu^{2+}	22, 173
Substrates	Reduced glutathione, L-cysteine, D-cysteine, and reduced and denatured proteins	22

enzyme prefers reduced glutathione as a substrate which, in the presence of oxygen, is converted into oxidized glutathione and hydrogen peroxide. It also can oxidize and reactivate reduced ribonuclease, which suggests that in cells it may function in the formation of the correct tertiary structure of proteins.[173] Its action on milk xanthine oxidase has been studied and it is able to convert the 'D' form (dehydrogenase) into the 'O' form (oxidase).[175] Its role in the formation of three-dimensional structures in proteins has recently been reviewed.[176]

Sulphydryl oxidase is of practical importance to the dairy industry mainly because of its potential use as a 'cooked' or 'burnt' flavour modifier in UHT milk.[22] A pilot scale immobilized enzyme reactor has been developed and conditions have been determined for effective regeneration of the enzyme, thus allowing for continuous use of the system.[22] Also, it has been suggested that it may be important with regard to flavour and oxidative stability of milk and dairy products.[176,177]

As sulphydryl oxidase survives pasteurization, heat-induced cooked flavours, produced as a result of SH group formation, are removed upon storage of milk. This enzymic oxidation of SH groups does not produce strong oxidative intermediates such as hydroxyl radicals and superoxide. This means that less lipid oxidation would occur in this case, in contrast to non-enzymic autoxidation of SH groups in which these oxidative intermediates are formed.[177] A mechanism of action for the enzyme has been proposed, and the enhancement of the reaction rate by peroxidase has been suggested to involve the direct transfer of an enzyme-bound hydroperoxy group to the active site of the peroxidase. The possibility, therefore, of lactoperoxidase interacting with sulphydryl oxidase in natural milk systems has been raised,[177] but the importance and likely implications of this have yet to be determined.

7. SUPEROXIDE DISMUTASE (EC 1.15.1.1)

The properties of bovine milk superoxide dismutase (SOD) and the consequences of its action in milk have been reviewed.[24] This enzyme is a common constituent of all microbial, plant and animal cells; it was originally known as erythrocuprein, and it was not until McCord and Fridovich[178] discovered the biological function of this protein that it was renamed superoxide dismutase because of its ability to convert superoxide anions into oxygen and hydrogen peroxide.

Superoxide anion (O_2^-) is produced by a number of biological systems that occur naturally in milk (e.g. xanthine oxidase,[179] lactoperoxidase[180]). The generation of superoxide is thought to lead to the formation of other oxidative species such as singlet oxygen and hydroxyl radicals. These species can cause lipid peroxidation,[181] destruction of biological membranes and degradation of macromolecules.[182]

Hill[23] first reported the presence of this enzyme in milk and found that it has the same molecular weight and similar chromatographic and electrophoretic properties to the enzyme from erythrocytes. Asada[183] and Korycka-Dahl et al.,[182] who also demonstrated the presence of SOD in bovine milk, confirmed these observations. The enzyme is a Cu^{2+}/Zn^{2+} metalloprotein, located wholly in the skim milk fraction of whole milk.[24,182] Its activity is unaffected by pasteurization and only about 50% inhibition occurs following 10 min at 76°C.[24,182,184]

Factors such as age of the cow, stage of lactation, and mastitis affect SOD levels in milk.[185] Only when milks have very high somatic cell counts (e.g. 10 000 000 cells ml^{-1}) does the level of SOD increase. The relationship between oxidation (TBA values) and SOD levels in milks has been examined but conflicting results have been obtained.[24,181] The discrepancies probably centre around the type of model system employed in the studies, and the absolute level of superoxide initially present in this system. Many factors, such as exposure to light and competing enzymic reactions involved in generating and destroying oxidative species, all appear to play important roles in determining the final oxidative stability of milk. It has been suggested that manipulating SOD levels in milk through slight variations in pasteurizing conditions may offer real possibilities for better control of oxidative problems in milk.[24] Marked improvements in the oxidative stability of milk high in linoleic acid were achieved by adding low levels of SOD.[186]

8. XANTHINE OXIDASE (EC 1.2.3.2)

Xanthine oxidase is a rather non-specific enzyme which catalyses the oxidation of several different compounds such as aldehydes, xanthine, hypoxanthine and NADH.[187,188] It occurs in milk in large quantities,[189] both in a soluble and a membrane-bound form.[190] It has been implicated in oxidized flavour problems in milk and dairy products by virtue of its production of superoxide during its oxidation of substrates.[181] However, no lipid peroxidation was observed in whole MFGM preparations containing high levels of XO in the presence of added hypoxanthine.[191] It may also be important in antimicrobial processes in milk owing to its ability to produce H_2O_2 and superoxide; the H_2O_2 produced could be used in the lactoperoxidase system.[192] More recently it has been studied because of the suggestion that it could be involved in initiation of atherosclerosis in humans who consume homogenized milk.[193] With regard to this proposal, extensive surveys of the levels of this enzyme in milk and a range of dairy products have been carried out.[194,195] The evidence for the involvement of xanthine oxidase in human heart disease has been examined, and it was concluded that the basis of the hypothesis is not well founded, and that more work needs to be carried out to investigate the relationship between this enzyme and coronary heart disease.[196]

Xanthine oxidase has been purified from milk by a variety of procedures.[197-203] The first step involves the release of the bound enzyme from membrane material. Protease treatment, solvent extraction procedures and detergent treatment have then been employed. More recently, procedures using techniques to inhibit intrinsic proteases[199,203] in the MFGM, followed by solvent extraction, electrofocusing and affinity chromatography, have been used. These and other purified preparations have indicated that the A_{280}/A_{450} ratio is in the range 4·35–5·4, that the activity/flavin ratio is 117–188, and that the specific activity against xanthine as substrate is 3–5 IU/mg protein.[203]

Xanthine oxidase purified from milk fat globule membrane by butan-1-ol extraction and ammonium sulphate precipitation can be separated into at least five distinct variants by electrofucusing techniques. These vary in isoelectric point from pH 6·9 to 7·6. However, if milk fat globule membrane material was only extracted with Triton X-100 prior to electrofocusing, some of the enzyme also focused at more acidic pH values (~4). These 'acidic' forms could be converted into forms with isoelectric points in the range pH 4·7–6·4 by treatment with neuraminidase. The isoelectric point of the major 'basic' form (pI ~7·2) was unchanged following neuraminidase treatment. Further electrophoretic analysis using both periodate/Schiff and Coomassie blue staining showed that xanthine oxidase itself is not a sialoglycoprotein, and the most likely explanation for the shift in the isoelectric point of the 'acidic' forms following neuraminidase treatment is that some of the xanthine oxidase in Triton X-100-solubilized extracts of MFGM exists as aggregates with a group of sialoglycoproteins which are immunologically unrelated to xanthine oxidase.[203]

A summary of the properties of xanthine oxidase is shown in Table 11. It is an important constituent of the MFGM and it has been estimated to account for up to 10% of the MFGM protein.[190,200] Mather et al.,[204] using preparative electrofocusing techniques, also showed that xanthine oxidase is the main component of MFGM polypeptide 3 (nomenclature of Mather and Keenan),[162] having a subunit molecular weight of 155 000.

Xanthine oxidase is not destroyed by pasteurization (72°C/15 s), and temperatures of around 77°C for up to 15 min are required to inactivate it fully. However, treatments that change the association of the enzyme

TABLE 11

PROPERTIES OF BOVINE MILK XANTHINE OXIDASE

Property	Comments	Ref.
pH optimum	8·0 — 9·0	200, 211, 214
Metal ion components	Mo, Fe	215, 216
Other cofactors	FAD	215, 216
Isoelectric point	6·9–7·6	203
Activity/flavin ratio	117–118	203
A_{280}/A_{450} ratio	4·8–5·4	199, 200, 203
Molecular weight		
Native	275 000–300 000	210, 216
Subunit	155 000	199, 200, 203
Substrates	Xanthine, hypoxanthine, aldehydes, $NADH_2$	1, 2
Acceptors	O_2, cytochrome c, NAD, ferricyanide	1, 2

with the MFGM (i.e. homogenization, cooling, heating, detergents) apparently activate the enzyme and subsequently change its heat stability.[195,205,206] Thus, the average heat inactivation achieved by processing pasteurized/homogenized milk is between 69% and 82%.[194,195]

Storage of whole milk at 5°C for 24 h results in an increase in the xanthine oxidase activity by 60%, with both membrane-bound and free enzyme showing increases.[190,207] It was suggested that the increase in activity may be caused by undetermined structural changes in the MFGM as a result of cold storage with a change in the accessibility of the substrate to the enzyme.

Xanthine oxidase can be converted into a NAD^+-dependent dehydrogenase by treatment with dithioerythritol or dihydrolipoic acid.[208] The NAD^+-dependent form has been purified from milk by pretreatment of the milk with dithioerythritol. About 94% of the purified enzyme was in this form but it reverted to the oxidase form if stored aerobically at 38°C. This conversion could be partially arrested by storage at 2°C aerobically or at 38°C anaerobically.[209] Also, the conversion from dehydrogenase into the oxidase form can be accomplished by sulphydryl oxidase or proteolytic enzymes.[210] Its association with MFGM also alters the ratio of its activity towards xanthine and NADH.[211]

Xanthine oxidase has been located in the mammary gland using immunofluorescence procedures; strong staining occurs in the apical regions of the cell.[191,212] Monoclonal antibodies to bovine milk XO have been obtained[213] and should provide more material for further immunological studies aimed at gaining a better understanding of the location and function of this enzyme in tissue cells.

9. LACTOPEROXIDASE (EC 1.11.1.7)

Lactoperoxidase catalyses the transfer of oxygen from hydrogen peroxide to other substrates such as thiocyanate. It is a glycoprotein containing a haem-iron prosthetic group which confers an additional non-enzymic activity upon this enzyme. Thus it has the ability to oxidize unsaturated fatty acids to form volatile products capable of contributing to oxidized flavours in dairy products.[217] Lactoperoxidase in the presence of the H_2O_2 and halide ion initiates lipid peroxidation and subsequently decolourizes β-carotene in model systems.[218–220] The functioning of this potentially detrimental system in milk could be of considerable interest considering the many practical uses suggested for this enzyme system.[221]

A more important physiological and technological function of lactoperoxidase is its ability, in the presence of H_2O_2 and thiocyanate (SCN^-), to inhibit the growth of both Gram-negative and Gram-positive bacteria.[221,222] Thus its activation in raw or heated milk by addition of sufficient thiocyanate and a H_2O_2-generating system such as glucose/glucose oxidase[223] or hypoxanthine/xanthine oxidase[192] results in a decrease in the total bacterial count which is followed by a period of bacteriostatis. The length of this inhibition is inversely related to the incubation temperature.[222] Practical use of this system has been demonstrated in the prevention of spoilage of milk when proper cooling cannot be achieved,[224] and in the destruction of *E. coli* in the abomasum of calves.[225] These antimicrobial aspects of the enzyme are discussed in more detail in Chapter 10 of this volume.

Lactoperoxidase has been purified from milk by many workers, and its properties have been studied in detail.[9,226–228] Hydrophobic chromatography has recently been used to obtain a purified preparation in high yield (10·5 mg/litre milk)[229] and an A_{412}/A_{280} ratio of 0·7 which is close to the accepted value of 0·9. The molecular weight of the enzyme obtained from sedimentation studies, iron content and amino acid composition is ~78 000.[228,230] It has been suggested that it consists of two almost identical subunits and that the *N*-terminal amino acid of one of the chains is blocked while that of the other is leucine. However, more recent studies have shown that lactoperoxidase consists of only one polypeptide chain whose *N*-terminal amino acid is indeed leucine.[231]

10. CATALASE (EC 1.11.1.6)

The level of catalase in bovine milk varies with the somatic cell count, and thus a measure of its activity in milk has been used to detect mastitis.[13,232] Catalase is found in both skim milk and cream, and it appears to be associated with membrane material in both of these fractions.[13] It can be partially released by detergent treatment of the membranes, and the free enzyme has a molecular weight of $\sim 250\,000$, i.e. similar to that of bovine liver catalase.[99,233,234] Ito[234] has purified catalase 18 000-fold from cream by a combination of churning, ammonium sulphate precipitation, butanol extraction, acetone precipitation and Sephadex G-200 chromatography. This partially purified preparation had a similar pH optimum and heat stability to those of bovine liver catalase but its K_m for H_2O_2 was double that of the liver catalase. The isoelectric point was pH 5·5 and heating for 1 h at 70°C was required to produce full inactivation. It is inhibited by heavy metals (Hg^{2+}, Fe^{2+}, Cu^{2+}), cyanide and nitrate. Catalase, being a haem-iron containing protein, may also be involved in lipid oxidation in a similar manner to that suggested for lactoperoxidase.[5,13] Further studies on the purification and properties of this enzyme have confirmed much of the earlier work.[235] In addition, the enzyme from bovine milk has now been crystallized.[235]

11. LYSOZYME (EC 3.2.1.17)

Lysozyme has been detected in the milk of many species;[2-4] bovine milk is one of the poorer sources of this enzyme,[236] while human and equine milks are extremely rich sources.[237,238] A figure of 79 mg/100 ml has been reported for equine milk,[238] while bovine milk has between 16 and 32 µg/100 ml, depending on the stage of lactation.[236]

Lysozyme has been purified from bovine milk by a combination of ion-exchange and gel-permeation chromatography, and the final purified protein was homogeneous by ultracentrifugation and electrophoretic analysis.[236] It has an isoelectric point of pH 9·5, which is in accordance with the known basic nature of the enzyme. The pH optimum for bovine milk lysozyme is 7·9 which is higher than that found for lysozymes from other sources. The molecular weight of bovine milk lysozyme has been reported to be 18 000, compared with a value of 15 000 for both human milk and egg white lysozymes.[239] Also, the amino acid composition of bovine milk lysozyme appears to be quite distinct from that found for other lysozymes.[240] The enzyme shows a pH-dependent temperature sensitivity with greater stability at lower pH values.[238,240] Reduction of bovine milk lysozyme with mercaptoethanol followed by re-oxidation results in a 3-fold increase in the specific activity. This has been explained as being the result of the formation of disulphide linkages different from those originally present, or to an altered tertiary structure due to different amino acid interactions.[241] The physiological role of lysozyme lies in its function as a bacteriocidal agent. This may be the case in human and equine milk but the extremely low levels in bovine milk probably have little, if any, effect on bacterial numbers or keeping quality of milk.

12. RIBONUCLEASE (EC 3.1.4.22)

Ribonuclease activity has been reported to be present in the milk of a variety of mammals.[242] It has been purified from both bovine and human milks.[243,244] Bovine milk ribonuclease was resolved into two components, A and B, by ion-exchange chromatography. The A form represents about 70% of the total protein and activity recovered, and appears to be identical to bovine pancreatic ribonuclease A by its similar electrophoretic and chromatographic properties and by immunological studies.[243] Similar results were found for human milk ribonuclease, in which case the two isoenzymes represent glycosylated (minor component) and unglycosylated forms of the enzyme.[244] The bovine milk enzyme has been crystallized.[245]

13. GLYCOSYLTRANSFERASES

Bovine milk and milks from some other mammals possess a wide range of sugar transferases which transfer the carbohydrate moiety of nucleoside diphosphate sugars to another carbohydrate acceptor which is either free or a part of a glycoprotein or glycolipid. The best known and best characterized of these is the galactosyltransferase (EC 2.4.1.22) which transfers galactose from UDP-galactose to a glycoprotein-bound N-acetylglucosamine or to free N-acetylglycosamine to form N-acetyl-lactosamine. The enzyme is physiologically important in the mammary gland because, upon interaction with α-lactalbumin, a change in its kinetic properties and specificity occurs. The presence of α-lactalbumin reduces its apparent K_m for glucose from 2M to 5 mM, resulting in the preferential transfer of the galactose moiety to glucose (instead of to N-acetylglucosamine), with the resultant formation of lactose.

The isolation of the enzyme from both human and bovine milks has been reported by a number of workers; the procedure usually involves a specific affinity chromatography step using covalently bound α-lactalbumin[246] or N-acetylglucosamine or UDP.[247]

The kinetic mechanism of the enzyme has been intensively studied, and it has been suggested that the mechanism involves an ordered sequential addition of the reactants, Mn^{2+}, UDP-galactose and N-acetylglucosamine, and an ordered release of the products N-acetyl-lactosamine and Mn^{2+}-UDP.[248,249] In lactose biosynthesis, α-lactalbumin binds after Mn^{2+} and UDP-galactose. Glucose is the last substrate to bind, and then lactose, α-lactalbumin and Mn^{2+}-UDP are sequentially released.[248,249] This mechanism holds for the fully intact enzyme species of apparent molecular weight 50 000 which occurs in bovine colostrum, but there are subtle changes in the mechanism if a partially degraded form of molecular weight 41 000 is used. Proteolytic attack on the colostral enzyme converts it into a form that has a more symmetrical shape and in which the nature of the binding of UDP-galactose to the enzyme–Mn^{2+} complex is altered.[250,251] Other workers have questioned the above ordered addition mechanism and suggest a random equilibrium addition of substrates and α-lactalbumin to the Mn^{2+}–enzyme complex.[252]

Galactosyltransferase has two metal-binding sites: site I is involved in maintaining the structural integrity of the protein while site II is associated with the UDP-galactose binding site.[253–255] Furthermore, chemical modification of the enzyme with reagents reacting with free sulphydryl groups has demonstrated the presence of an SH group at or near the UDP-galactose binding site.[256–258] By immobilizing galactosyltransferase through one of its free SH groups to Sepharose beads (previously converted into a 5-nitropyridyl disulphide), it was demonstrated that some N-acetyl-lactosamine and lactose synthetase activity was retained. It was inferred that the SH group which was bound to the gel is in the region of the protein that is distal from the α-lactalbumin and substrate binding site.[259] The implications of these findings still need to be examined further in relation to the kinetic and physical studies which do not entirely support this view. Other studies using inhibitors directed towards tyrosine residues have shown the importance of this

TABLE 12

VARIOUS GLYCOSYLTRANSFERASES OCCURRING IN MILK

Enzyme	Donor	Acceptor	Ref.
Galactosyltransferase	UDP-galactose	*N*-Acetylglucosamine	14
Galactosyltransferase	UDP-galactose	Ceramides, gangliosides	261
Fucosyltransferase	GDP-fucose	*N*-Acetyl-lactosamine	262
Sialyltransferase	CMP-*N*-acetyl-neuraminic acid	*N*-Acetyl-lactosamine	263, 264
N-Acetylglucosamine transferase (I and II)	UDP-*N*-acetyl-glucosamine	Mannose or *N*-acetylglucosamine	265

residue in and around the substrate binding site of the enzyme.[260]

Table 12 lists the different types of sugar transferases that have been found in milk. Some of these have been purified and their properties studied in detail.

14. α-AMYLASE (EC 3.2.1.1)

α-Amylase has been detected in cow's milk,[266] and a measure of its residual activity was suggested as an index of pasteurization efficiency.[267] Human milk also contains α-amylase activity which probably contributes to the infant's ability to digest starch.[268] The bovine milk enzyme has been partially purified by batch-wise adsorption on rice starch followed by specific elution with $CaSO_4$ solution. This procedure essentially separates α-amylase from β-amylase. The α-amylase has a pH optimum of 7·4 at 34°C and requires both Ca^{2+} and Cl^- for activity. Heat treatment at 60°C for 30 min results in 80–90% reduction in activity.[266]

15. *N*-ACETYL-β-D-GLUCOSAMINIDASE (EC 3.2.1.30)

This enzyme has been detected in both bovine[15,269] and human milks.[270] Its significance in the dairy industry lies in its ability to be used as an indicator of mastitis.[15] The elevated activity in milk from mastitic animals has been suggested to be the result of tissue damage in the udder caused by the pathogenic bacteria.[271,272] Rapid and simple diagnostic tests have been developed for its routine use in mastitis monitoring programs. Bovine milk probably contains at least two pools of N-acetyl-β-D-glucosaminidase (NAGase), one from the leucocytes and the other from the mammary gland tissue cells. Mellors[269] has examined the properties of the enzyme from separator bowl sediment and which is more than likely the leucocyte enzyme. In normal and mastitic milks, leucocytes contribute only 10–15% of the total NAGase activity, while the enzyme originating from mammary gland tissue accounts for the major portion of this activity.[271] The properties of both NAGases are very similar with pH optima in the range 4·2–4·5 and similar K_m values (i.e. 1–1·25 mM) for p-nitrophenyl-N-acetyl-β-D-glucosaminide.

The major portion of the total NAGase in bovine milk occurs in the soluble whey fraction,[271] and this enzyme shows similar gel filtration and electrophoretic behaviour to that of mammary gland tissue NAGase (Kitchen, unpublished). Mammary gland NAGase has been purified by ion-exchange chromatography, gel filtration, affinity chromatography and preparative isoelectric focusing.[273] Two forms of activity, designated as A and B forms, were obtained which differed in charge properties, molecular weights and heat stability, the A form being more acidic, of lower molecular weight and less heat-stable. The native molecular weights of the A and B forms are 120 000 and 240 000, respectively. Each form has a major and a minor subunit of 55 000 and 25 000, respectively, after treatment with mercaptoethanol and sodium dodecyl sulphate.

Milk NAGase activity is high just following calving, but it rapidly falls to normal levels within 7–10 days if the udder is free of infection (Marschke and Kitchen, unpublished). Similar observations have been reported by Bertoni *et al.*[274]

16. OTHER ENZYMES

Many other enzymes have been reported to be present in milk. These include aldolase[13] (EC 4.1.2.13), lactate dehydrogenase[16,275] (EC 1.1.1.27), α-mannosidase[15,276] (EC 3.2.1.24), β-glucuronidase[15,277] (EC 3.2.1.31) and arylsulphatase[15] (EC 3.1.6.1). They have mainly been examined for their possible variation in milk as a result of mastitis. Little is known about their molecular properties, or their physiological and technological significance in milk.

REFERENCES

1. WHITNEY, R. McL., *J. Dairy Sci.*, 1958, **41**, 1303.
2. SHAHANI, K. M., *J. Dairy Sci.*, 1966, **49**, 907.
3. SHAHANI, K. M., HARPER, W. J., JENSEN, R. G., PARRY, R. M. and ZITTLE, C. A., *J. Dairy Sci.*, 1973, **56**, 531.
4. GROVES, M. L., In: *Milk Proteins: Chemistry and Molecular Biology*, H. A. McKenzie (ed.), Vol. II, 1971, Academic Press, New York, p. 367.
5. FOX, P. F. and MORRISSEY, P. A., In: *Enzymes and Food Processing*, G. G. Birch, N. Blakenbrough and K. J. Parker (eds), 1981, Applied Science Publishers, London, p. 213.
6. BLANC, B., *Lait*, 1982, **62**, 350.
7. MORTON, R. K., *Biochem. J.*, 1953, **55**, 795.
8. MORTON, R. K., *Biochem. J.*, 1953, **55**, 786.
9. POLIS, B. D. and SHMUKLER, H. W., *J. Biol. Chem.*, 1953, **201**, 475.
10. HERRINGTON, B. L., *J. Dairy Sci.*, 1954, **38**, 775.
11. BALL, E. G., *J. Biol. Chem.*, 1939, **128**, 51.
12. WARNER, R. C. and POLIS, E., *J. Am. Chem. Soc.*, 1945, **67**, 529.
13. KITCHEN, B. J., TAYLOR, G. C. and WHITE, I. C., *J. Dairy Res.*, 1970, **37**, 279.
14. EBNER, K. E. and SCHANBACHER, F. L., In: *Lactation: A Comprehensive Treatise*, B. L. Larson and V. R. Smith (eds), Vol. II, 1974, Academic Press, New York, p. 77.
15. KITCHEN, B. J., *J. Dairy Res.*, 1976, **43**, 251.
16. KITCHEN, B. J., *J. Dairy Res.*, 1981, **48**, 167.
17. SHAHANI, K. M., KWAN, A. J. and FRIEND, B. A., *Am. J. Clin. Nutr.*, 1980, **33**, 1861.
18. SUHREN, G., *IDF-Doc. 157*, 1983, p. 17.
19. DOWNEY, W. K., *IDF-Doc. 86*, 1975.
20. ASCHAFFENBURG, R. and MULLEN, J. E. C., *J. Dairy Res.*, 1949, **16**, 58.
21. BAUMRUCKER, C. R., *J. Dairy Sci.*, 1980, **63**, 49.
22. SWAISGOOD, H. E., *Enzyme Microb. Technol.*, 1980, **2**, 265.
23. HILL, R. D., *Aust. J. Dairy Technol.*, 1975, **30**, 26.
24. HICKS, C. L., *J. Dairy Sci.*, 1980, **63**, 1199.
25. AURAND, L. W., CHU, T. M., SINGLETON, J. A. and SHEN, R., *J. Dairy Sci.*, 1967, **50**, 465.
26. BJORCK, L., CLAESSON, O. and SCHULTHESS, W., *Milchwissenschaft*, 1979, **34**, 726.
27. HUMBERT, G. and ALAIS, C., *J. Dairy Res.*, 1979, **46**, 559.
28. VISSER, S., *Neth. Milk Dairy J.*, 1981, **35**, 65.
29. REIMERDES, E. H., In: *Developments in Dairy Chemistry*, P. F. Fox (ed.), Vol. 1 (Proteins), 1983, Applied Science Publishers, London, p. 271.
30. SNOEREN, T. H. M., VAN DER SPEK, C. A., DEKKER, R. and BOTH, P., *Neth. Milk Dairy J.*, 1979, **33**, 31.
31. NOOMEN, A., *Neth. Milk Dairy J.*, 1975, **29**, 153.
32. VISSER, F. M. W., *Neth. Milk Dairy J.*, 1977, **31**, 210.
33. KAMINOGAWA, S., MIZOBUCHI, H. and YAMAUCHI, K., *Agric. Biol. Chem.*, 1972, **36**, 2163.

34. HALPAAP, I., REIMERDES, E. H. and KLOSTERMEYER, H., *Milchwissenschaft*, 1977, **32**, 341.
35. EIGEL, W. N., HOFMANN, C. J., CHIBBER, B. A. K., TOMICH, J. M., KEENAN, T. W. and MERTZ, E. T., *Proc. Nat. Acad. Sci. USA*, 1979, **76**, 2244.
36. RICHARDSON, B. C. and PEARCE, K. N., *N.Z. J. Dairy Sci. Technol.*, 1981, **16**, 209.
37. KORYCKA-DAHL, M., DUMAS, B. R., CHENE, N. and MARTAL, J., *J. Dairy Sci.*, 1983, **66**, 704.
38. ASTRUP, T. and STERNDORFF, I., *Proc. Soc. Exp. Biol. Med.*, 1953, **84**, 605.
39. KIERMEIER, F. and SEMPER, G., *Z. Lebensm.-Forsch.*, 1960, **111**, 282.
40. REIMERDES, E. H., KLOSTERMEYER, H. and SAYK, E., *Milchwissenschaft*, 1976, **31**, 329.
41. ROLLEMA, H. S., VISSER, S. and POLL, J. K., *Milchwissenschaft*, 1983, **38**, 214.
42. PULVERER, G., WEGRZYNOWICZ, Z., KO, H. L. and JELJASZEWICZ, J., *Zbl. Bakt., Hyg., I. Abt. Orig.*, 1980, **A249**, 99.
43. LEYTUS, S. P., BOWLES, L. K., KONISKY, J. and MANGEL, W. F., *Proc. Nat. Acad. Sci. USA*, 1981, **78**, 1485.
44. NOOMEN, A., *Neth. Milk Dairy J.*, 1972, **29**, 153.
45. DULLEY, J. R., *J. Dairy Res.*, 1972, **39**, 1.
46. ROBBINS, K. C. and SUMMARIA, L., In: *The Enzymes*, G. E. Perlmann and L. Lorand (eds), Vol. 19, 1970, Academic Press, New York, p. 184.
47. EIGEL, W. N., *Int. J. Biochem.*, 1977, **8**, 187.
48. REIMERDES, E. H., PETERSEN, F. and KIELWEIN, G., *Milchwissenschaft*, 1979, **34**, 548.
49. ANDREWS, A. T., *Eur. J. Biochem.*, 1978, **90**, 59.
50. ANDREWS, A. T., *Eur. J. Biochem.*, 1978, **90**, 67.
51. EIGEL, W. N. and KEENAN, T. W., *Int. J. Biochem.*, 1979, **10**, 529.
52. EIGEL, W. N., *Int. J. Biochem.*, 1981, **13**, 1081.
53. ANDREWS, A. T., *J. Dairy Res.*, 1983, **50**, 45.
54. ANDREWS, A. T., *J. Dairy Res.*, 1983, **50**, 57.
55. SNOEREN, T. H. M. and VAN RIEL, J. A. M., *Milchwissenschaft*, 1979, **34**, 528.
56. AIMUTIS, W. R. and EIGEL, W. N., *J. Dairy Sci.*, 1982, **65**, 175.
57. HOFMANN, C. J., KEENAN, T. W. and EIGEL, W. N., *Int. J. Biochem.*, 1979, **10**, 909.
58. REIMERDES, E. H., HALPAAP, I. and KLOSTERMEYER, H., *Milchwissenschaft*, 1981, **36**, 73.
59. KAMINOGAWA, S. and YAMAUCHI, K., *Agric. Biol. Chem.*, 1972, **36**, 255.
60. KAMINOGAWA, S. and YAMAUCHI, K., *Agric. Biol. Chem.*, 1972, **36**, 2351.
61. KAMINOGAWA, S., YAMAUCHI, K., MIYAZAWA, S. and KOGA, Y., *J. Dairy Sci.*, 1980, **63**, 701.
62. MELLORS, A., *Can. J. Biochem.*, 1969, **47**, 173.
63. GRUNDIG, C. A. and HANSON, H., *Hoppe-Seyler's Z. Physiol. Chem.*, 1973, **354**, 487.
64. OLIVECRONA, T., EGELRUD, T., HERNELL, O., CASTBERG, H. and SOLBERG, P., *IDF-Doc. 86*, 1975.
65. DOWNEY, W. K. and MURPHY, R. F., *IDF-Doc. 86*, 1975.

66. JENSEN, R. G. and PITAS, R. E., *J. Dairy Sci.*, 1976, **59**, 1203.
67. OLIVECRONA, T., *IDF-Doc. 118*, 1980.
68. ANDERSON, M., *J. Soc. Dairy Technol.*, 1983, **36**, 3.
69. DEETH, H. C. and FITZ-GERALD, C. H., In: *Developments in Dairy Chemistry*, P. F. Fox (ed.), Vol. 2 (Lipids), 1983, Applied Science Publishers, London, p. 195.
70. TARASSUK, N. R. and FRANKEL, E. N., *J. Dairy Sci.*, 1957, **40**, 418.
71. DOWNEY, W. K. and ANDREWS, P., *Biochem. J.*, 1969, **112**, 559.
72. EGELRUD, T. and OLIVECRONA, T., *J. Biol. Chem.*, 1972, **247**, 6212.
73. IVERIUS, P. H. and OSTLUND-LINDQVIST, A.-M., *J. Biol. Chem.*, 1976, **251**, 7791.
74. OLIVECRONA, T., BENGTSSON, G. and OSBORNE, J. C., *Eur. J. Biochem.*, 1982, **124**, 629.
75. CHANDAN, R. C. and SHAHANI, K. M., *J. Dairy Sci.*, 1963, **46**, 275.
76. EGELRUD, T. and OLIVECRONA, T., *Biochim. Biophys. Acta*, 1973, **306**, 115.
77. SCOW, R. O. and EGELRUD, T., *Biochim. Biophys. Acta*, 1976, **431**, 538.
78. RAPP, D. and OLIVECRONA, T., *Eur. J. Biochem.*, 1978, **91**, 379.
79. BENGTSSON, G. and OLIVECRONA, T., *FEBS Lett.*, 1979, **106**, 345.
80. BENGTSSON, G. and OLIVECRONA, T., *Eur. J. Biochem.*, 1980, **106**, 549.
81. SHIRAI, K., WISNER, D. A., JOHNSON, J. D., SRIVASTAVA, L. S. and JACKSON, R. L., *Biochim. Biophys. Acta*, 1982, **712**, 10.
82. BENGTSSON, G. and OLIVECRONA, T., *FEBS Lett.*, 1982, **147**, 183.
83. VAINIO, P., VIRTANEN, J. A. and KINNUNEN, P. K. J., *Biochim. Biophys. Acta*, 1982, **711**, 386.
84. BENGTSSON, G. and OLIVECRONA, T., *Biochim. Biophys. Acta*, 1982, **712**, 196.
85. BENGTSSON, G. and OLIVECRONA, T., *Biochim. Biophys. Acta*, 1983, **751**, 254.
86. FORSTER, T., BENDIXEN, H. A. and MONTGOMERY, M. W., *J. Dairy Sci.*, 1959, **42**, 1903.
87. KITCHEN, B. J., *J. Dairy Res.*, 1971, **38**, 171.
88. NAKANISHI, T. and TAGATA, Y., *Jap. J. Dairy Sci.*, 1972, **21**, A207.
89. ARBABI, P., *20th Int. Dairy Congr.*, 1978, **E**, 319.
90. DEETH, H. C., *20th Int. Dairy Congr.*, 1978, **E**, 364.
91. PURR, A., *Nährung*, 1965, **9**, 445.
92. PURR, A., MATHIES, P. and KOTTER, L., *Dte. Lebensm. Rdsh.*, 1969, **65**, 105.
93. DOWNEY, W. K. and ANDREWS, P., *Biochem. J.*, 1965, **96**, 21.
94. HOLMES, R. S. and MASTERS, C. J., *Biochim. Biophys. Acta*, 1967, **132**, 239.
95. MARQUARDT, R. R. and FORSTER, T. L., *J. Dairy Sci.*, 1965, **48**, 1526.
96. FORSTER, T. L., MONTGOMERY, M. W. and MONTOURE, J. E., *J. Dairy Sci.*, 1961, **44**, 1420.
97. KITCHEN, B. J., MIDDLETON, G., DURWARD, I. G., ANDREWS, R. J. and SALMON, M. C., *J. Dairy Sci.*, 1980, **63**, 978.
98. KITCHEN, B. J., MASTERS, C. J. and WINZOR, D. J., *Biochem. J.*, 1973, **135**, 93.
99. KITCHEN, B. J., MSc Thesis, University of Queensland, 1971.
100. SUNDHEIM, G., ZIMMER, T. L. and ASTRUP, H. N., *J. Dairy Sci.*, 1983, **66**, 400.

101. VAINIO, P., VIRTANEN, J. A., SPARROW, J. T., GOTTO, A. M. and KINNUNEN, P. K. J., *Chem. Phys. Lipids*, 1983, **33**, 21.
102. PEEREBOOM, J. W. C., *Ernährungsindustrie*, 1970, **72**, 299.
103. PEEREBOOM, J. W. C., *Milchwissenschaft*, 1969, **24**, 266.
104. STANNARD, D. J., *J. Dairy Res.*, 1975, **42**, 241.
105. MORTON, R. K., *Biochem. J.*, 1955, **61**, 232.
106. ZITTLE, C. A. and BINGHAM, E. A., *J. Dairy Sci.*, 1959, **42**, 1772.
107. LORRIENT, D. and LINDEN, G., *J. Dairy Res.*, 1976, **43**, 29.
108. LINDEN, G. and ALAIS, C., *Biochim. Biophys. Acta*, 1976, **429**, 205.
109. FARREL, H. M. and THOMPSON, M. P., *J. Dairy Sci.*, 1971, **54**, 1219.
110. ZITTLE, C. A. and DELLA MONICA, E. S., *Arch. Biochem. Biophys.*, 1952, **35**, 321.
111. LYSTER, R. L. J. and ASCHAFFENBURG, R., *J. Dairy Res.*, 1962, **29**, 21.
112. ANDREWS, P., *Biochem. J.*, 1965, **96**, 595.
113. MORTON, R. K., In: *Methods in Enzymology*, S. P. Colowick and N. O. Kaplan (eds), Vol. II, 1955, Academic Press, New York, p. 533.
114. SHARMA, R. S. and GANGULI, N. C., *Indian J. Dairy Sci.*, 1969, **22**, 272.
115. PEEREBOOM, J. W. C., *Neth. Milk Dairy J.*, 1968, **22**, 137.
116. BARMAN, T. E. and GUTFREUND, H., *Biochem. J.*, 1966, **101**, 460.
117. LINDEN, G., MAZERON, P., MICHALOWSKI, J. B. and ALAIS, C., *Biochim. Biophys. Acta*, 1974, **358**, 82.
118. BURVIANA, L. M. and DEMA, A., *Rev. Roum. Biochim.*, 1969, **6**, 275.
119. O'KEEFE, R. B. and KINSELLA, J. E., *Int. J. Biochem.*, 1979, **10**, 125.
120. KUZUYA, Y., KANAMARU, Y. and TANAHASHI, T., *Jap. J. Zootech. Sci.*, 1980, **51**, 501.
121. KUZUYA, Y., KANAMARU, Y. and TANAHASHI, T., *Jap. J. Zootech. Sci.*, 1982, **53**, 45.
122. LINDEN, G. and MICHALOWSKI, J. B., *Milchwissenschaft*, 1976, **31**, 78.
123. DIAZ-MAURINO, T. and NIETO, M., *Biochim. Biophys. Acta*, 1976, **448**, 234.
124. LINDEN, G., CHAPPELET-TORDO, D. and LAZDUNSKI, M., *Biochim. Biophys. Acta*, 1977, **483**, 100.
125. KANNAN, A. and BASU, K. P., *Indian J. Dairy Sci.*, 1949, **2**, 51.
126. BINGHAM, E. W. and ZITTLE, C. A., *Arch. Biochem. Biophys.*, 1963, **101**, 471.
127. KITCHEN, B. J., *Biochim. Biophys. Acta*, 1974, **356**, 257.
128. MCPHERSON, A. V. and KITCHEN, B. J., *21st Int. Dairy Congr.*, 1982, **1** (book 2), 222.
129. ANDREWS, A. T. and PALLAVICINI, C., *Biochim. Biophys. Acta*, 1973, **321**, 197.
130. BINGHAM, E. A., FARRELL, H. M. and CARROLL, R. J., *Biochemistry*, 1972, **11**, 2450.
131. KNOOP, A. M. and PETERS, R. H., *Milchwissenschaft*, 1975, **30**, 680.
132. DULLEY, J. R. and KITCHEN, B. J., *Aust. J. Dairy Technol.*, 1973, **28**, 114.
133. ANDREWS, A. T. and ALICHANIDIS, E., *J. Dairy Res.*, 1975, **42**, 327.
134. SCHORMULLER, J., LEHMANN, K. and BELITZ, H. D., *Z. Lebensm.-Forsch.*, 1960, **111**, 180.
135. MULLEN, J. E. C., *J. Dairy Res.*, 1950, **17**, 288.
136. BINGHAM, E. W., JASEWICZ, L. and ZITTLE, C. A., *J. Dairy Sci.*, 1961, **44**, 1247.

137. KUZUYA, Y., *Res. Bull. Fac. Agr. Gifu Univ.*, 1976, **39**, 263.
138. ANDREWS, A. T., *Biochim. Biophys. Acta*, 1976, **434**, 345.
139. KUZUYA, Y., WAGNER, F. W., KILARA, A. and SHAHANI, K. M., *Milchwissenschaft*, 1976, **31**, 358.
140. KUZUYA, Y., *Res. Bull. Fac. Agr. Gifu Univ.*, 1976, **39**, 269.
141. ANDREWS, A. T. and ALICHANIDIS, E., *J. Dairy Res.*, 1975, **42**, 391.
142. ANDERSON, M., BROOKER, B. E., ANDREWS, A. T. and ALICHANIDIS, E., *J. Dairy Res.*, 1975, **42**, 401.
143. ANDREWS, A. T., *J. Dairy Res.*, 1976, **43**, 127.
144. HUANG, C. M. and KEENAN, T. W., *Biochim. Biophys. Acta*, 1972, **274**, 246.
145. CAULINI, G., BINOTTI, I., ALESSANDRINI, M., IPATA, P. L. and MAGNI, G., *Soc. Ital. Biol. Sperim.*, 1972, **48**, 890.
146. DOSS, R. C., CAROTHERS CARRAWAY, C. A. and CARRAWAY, K. L., *Biochim. Biophys. Acta*, 1979, **570**, 96.
147. BHAVADASAN, M. K. and GANGULI, N. C., *Indian J. Biochem. Biophys.*, 1976, **13**, 255.
148. SNOW, L. D., DOSS, R. C. and CARRAWAY, K. L., *Biochim. Biophys. Acta*, 1980, **611**, 333.
149. McPHERSON, A. V. and KITCHEN, B. J., *J. Dairy Res.*, 1983, **50**, 107.
150. HUANG, C. M. and KEENAN, T. W., *Comp. Biochem. Physiol.*, 1972, **43**, 277.
151. GILL, D. S., AHUJA, S. P. and BHATIA, I. S., *Indian J. Biochem. Biophys.*, 1980, **17**, 282.
152. LINZELL, J. L. and PEAKER, M., *Physiol. Rev.*, 1971, **51**, 564.
153. DOWBEN, R. M., BRUNNER, J. R. and PHILPOTT, D. E., *Biochim. Biophys. Acta*, 1967, **135**, 1.
154. MATSUSHITA, S., IBUKI, F., MORI, T. and HATA, T., *Agric. Biol. Chem.*, 1965, **29**, 436.
155. BAUMRUCKER, C. R., POCIUS, P. A. and RISS, T. L., *J. Cell Biol.*, 1980, **87**, 297.
156. MAJUMDER, G. C. and GANGULI, N. C., *Milchwissenschaft*, 1971, **26**, 668.
157. MUJUMDER, G. C. and GANGULI, N. C., *Milchwissenschaft*, 1972, **27**, 296.
158. MUJUMDER, G. C. and GANGULI, N. C., *Milchwissenschaft*, 1972, **27**, 486.
159. SOBIECH, K. A., ZIOMEK, E. and SZEWCZUK, A., *Arch. Immun. Therap. Exper.*, 1974, **22**, 645.
160. YASUMOTO, K., IWAMI, K., FUSHIKI, T. and MITSUDA, H., *J. Biochem.*, 1978, **84**, 1227.
161. BAUMRUCKER, C. R., *J. Dairy Sci.*, 1979, **62**, 253.
162. MATHER, I. H. and KEENAN, T. W., *J. Membr. Biol.*, 1975, **21**, 65.
163. TATE, S. S. and ORLANDO, J., *J. Biol. Chem.*, 1979, **254**, 5573.
164. IWAMI, K., FUSHIKI, T., YASUMOTO, K. and IWAI, K., *Agric. Biol. Chem.*, 1981, **45**, 307.
165. SLIWKOWSKI, M. B., SLIWKOWSKI, M. X., SWAISGOOD, H. E. and HORTON, H. R., *Arch. Biochem. Biophys.*, 1981, **211**, 731.
166. SCHMELZER, C. H., SWAISGOOD, H. E. and HORTON, H. R., *Biochem. Biophys. Res. Comm.*, 1982, **107**, 196.
167. YASUMOTO, K., IWAMI, K., FUSHIKI, T. and MITSUDA, H., *FEBS Lett.*, 1976, **67**, 328.
168. PATIL, K. P. and RANGNEKAR, N. R., *Clin. Chem.*, 1982, **28**, 1724.

169. FUSHIKI, T., IWAMI, K., YASUMOTO, K. and IWAI, K., *J. Biochem.*, 1983, **93**, 795.

170. KIERMEIER, F. and PETZ, E., *Z. Lebensm.-Forsch.*, 1967, **132**, 342.

171. KIERMEIER, F. and PETZ, E., *Z. Lebensm.-Forsch.*, 1967, **134**, 149.

172. KIERMEIER, F. and RANFFT, K., *Z. Lebensm.-Forsch.*, 1970, **143**, 11.

173. JANOLINO, V. G. and SWAISGOOD, H. E., *J. Biol. Chem.*, 1975, **250**, 2532.

174. JANOLINO, V. G., CLARE, D. A. and SWAISGOOD, H. E., *Biochim. Biophys. Acta*, 1981, **658**, 406.

175. CLARE, D. A., BLAKISTONE, B. A., SWAISGOOD, H. E. and HORTON, H. R., *Arch. Biochem. Biophys.*, 1981, **211**, 44.

176. SWAISGOOD, H. E. and HORTON, H. R., *CIBA Foundation Symposium 72: Sulphur in Biology*, 1980, p. 205.

177. SWAISGOOD, H. E. and ABRAHAM, P., *J. Dairy Sci.*, 1980, **63**, 1205.

178. McCORD, J. M. and FRIDOVICH, I., *J. Biol. Chem.*, 1969, **244**, 6049.

179. FRIDOVICH, I., In: *Free Radicles in Biology*, W. A. Prior (ed.), Vol. I, 1976, Academic Press, New York, p. 238.

180. HILL, R. D., *N.Z. J. Dairy Sci. Technol.*, 1977, **12**, 37.

181. AURAND, L. W., BOONE, N. H. and GIDDINGS, G. G., *J. Dairy Sci.*, 1977, **60**, 363.

182. KORYCKA-DAHL, M., RICHARDSON, T. and HICKS, C. L., *J. Food Prot.*, 1979, **42**, 867.

183. ASADA, K., *Agric. Biol. Chem.*, 1976, **40**, 1659.

184. HICKS, C. L., BUCY, J. and STOFER, W., *J. Dairy Sci.*, 1979, **62**, 529.

185. HOLBROOK, J. and HICKS, C. L., *J. Dairy Sci.*, 1978, **61**, 1072.

186. HILL, R. D., VAN LEEUWEN, V. and WILKINSON, R. A., *N.Z. J. Dairy Sci. Technol.*, 1977, **12**, 69.

187. ZITTLE, C. A., *J. Dairy Sci.*, 1964, **47**, 202.

188. BRAY, R. C., In: *The Enzymes*, P. D. Boyer (ed.), Vol. 12, 1975, Academic Press, New York, p. 299.

189. CORRAN, H. S., DEWAN, J. G., GORDON, A. H. and GREEN, D. E., *Biochem. J.*, 1939, **33**, 1694.

190. BRILEY, M. S. and EISENTHAL, R., *Biochem. J.*, 1974, **143**, 149.

191. BRUDER, G., HEID, H., JARASCH, E-D., KEENAN, T. W. and MATHER, I. H., *Biochim. Biophys. Acta*, 1982, **701**, 357.

192. BJORCK, L. and CLAESSON, O., *J. Dairy Sci.*, 1979, **62**, 1211.

193. OSTER, K. A., *Am. J. Clin. Res.*, 1971, **2**, 30.

194. CERBULIS, J. and FARRELL, H. M., *J. Dairy Sci.*, 1977, **60**, 170.

195. ZIKAKIS, J. P. and WOOTERS, S. C., *J. Dairy Sci.*, 1980, **63**, 893.

196. DEETH, H. C., *J. Dairy Sci.*, 1983, **66**, 1419.

197. GILBERT, D. A. and BERGEL, F., *Biochem. J.*, 1964, **90**, 350.

198. EDMONDSON, D., MASSEY, V., PALMER, G., BEACHAM, L. M. and ELION, G. B., *J. Biol. Chem.*, 1972, **247**, 1597.

199. NATHANS, G. R. and HADE, E. P. K., *Biochem. Biophys. Res. Comm.*, 1975, **66**, 108.

200. MANGINO, M. E. and BRUNNER, J. R., *J. Dairy Sci.*, 1977, **60**, 841.

201. NATHANS, G. R. and HADE, E. P. K., *Biochim. Biophys. Acta*, 1978, **526**, 328.

202. NISHINO, T., NISHINO, T. and TSUSHIMA, K., *FEBS Lett.*, 1981, **131**, 369.

203. SULLIVAN, C. H., MATHER, I. H., GREENWALT, D. E. and MADARA, P. J., *Mol. Cell Biochem.*, 1982, **44**, 13.
204. MATHER, I. H., TAMPLIN, C. B. and IRVING, M. G., *Eur. J. Biochem.*, 1980, **110**, 327.
205. ROBERTS, I. and POLONOVSKI, J., *Discuss. Faraday Soc.*, 1955, **25**, 54.
206. GREENBANK, G. R. and PALLANSCH, M. J., *J. Dairy Sci.*, 1962, **45**, 958.
207. BHAVADASAN, M. K. and GANGULI, N. C., *J. Dairy Sci.*, 1980, **63**, 362.
208. BATELLI, M. G., LORENZONI, E. and STIRPE, F., *Biochem. J.*, 1973, **131**, 191.
209. NAKAMURA, M. and YAMAZAKI, I., *J. Biochem.*, 1982, **92**, 1279.
210. WAUD, W. R., BRADY, F. O., WILEY, R. D. and RAJAGOPALAN, K. V., *Arch. Biochem. Biophys.*, 1975, **169**, 695.
211. BRILEY, M. S. and EISENTHAL, R., *Biochem. J.*, 1975, **147**, 417.
212. JARASCH, E-D., GRUND, C., BRUDER, G., HEID, H. W., KEENAN, T. W. and FRANKE, W. W., *Cell*, 1981, **25**, 67.
213. MATHER, I. H., NACE, C. S., JOHNSON, V. G. and GOLDSBY, R. A., *Biochem. J.*, 1980, **188**, 925.
214. PALMER, G., BRAY, R. C. and BEINERT, H., *J. Biol. Chem.*, 1964, **239**, 2657.
215. AVIS, P. G., BERGEL, F. and BRAY, R. C., *J. Chem. Soc.*, 1955, 1100.
216. HART, L. I., MCGARTOLL, M. A., CHAPMAN, H. R. and BRAY, R. C., *Biochem. J.*, 1970, **116**, 851.
217. ERIKSSON, C. E., *J. Dairy Sci.*, 1969, **53**, 1649.
218. KANNER, J. and KINSELLA, J. E., *Lipids*, 1983, **18**, 198.
219. KANNER, J. and KINSELLA, J. E., *Lipids*, 1983, **18**, 204.
220. KANNER, J. and KINSELLA, J. E., *J. Agric. Food Chem.*, 1983, **31**, 370.
221. BJORCK, L., ROSEN, C-G., MARSHALL, V. and REITER, B., *Appl. Microbiol.*, 1975, **30**, 199.
222. BJORCK, L., *IDF Symposium on Bacteriological Quality of Raw Milk*, 1981, Part II, p. 5.
223. REITER, B. and HARNULV, B. G., *IDF Symposium on Bacteriological Quality of Raw Milk*, 1981, Part II, p. 50.
224. HARNULV, B. G. and KANDASAMY, C., *IDF Symposium on Bacteriological Quality of Raw Milk*, 1981, Part II, p. 47.
225. REITER, B., MARSHALL, V. M. and PHILIPS, S. M., *Res. Vet. Sci.*, 1980, **28**, 116.
226. MORRISON, M. and HULTQUIST, D. E., *J. Biol. Chem.*, 1963, **228**, 767.
227. CARLSTROM, A., *Acta Chem. Scand.*, 1965, **19**, 2387.
228. CARLSTROM, A., *Acta Chem. Scand.*, 1969, **23**, 185.
229. PAUL, K. G., OHLSSON, P. I. and HENRIKSSON, A., *FEBS Lett.*, 1980, **110**, 200.
230. ROMBAUTS, W. A., SCHROEDER, W. A. and MORRISON, M., *Biochemistry*, 1967, **6**, 2965.
231. SIEVERS, G., *FEBS Lett.*, 1981, **127**, 253.
232. SPENCER, G. R. and SIMON, J., *Am. J. Vet. Res.*, 1960, **21**, 578.
233. ITO, O., *Bull. Nippon Vet. Zootech. Coll.*, 1972, **21**, 74.
234. ITO, O., *Bull. Nippon Vet. Zootech. Coll.*, 1972, **22**, 47.
235. ITO, O. and AKUZAWA, R., *J. Dairy Sci.*, 1983, **66**, 967.
236. CHANDAN, R. C., PARRY, R. M. and SHAHANI, K. M., *Biochim. Biophys. Acta*, 1965, **110**, 389.

237. PARRY, R. M., CHANDAN, R. C. and SHAHANI, K. M., *Arch. Biochem. Biophys.*, 1969, **103**, 59.
238. JAUREGUI-ADELL, J., *J. Dairy Sci.*, 1974, **58**, 835.
239. EITENMILLER, R. R., FRIEND, B. A. and SHAHANI, K. M., *J. Dairy Sci.*, 1971, **54**, 762.
240. EITENMILLER, R. R., FRIEND, B. A. and SHAHANI, K. M., *J. Dairy Sci.*, 1976, **59**, 834.
241. FRIEND, B. A., EITENMILLER, R. R. and SHAHANI, K. M., *Arch. Biochem. Biophys.*, 1972, **149**, 435.
242. CHANDAN, R. C., PARRY, R. M. and SHAHANI, K. M., *J. Dairy Sci.*, 1968, **51**, 606.
243. BINGHAM, E. W. and ZITTLE, C. A., *Arch. Biochem. Biophys.*, 1964, **106**, 235.
244. DALALY, B. K., EITENMILLER, R. R., FRIEND, B. A. and SHAHANI, K. M., *Biochim. Biophys. Acta*, 1980, **615**, 381.
245. GROVES, M. L., *J. Dairy Sci.*, 1966, **49**, 204.
246. ANDREWS, P., *FEBS Lett.*, 1970, **9**, 297.
247. BARKER, R., OLSEN, K. W., SHAPER, J. H. and HILL, R. L., *J. Biol. Chem.*, 1972, **247**, 7135.
248. MORRISON, J. F. and EBNER, K. E., *J. Biol. Chem.*, 1971, **246**, 3977.
249. MORRISON, J. F. and EBNER, K. E., *J. Biol. Chem.*, 1971, **246**, 3992.
250. POWELL, J. T. and BREW, K., *Eur. J. Biochem.*, 1974, **48**, 217.
251. PRIEELS, J. P., MAES, E., DOLMANS, M. and LEONIS, J., *Eur. J. Biochem.*, 1975, **60**, 525.
252. BELL, J. E., BEYER, T. A. and HILL, R. L., *J. Biol. Chem.*, 1976, **251**, 3003.
253. POWELL, J. T. and BREW, K., *J. Biol. Chem.*, 1976, **251**, 3645.
254. ANDREE, P. J. and BERLINER, L. J., *Biochemistry*, 1980, **19**, 929.
255. O'KEEFFE, E. T., HILL, R. L. and BELL, J. E., *Biochemistry*, 1980, **19**, 4954.
256. KITCHEN, B. J. and ANDREWS, P., *Biochem. J.*, 1974, **141**, 173.
257. KITCHEN, B. J. and ANDREWS, P., *Biochem. J.*, 1974, **143**, 587.
258. MAGEE, S. C. and EBNER, K. E., *J. Biol. Chem.*, 1974, **249**, 6992.
259. GRUNWALD, J. and BERLINER, L. J., *Biochim. Biophys. Acta*, 1978, **523**, 53.
260. CHANDLER, D. K., SILVIA, J. C. and EBNER, K. E., *Biochim. Biophys. Acta*, 1980, **616**, 179.
261. BUSHWAY, A. A. and KEENAN, T. W., *Biochim. Biophys. Acta*, 1979, **572**, 146.
262. PRIEELS, J. P. and BEYER, T. A., *Fed. Proc.*, 1979, **38**, 631.
263. PAULSON, J. C., BERANEK, W. E. and HILL, R. L., *J. Biol. Chem.*, 1977, **252**, 2356.
264. PAULSON, J. C., REARICK, J. I. and HILL, R. L., *J. Biol. Chem.*, 1977, **252**, 2363.
265. HARPAZ, N. and SCHACHTER, H., *J. Biol. Chem.*, 1980, **255**, 4885.
266. GUY, E. J. and JENNESS, R., *J. Dairy Sci.*, 1958, **41**, 13.
267. GOULD, B. S., *J. Dairy Sci.*, 1932, **15**, 230.
268. LINDBERG, T. and SKUDE, G., *Pediatrics*, 1982, **70**, 235.
269. MELLORS, A., *Can. J. Biochem.*, 1968, **46**, 451.
270. OBERKOTTER, L. V., KOLDOVSKY, O. and TENORE, A., *Int. J. Biochem.*, 1982, **14**, 151.

271. KITCHEN, B. J., MIDDLETON, G. and SALMON, M., *J. Dairy Res.*, 1978, **45**, 15.
272. KITCHEN, B. J. and MIDDLETON, G., *J. Dairy Res.*, 1976, **43**, 491.
273. KITCHEN, B. J., MASTERS, C. J. and WINZOR, D. J., *Proc. Aust. Biochem. Soc.*, 1979, **12**, 26.
274. BERTONI, G., BANI, P. and SPERONI, A., *Atti Soc. Ital. Sci. Vet.*, 1981, **35**, 537.
275. BOGIN, E., ZIV, G., AVIDAR, J., RIVETZ, B., GORDIN, S. and SARAN, A., *Res. Vet. Sci.*, 1977, **22**, 198.
276. MELLORS, A. and HARWALKAR, V. R., *Can. J. Biochem.*, 1968, **46**, 1351.
277. KIERMEIER, F. and GULL, J., *Z. Lebensm.-Forsch.*, 1968, **138**, 205.

Chapter 10

THE BIOLOGICAL SIGNIFICANCE OF THE NON-IMMUNOGLOBULIN PROTECTIVE PROTEINS IN MILK: LYSOZYME, LACTOFERRIN, LACTOPEROXIDASE

BRUNO REITER*

*Formerly National Institute for Research in Dairying,
University of Reading, UK*

*Hon. Research Fellow, Department of Paediatrics,
University of Oxford, UK*

1. INTRODUCTION

'... apart from phagocytosis and the bacteriocidal power of blood fluids, the tissues and secretions have also primary antiseptic properties'; '... its (lysozyme) importance in connection with natural immunity does not seem to be generally appreciated'.[1,2]

The above quotations refer to the first defined non-immunoglobulin antibacterial agent and express the disappointment with the poor response of the medical profession. Today, so many years later, the general attitude has changed only marginally.

The bacteriocidal and bacteriostatic properties of milk were observed at about the same time as those of blood (e.g. Ref. 3) but, naturally, have not attracted the same degree of attention. However, medical workers were interested in the spread of diseases such as cholera, salmonellosis,

*Present address: 23 Brampton Court, Ray Park Avenue, Maidenhead, Berkshire SL6 8EA, UK.

and 'scarlet fever' (*Streptococcus pyogenes*) through consumption of contaminated milk (e.g. Ref. 4). Veterinarians investigated the anti-streptococcal activity of milk in relation to inflammation of the bovine udder (e.g. Ref. 5); dairy bacteriologists were concerned with the sporadic cessation of lactic acid production by starter streptococci in cheese making (other than that caused by bacteriophage) and the keeping quality of milk.[6,7]

Those inhibitors were referred to by the trivial name, 'lactenins' until they were partly identified as agglutinins (specific antibodies to streptococci) or associated with lactoperoxidase.[8,9] Eventually, the complete 'lactoperoxidase system' was demonstrated to include H_2O_2[10] and thiocyanate.[11]

Lysozyme, 'the extraordinary bacteriolytic agent', was demonstrated to occur in tissues, secretions and egg white,[1,2] and in human but not in bovine milk.[12] Fleming[2] used a more sensitive assay organism (*Micrococcus lysodeicticus*) than Bordet[12] and detected low concentrations of lysozyme in bovine milk also, particularly milk from Jersey cows. In the same paper, Fleming[2] cites the findings of Rosenthal and Lieberman[13] that lysozyme could be detected in the faeces of breast-fed but not of 'artificially fed' infants. He concurred with the suggestion that the lysozyme of human milk has 'a marked influence on the bacterial flora of the intestine of infants'. This appears to be only the second instance of maternal milk being considered beneficial to the newborn, other than as a nutrient. Ehrlich[14] was the first to suggest that mouse milk could transfer passive immunity (against ricin) to the offspring, and that antibodies may be derived from mammalian tissue and pass unaltered through the digestive tract. The possible *in vivo* role of lysozyme remained largely ignored and the enzyme became a tool in biophysics, chemistry, physiology and clinical medicine. It is now time to reassess its biological role.

While the lytic activity of lysozyme was described first and the enzyme purified much later,[15] lactoferrin was first isolated as an iron-containing red protein from human milk by several workers simultaneously[16–18] and later from other secretions and leucocytes (see Ref. 19). The antibacterial activity of lactoferrin was demonstrated simultaneously in bronchial mucus against '*Staphylococcus albus*'[20] and in bovine milk and the secretion of the non-lactating bovine udder against *Bacillus stearothermophilus*.[21,22]

Lactoperoxidase was purified from bovine milk[23] long before the lactoperoxidase/H_2O_2/thiocyanate system was recognized. The peroxidative activity of human milk is much lower than that of cow's milk and is now known to be derived partly from polymorphonuclear leucocytes as myeloperoxidase and salivary peroxidase[213] which is more akin to lacto-peroxidase[25,26] (also K. M. Pruitt, University of Birmingham, Alabama, USA, pers. comm.). It is very likely that the low peroxidative activity of human milk is supplemented by salivary peroxidase. Since bovine milk lacks high concentrations of lactoferrin and lysozyme but is rich in lactoperoxidase, it is not surprising that the role of the lactoperoxidase system can be investigated best in calves.

Summing up, there is little doubt that the total antibacterial effect of the non-immunoglobulin proteins and immunoglobins is greater than the sum of each acting individually because of their interaction (see Section 5 on synergism). These proteins are not restricted to milk but occur in all secretions bathing mucous membranes: saliva, tears, bronchial and nasal secretions, hepatic bile, pancreatic fluid, etc; also, the same proteins are involved in the intracellular killing of bacteria by leucocytes.

The functions of the protective milk proteins can be two-fold: protection against infection of the mammary gland and/or protection of the newborn. The latter appears to be more important, or at least we have more evidence for their *in vivo* activity. (For historic accounts and reviews see Refs 27–38.)

2. LYSOZYME (MURAMIDASE) (*N*-ACETYLMURAMYL HYDROLASE, EC 3.2.1.17)

Lysozyme has recently attracted renewed interest as a component of the antibacterial systems of milk which influence the intestinal bacterial flora of the neonate, possibly also affecting the general immune system. Lysozyme also inhibits the outgrowth of spores and their vegetative cells. Although the concentration of native lysozyme in bovine milk is too low for practical effects, the addition of egg white lysozyme to dairy products prevents the blowing of semi-hard cheeses caused by clostridia (for review see Ref. 39).

2.1. Mode of Action

Lysozyme is a small basic protein (mol. wt. $\sim 15\,000$) which attaches itself to the basic bacterial surface in the presence of electrolytes and has a low energy of activation (appreciable rate of activity at low temperatures).[40] Its activity depends not only on the concentration of the enzyme but also on its origin; lysozyme isolated from human or bovine milk has a far greater lytic activity on the most sensitive assay organism—*Micrococcus lysodeicticus*—than that isolated from egg white (Table 1).[41] In the past, this fact has been generally neglected with experimental results both *in vitro* and *in vivo*. Lysis is not the only criterion of the activity of the enzyme. At low ionic concentrations, viability through disruption of the cellular metabolism can be rapidly reduced without loss of cellular integrity (no leakage of protein, DNA or RNA) although

TABLE 1

THE LYTIC EFFECT OF BOVINE MILK, HUMAN MILK AND EGG WHITE LYSOZYME ON
LIVE *Micrococcus lysodeicticus*[41]

Source of lysozyme	Rate of lysis ($\Delta\%\ T\ min^{-1}$)[a]
Bovine milk	1·82
Human milk	1·70
Egg white	0·63

[a]Rate of lysis was determined in a spectrophotomer at 540 μm.
Phosphate buffer, pH 6·2.

the cell wall is hydrolysed (release of N-acetylhexosamine). Also, lysis depends not only on the species used but also on the strain, the growth phase and the medium in which the culture was grown.[42,43] Recently it has been shown that the activity of the enzyme depends not only on the concentration of the electrolytes but also on the nature of the anion (Fig. 1); SCN^- and HCO_3^- are the best promoters, both being more effective than Cl^- or F^-.[44,45]

The enzyme is stable at low pH and lysis is also promoted when it is transferred from a low pH to a high pH.[46,47] Considering the low pH in the stomach (particularly during fasting when the stomach is empty), the active secretion of SCN^- (see Section 4 on lactoperoxidase), and the high pH in the intestinal fluid buffered by HCO_3^-, it is surprising that the *in vitro* results have so far not been investigated sufficiently and related to the *in vivo* situation.

2.2. Sources

Bovine milk contains an average of only 13 μg lysozyme/100 ml but human milk contains 10 mg/100 ml.[30,41] The enzyme is generally considered to be synthesized in the mammary gland but *in vitro* synthesis by mammary tissue has not yet been attempted. Tissues of the gastrointestinal tract, of the normal and inflamed respiratory tract, monocytes and macrophages incorporate ^{14}C-labelled amino acids thus proving *in vitro* synthesis.[48] Gastric mucosa appears to synthesize both secretory component (for the transport of circulating blood dimeric IgA) and lysozyme.[49] The Paneth cells appear to be the main source of lysozyme in the crypt lumen.[50]

The so-called β-lysin[51] of blood serum, which has been known to be bacteriocidal for some Gram-positive organisms, is now recognized to be identical with lysozyme.[52] We must therefore also consider whether the enzyme diffuses into the mammary gland or intestinal tract when tissues become inflamed. The faeces of infants with diarrhoea, for instance, contain more lysozyme than those of healthy infants.[53,54] This increase can be ascribed to greater synthesis of lysozyme, to infiltration of leucocytes or to direct diffusion from the blood. No comparable data are available for animals. However, infected udders secrete milk containing increased numbers of leucocytes and, according to Korhonen,[55] increased levels of lysozyme. Since bovine leucocytes lack lysozyme,[56] they cannot be a source of it. The inflamed udder therefore either increases the synthesis of lysozyme by analogy with lactoferrin (see later) or lysozyme (β-lysin) diffuses from the blood like other blood proteins.

2.3. Possible Biological Significance

In vivo we have, so far, no direct evidence that either lysozyme of ingested milk or intestinally secreted lysozyme affects the intestinal bacteria, particularly Gram-negative potential pathogens. Claims have been made that, in rats, the addition of egg white lysozyme to formula feeds reduces the pH of their faeces and increases the number of bifidobacteria, thereby bestowing a 'clinical' benefit.[57] Numerous attempts to detect any benefit from feeding lysozyme to infants are even more difficult to interpret.

Theoretically it is possible that hydrolysates of cell walls, *N*-acetylglucosamine and *N*-acetylmuramic acid, stimulate the growth of bifidobacteria, e.g. *Lactobacillus bifidus* var. *pennsylvania* which is unable to synthesize peptidoglycan in the absence of *N*-sugars (which occur only in human milk at adequate levels). The organisms, for instance, grew adequately in a synthetic medium in which *M. lysodeicticus* had previously been lysed by lysozyme (unpublished). Recently Spik *et al.*[58] have demonstrated that ammonium sulphate-precipitated fractions (0·50–0·75 sat.) of human milk yield growth promoting factor(s) for bifidobacteria *in vitro*; their nature has not, as yet, been established.

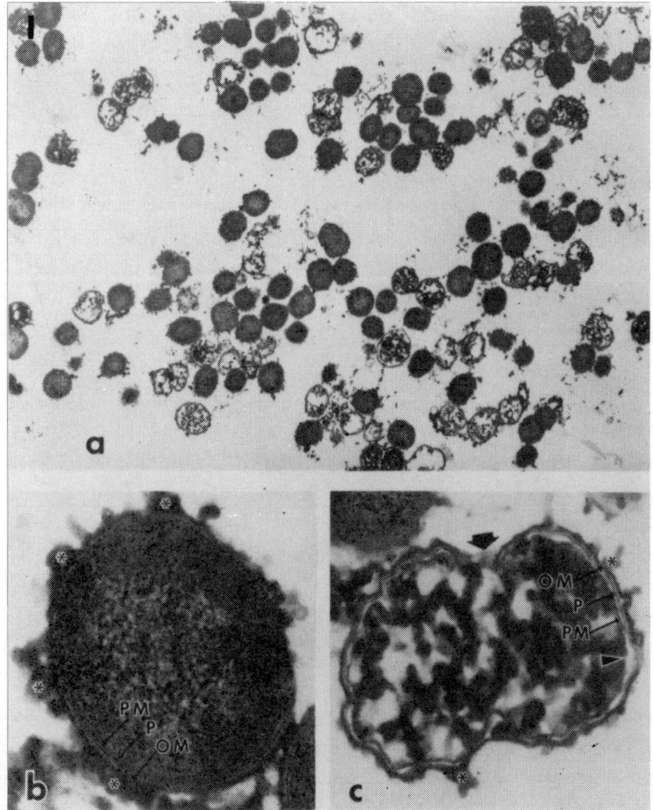

FIG. 1. (I) Electron micrographs of *Vibrio alcalescens* treated with human lysozyme alone: (a) ×13 000, lysed and unlysed cells; (b) ×90 000, unlysed cell but showing 'blibbing'—dense material attached to outer membrane (compare with Fig. 6, damage to *E. coli* by lactoperoxidase system); (c) ×61 000, lysed cell. ▶, Septum, damaged outer membrane (OM), peptidoglycan (P) and plasma membrane (PM) missing; ➡ , separation of septum.

An entirely new hypothesis for the function of lysozyme was proposed by Jollès,[59] who suggested that the hydrolysis products of peptidoglycan may act as an adjuvant or immunomodulator. Indeed, without apparent knowledge of this hypothesis, Lodinova and Jouja[60] reported that feeding lysozyme to infants increased the immunoglobulin (secretory IgA) level in the faeces compared with the faeces of infants given formula feed only; the immunoglobulin level in the blood was not affected. Other workers,[61] acknowledging the Jollès hypothesis, showed that injection of

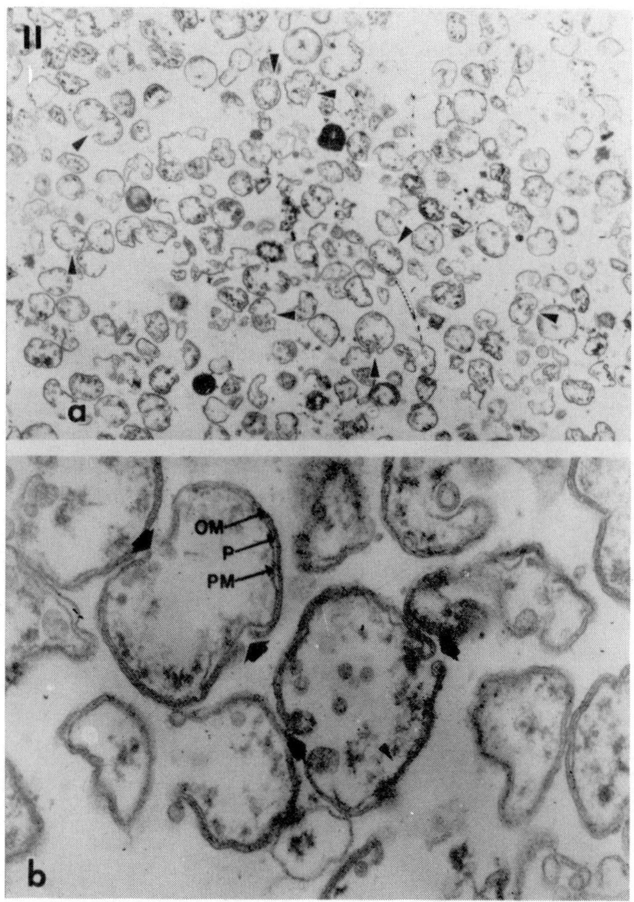

Fig. 1—*contd.* (II) Electron micrographs of *Vibrio alcalescens* treated with human lysozyme followed by $NaHCO_3$: (a) ×13 000, nearly all cells lysed; (b) ×13 000, lysed cells—OM, P and PM still remain or are missing. Symbols as for Fig. 1(I) (from Ref. 45).

hepatitis B antigen into the foot pads of guinea pigs increased the circulating hepatitis antibodies in the blood and stimulated cellular immunity (skin and corneal tests for hypersensitivity). Digests of bacterial cell walls by lysozyme and pronase or *N*-acetylmuramyl dipeptide also increased the serum and blood antibody levels but not as efficiently as Freund's complete adjuvant.

A novel application for lysozyme is now being established in the cheese industry. Preparations of lactoferrin isolated from secretions of the non-lactating udder were shown to inhibit vegetative cells and the germination of *Bacillus stearothermophilus* spores, and inhibition could be reversed by iron.[22] Subsequently it was discovered that the lactoferrin preparation had been contaminated with lysozyme, which was responsible for the inhibition. Egg white lysozyme was also found to be effective; furthermore, iron reversed the inhibition when the spores were suspended in phosphate but not citrate buffer, hence the previous erroneous conclusion that lactoferrin was involved becomes understandable.[31] It has since been shown[58,185] that lactoferrin and lactoperoxidase form a strong complex with an electrophoretic mobility distinct from that of lactoferrin and lysozyme. It is therefore not surprising that the 'purest' preparation of bovine lactoferrin contains traces of lysozyme (and lactoperoxidase).[180]

Egg white lysozyme has been used for some time to prevent 'butyric acid blowing' of cheese (Pulay and Krasz, 1967, cited in Carini and Lodi[62]) and it appears to be a feasible alternative to the addition of toxicologically undesirable nitrates.[63-65] It appears[63,64] that lysozyme inhibits the overall outgrowth of spores of *Clostridium tyrobutyricum* into vegetative cells but without affecting the early stages of germination, such as changes in optical refractibility and heat sensitivity. Considering the greater activity of milk lysozyme (see before), it would be of interest to determine its effect on various species of spores.

In this context it is interesting to recall that peasants traditionally fed raw eggs to piglets and calves to relieve scouring. Recently it has been reported that the count of clostridia is markedly reduced in the faeces of breast fed compared with artificially fed infants.[66] This could be attributed to the high lysozyme content of human milk (some clostridia are, of course, known to cause diarrhoea in man and animals).

Finally, it can be speculated that lysozyme promotes protection through its effect on leucocytes. Human lysozyme at concentrations of 10–100 mg ml^{-1} significantly ($p < 0.001$) stimulates phagocytosis of yeast cells by leucocytes in the absence of any serum factors (opsonins, complements); egg white lysozyme is ineffective.[67]

3. LACTOFERRIN

It is now well established that iron-chelating proteins such as transferrin, lactoferrin and ovotransferrin have antibacterial activity. While transferrin is mainly a blood and extravascular protein, although it occurs also in milk, lactoferrin is exclusively a secretory protein. It occurs in milk, saliva, tears, nasal and intestinal secretions, pancreatic juice and seminal fluid as well as in the secondary granules of neutrophils. Ovotransferrin, also referred to as conalbumin, is exclusively an avian egg white (albumin) protein.

3.1. Mode of Action

Lactoferrin is a single glycoprotein (mol. wt. 76 500) to which two carbohydrate groups are attached. The molecular, functional and evolutionary comparisons with transferrin and ovotransferrin have been reviewed.[58,68,69a,69b,70–72] The three proteins possess two metal-binding sites, each of which can bind a ferric ion (Fe^{3+}) together with a bicarbonate ion (HCO_3^-). In the native state, lactoferrin is only partly saturated with iron (8–30%), which is physiologically important because iron can be chelated and thus inhibit bacteria by depriving them of iron essential for growth. Iron-saturated proteins are devoid of antibacterial activity (Fig. 2).[20–22] The capacity of lactoferrin to bind iron depends on the presence of bicarbonate, mole per mole,[73] but is inversely related to citrate concentration[74,75,75a] (Table 2). Citrate can exchange iron chelated by lactoferrin (and conalbumin), and iron citrate is then actively taken up by microorganisms.[76] Bovine and human milks contain

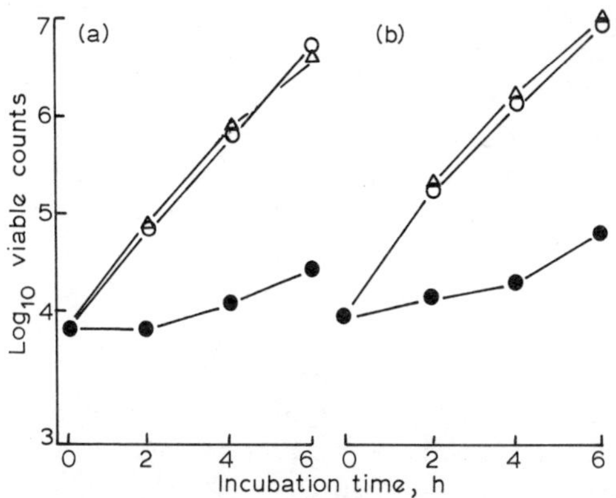

FIG. 2. Inhibition of (a) *Escherichia coli* serotype 0101 (serum resistant) and (b) *E. coli* NCTC 9703, serotype 0111 (serum susceptible) in carbon–ammonium salts medium by lactoferrin isolated from bovine whey. O——O, Control (no added lactoferrin or Fe^{3+}); ●——●, medium containing 2·5 mg ml^{-1} lactoferrin; △—△, medium containing 2·5 mg ml^{-1} lactoferrin saturated with Fe^{3+} (from Ref. 180).

TABLE 2

EFFECT OF CITRATE AND BICARBONATE ON THE BACTERIOSTATIC ACTIVITY OF DIALYSED COLOSTRUM[74]

HCO_3^- (μM ml^{-1})	Citrate (μM ml^{-1})			
	0	*0·1*	*1*	*10*
	Increase in log_{10} viable count after 6 h incubation			
0	0·7	1·9	2·6	2·4
0·7	0·7	0·9	2·5	2·5
7·0	1·1	1·1	1·1	2·4
70·0	1·0	0·7	0·7	0·8

4–8 mM and 2–3 mM citrate, respectively; thus, iron citrate can counteract the bacteriostatic activity of lactoferrin unless the bicarbonate concentration is high, as in the intestinal fluid where bicarbonate is the main buffer. Also, it was reported *in vivo* (in calves) that citrate is rapidly absorbed in the upper duodenum.[33] Lactoferrin (and ovotransferrin) differs from transferrin in its ability to retain iron at very low pH (down to 2·2); it therefore passes through the acid gastric fluid unharmed.

Bacteriostasis is only temporary because some Gram-negative pathogens adapt to iron-restrictive conditions *in vitro* and *in vivo* by synthesizing low molecular weight iron chelators—siderophores (sidochromes)—that can remove iron from lactoferrin (or transferrin). The iron-laden siderophores are taken up by newly synthesized outer membrane protein receptors and the iron released by enzymes. The iron–citrate complex is also taken up by new outer membrane proteins, part of the citrate-dependent iron-transport system. *E. coli* produces two types of siderophores: enterocholin (gene-controlled) and aerobactin (plasmid-controlled).[69,69a,69b,69c,77] (It is of some interest that the outer membrane protein receptors act also as bacteriophage and colicin receptors.)

It was generally assumed that lactoferrin inhibits only bacteria with high iron requirements (e.g. coliforms) while lactic bacteria, with low iron requirements, remain unaffected.[78] This explanation appears to be insufficient because it has now been shown[79–82] that apolactoferrin (iron-free) has a bacteriocidal effect against a wide range of microorganisms: Gram-positive, Gram-negative, aerobic and anaerobic bacteria, as well as yeasts. This effect appears to be strain-specific because the strains of species previously shown to be inhibited only are not killed. It is necessary to demonstrate that the bacteriocidal effect is valid for 'native' lactoferrin, which is partly saturated, and tested under physiological conditions, since the bacteriocidal effect was observed only with suspensions of bacteria in distilled water or saline; under conditions of growth, viability was maintained but the uptake of glucose, production of lactic acid and the subsequent synthesis of DNA, RNA and protein were impaired. The enhancement of the antibacterial activity of lactoferrin by antibodies will be discussed in Section 5 on synergism.

3.2. Sources

Lactoferrin is synthesized *in situ* in the mammary gland, intestinal tract, salivary and haderian glands, and neutrophils. Blood contains only traces of lactoferrin so it is unlikely to contribute to the levels in the mammary gland. Artificially infected bovine udders produce milk with increased lactoferrin concentration which appears to be a direct response to infection because it was calculated that the increased number of leucocytes do not contain sufficient lactoferrin.[83] So far, a possible increase in the level of transferrin by diffusion from the blood has not been investigated. This could be of particular importance in the case of coliform infections, which are known to cause the rapid appearance of blood proteins in milk; this may be of some importance for iron-requiring bacteria which are haemolytic because they can utilize free haem as a source of iron.

Bovine milk contains between 0·02 and 0·35 mg ml^{-1} of lactoferrin, depending on the period of lactation, while human milk contains 2–4 mg ml^{-1}. The relatively high levels of lactoferrin in human colostra (6–8 mg ml^{-1}) and bovine colostra ($\sim$ 1 mg ml^{-1}) decline rapidly within the first few weeks of lactation.[30,84,85]

3.3. Biological Significance

It is unlikely that lactoferrin plays a significant role in the defence of the bovine mammary gland against infections during lactation; its concentration is too low and the citrate level too high. In the non-lactating udder, however, the conditions for antibacterial acitivity are very favourable; the lactoferrin concentration increases up to 70 mg ml^{-1}, the citrate is rapidly reabsorbed after lactation ceases, and, at the same time, bicarbonate diffuses from the blood. Indeed, the secretion of the non-lactating udder is very inhibitory to *E. coli* and can be completely reversed by the addition of iron. *In vivo* experiments showed that the non-lactating bovine udder is exceptionally resistant to *E. coli*.[86]

There is some evidence that lactoferrin has a role in the protection of the newborn; the breast-fed human infant ingests $\sim$3000 mg lactoferrin per day during the first week of life, declining to $\sim$1000 mg during the following three weeks.[30] Unfortunately we have no data for the calf but, assuming that the calf drinks 1–2 litres of colostrum and 4 litres of milk per day, it can be calculated that a calf ingests 1000–2000 mg lactoferrin per day while consuming colostrum, declining to 80–1400 mg per day when consuming 'normal' milk (related to relative body weights, the calf ingests only 10% as much lactoferrin as the human infant).

Whatever the intake of lactoferrin, it is important to know how well the protein resists digestion and survives passage through the intestinal tract. *In vitro* experiments show that purified lactoferrin and the lactoferrin in milk resist digestion by trypsin and chymotrypsin but not by pepsin.[87,88] Lactoferrin can be recovered from the faeces of breast-fed infants and is still able to bind iron.[89,90] The faeces of artificially fed infants contain no lactoferrin unless the feed based on cow's milk is supplemented with human or bovine lactoferrin; the latter resists digestion even better than the former. Lactoferrin, besides its bacteriostatic activity, may be involved also in the absorption of iron by mucosal cells. There is some evidence for specific lactoferrin receptors in the human intestine, and *in vitro* experiments have shown that iron is transported across the duodenal brush border.[91] However, based on *in vivo* results, some workers dispute that lactoferrin is involved in iron transport.[92,93] Recent work has confirmed the *in vitro* results of Cox *et al.*[91] In piglets and mice anaemia was remedied by feeding either iron–chloride or lactoferrin. Although lactoferrin was not superior, it would have the advantage of preventing iron becoming available for the growth of pathogens.[91a,91b]

3.4. *In Vivo* Experiments

While the antibacterial activity of lactoferrin *in vitro* is well documented, only a few animal experiments have been performed.

The non-lactating bovine udder was chosen for a limited investigation[86] to test the effect of lactoferrin *in vivo*. A strain of *E. coli* was selected which was known to cause severe mastitis in lactating udders or even death of the cow if not treated in time. Out of 17 quarters, 13 became infected when small numbers of microorganisms (250–300 cfu/quarter) were infused during lactation; in contrast, none of 14 dry quarters became infected. However, when 2 quarters were infused immediately before parturition, both became infected (Table 3). Although the latter experiment ought to be repeated with a larger number of

TABLE 3

RESPONSE OF LACTATING AND NON-LACTATING UDDERS TO THE INFUSION OF *E. coli*[86]

| | Number of quarters | | |
State of quarter	Infused	Infected	% Infection
Lactating	17	13	76
Non-lactating	14	0	0
Two days pre-calving	2	2	100

Infusion of 250–300 cfu of *E. coli* per quarter.

quarters, the observation agrees with the well-known fact that coli-infections do occur frequently after calving. Two dry quarters infused with both iron and *E. coli* became infected, indicating that lactoferrin is one of the systems that may play a part in preventing the multiplication of pathogens. The failure of lactoferrin to prevent infection in the quarters prior to parturition can be explained by the changed composition of the secretion; according to Peaker and Linzell,[94] citrate, which they regarded as the harbinger of parturition, reappears during the formation of colostrum. We may also assume that the level of bicarbonate decreases at that time. The favourable conditions for the antibacterial activity of lactoferrin in the dry udder are therefore reversed and any 'dormant' bacteria can multiply, causing mastitis.

Bullen *et al.*[95] investigated the effect of lactoferrin on the intestinal bacterial flora of sucking guinea piglets which were chosen because guinea pig milk is as rich as human milk in lactoferrin. Shortly after birth, the piglets were dosed with *E. coli* 0111 but developed a normal intestinal flora within 6 days; lactobacilli became the dominant flora in the small and large intestines and *E. coli* counts declined to $\sim 10^2$ per gram of intestine within 2 days. When guinea piglets were fed an artificial diet, coliforms remained the dominant microorganisms; lactobacilli appeared only after 3 days and remained well below the coliform count. When sucking guinea piglets were also given haematin, as a source of iron, the coliform count increased 10 000-fold in the small intestine compared with piglets without a source of iron; the increase was smaller in the large intestine (only 100-fold) but still significant. Stephens *et al.*[96] showed that guinea pig milk is not only rich in lactoferrin but also in lactoperoxidase (~ 20-fold higher than in bovine milk). They found that guinea pig milk 3 days post-partum inhibited *E. coli* 0111 and that the inhibition could be reversed by iron. However, milk obtained 7 days post-partum killed *E.*

coli, and this bacteriocidal effect could not be reversed by iron but only by reducing agents such as cysteine or penicillamine which indicates that the lactoperoxidase system was involved (see later). Unfortunately, although lactoperoxidase was assayed, thiocyanate was not, and there is no indication of a source of H_2O_2 which is required to activate the lactoperoxidase system. However, there is a possible explanation. It is well known that guinea pigs have a massive requirement for ascorbic acid when fed artificially, and it can therefore be assumed that the milk contains naturally high levels of ascorbic acid which is oxidized to dehydroascorbic acid in the presence of copper, thus producing H_2O_2. Ascorbic acid was considered[97] as an alternative source of H_2O_2 in bovine milk, and it has since been proven that physiological levels in the presence of copper effectively activate the lactoperoxidase system in milk *in vitro* (unpublished).

The problem of inhibitors other than lactoferrin which may interfere with the interpretation of *in vivo* experiments was overcome by using a preparation of conalbumin (Ricordati, Milano*) as an alternative source of an iron-chelating protein. Conalbumin can replace lactoferrin in the presence of bicarbonate *in vitro* and *in vivo*.[98–101] When newborn guinea piglets were orally infected with *E. coli* 0111 and fed milk powder plus conalbumin, a similar protection was provided as when suckling the dam. These favourable results encouraged the planning of large-scale trials in which infants with acute enteritis were treated in about thirty hospitals. The preliminary results appear to be favourable (Ref. 102, and to be published); it was found that infants given conalbumin in a cow's milk formula feed improved their general status and normalized the alvus sooner than the 'controls' ($P < 0.01$).

Although we have no direct evidence that lactoferrin contributes to the suppression of *E. coli* in the human infant, it seems likely (see also Ref. 58).

4. THE LACTOPEROXIDASE SYSTEM

Wright and Tramer[9] associated the inhibition of some starter strains with lactoperoxidase. They observed a high correlation between the inactivation of the enzyme by heat, at low pH or by sodium azide and

---

*British Patent 1463327 (Method of isolating transferrins from biological materials).

the inhibitory activity of raw milk against some strains of lactic acid streptococci. Since reducing agents such as cysteine or sodium sulphite reversed the inhibition, they suggested the formation by lactoperoxidase and H_2O_2 of an inhibitory oxidation product having a quinonoid structure. The role of lactoperoxidase was rapidly confirmed when it was shown that purified lactoperoxidase added to milk, heated to destroy the native enzyme, restored the original inhibition of susceptible strains.[103] Next, the necessity of H_2O_2 was established[10] and the oxidizable substrate identified as thiocyanate.[11] Inhibition was also confirmed in a synthetic medium, provided that all three components—lactoperoxidase, H_2O_2, thiocyanate—were present.[104]

The inhibition or killing of many bacterial species *in vitro* by the lactoperoxidase system is now well established (for reviews see Refs 105–108a, b).

4.1. Distribution of the Components

4.1.1. Lactoperoxidase

Bovine milk contains $\sim 30\,\mu g\,ml^{-1}$ of this enzyme.[23] The values of unit activity per ml reported in the literature are difficult to compare because different substrates, e.g. phenylenediamine, *o*-dianisidine or pyrogallol, have been used to measure colour formation catalysed by lactoperoxidase/H_2O_2. In a large-scale survey, Kiermeier and Kayser[109] reported that mid-lactation milk contains 1·7–17·6 'phenylenediamine' $U\,ml^{-1}$. Peroxidase activity is very low in colostrum, reaches a peak at 5 days post-partum and declines thereafter. The enzyme level also depends on the sexual cycle of the cow and varies according to summer or winter season, feeding regime and breed.[110] Korhonen,[55] also using phenylenediamine but under different conditions, reported average values of 0·46 (range 0·3–0·6) $U\,ml^{-1}$ in samples of milk which contained less than 5×10^6 leucocytes ml^{-1} and hence were assumed to be from healthy udders; the values were slightly higher in milks with higher cell counts. Since we now know that leucocytes suspended in milk phagocyte (ingest) casein micelles and fat globules,[111–113] it is likely that the increased peroxidative values were caused by leakage of myeloperoxidase, the peroxidase of polymorphonucleated leucocytes; phagocytosis induces a metabolic burst, generating H_2O_2 and diffusing H_2O_2 as well as myeloperoxidase (Korhonen and Reiter, cited in Ref. 107). Recently a more sensitive assay has come into general use employing 2,2-azino-di-(3-ethylbenzthiazoline-6-sulphonic acid) (ABTS) as substrate.[114] According to Stephens *et al.*,[96] bovine milk contains, on average, 1422 (range 738–3889) ABTS mU ml^{-1}. It is important that all the peroxidative values for bovine milk quoted are far above the minimum requirements for the lactoperoxidase/H_2O_2/SCN system. For instance, it was found that as little as 20 ABTS mU ml^{-1} in human milk are sufficient to kill *E. coli* in the presence of thiocyanate and H_2O_2 (Fig. 3).[36] However, all the

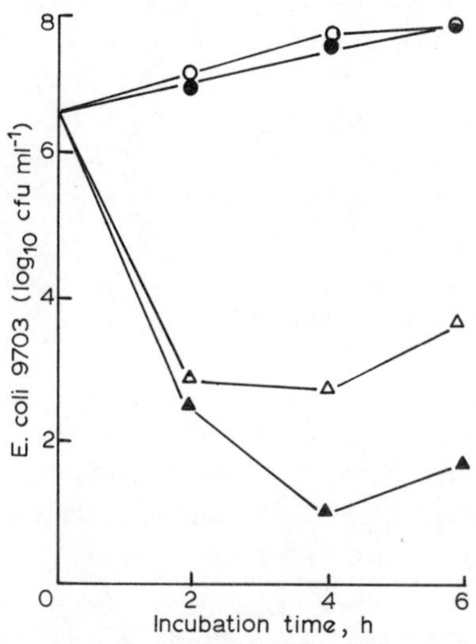

FIG. 3. Bactericidal effect on *E. coli* 9703 of human milk (10 days post-partum) containing 20 mU ml^{-1} lactoperoxidase. Control milk (○); milk plus 0·225 mM SCN$^-$ and 0·225 mM H_2O_2 (△); milk plus 0·225 mM SCN$^-$, 0·225 mM H_2O_2 and 200 mU ml^{-1} lactoperoxidase (▲); milk plus 0·225 mM SCN$^-$, 0·225 mM H_2O_2 and 1 mM $Na_2S_2O_4$ (●) (from Ref. 36).

peroxidative activity in human milk was supposed to be derived from leucocytes and hence is due to myeloperoxidase and not lactoperoxidase.[213] This appeared to explain why the milks of Gambian mothers were found to contain persistently, for several months, high levels of lactoperoxidase in contrast to the milks of women obtained from local hospitals;[36] the breasts may have been infected, and the milk may have contained high numbers of cells which were lysed after freezing for transport and thawing for sampling.

A more recent report seemed to contradict this explanation. Partially purified lactoperoxidase from human colostrum was demonstrated[24] to have stability, chromatographic and immunoreactive properties similar to those of bovine lactoperoxidase and not myeloperoxidase. The preparation failed to be adsorbed on immunoadsorbent columns prepared with high-titre polyclonal antibodies raised against human myeloperoxidase and human eosinophile peroxidase. This latest concept is under intensive investigation by R. M. Pruitt and his collaborators[213] (also R. M. Pruitt, University of Alabama, Birmingham, USA, pers. comm.). It now appears that human milk can contain varying levels of myeloperoxidase as well as peroxidase with the characteristics of salivary peroxidase.[25] Since lactoperoxidase in contrast to myeloperoxidase cannot oxidize Cl^-, a useful assay to distinguish the enzymes has helped to demonstrate their relative proportions in human milk. These latest results do not invalidate the bacteriocidal effect of human milk, as long as a minimum peroxidative value of 20 mU (ABTS) is present[36] (Fig. 3).

The general distribution of lactoperoxidase in various secretions is similar to all the other protective proteins (including immunoglobulins) but it is not clear whether its occurrence in intestinal mucosal scrapings is due to *in situ* synthesis or is derived from eosinophiles.[115] The contribution of human salivary lactoperoxidase to the suppression of lactic acid bacteria, and hence of caries-producing oral streptococci, is now well documented.[116,117]

Human salivary peroxidase may have yet another function, in particular for the newborn infant; it may compensate for the lack or paucity of lactoperoxidase in human milk. The concentration of salivary peroxidase varies greatly, from nil to 2353 mU ml^{-1}, amongst infants and varies throughout the day. The enzyme level is very low in calf's saliva up to 1–2 weeks after birth, ~ 265 mU ml^{-1} up to 3 days, but rises to ~ 5000 mU ml^{-1}.[118] The gastric juice of the fasting calf contains no lactoperoxidase because saliva normally enters the rumen and enters the abomasum only during feeding.[119] This is, of course, in contrast to the human infant whose saliva enters the stomach all the time. It has been known for some time that lactoperoxidase is resistant to low pH and proteolysis[8,118,119] and it was therefore no surprise that the enzyme could be demonstrated in calf's abomasum after ingestion of raw milk[119] (Table 4). It has now been shown (J. Carlsson and L. Gothefors, University of Umea, Sweden, pers. comm.) that the gastric juice of infants contains appreciable levels of peroxidative activity derived from the ingested breast milk and from the saliva of the infant and SCN^- secreted in the stomach.

4.1.2. Thiocyanate (SCN^-)

This anion, which permeates freely, occurs ubiquitously in animal tissues

TABLE 4

CONCENTRATION OF THIOCYANATE AND LACTOPEROXIDASE IN CALF ABOMASAL FLUID BEFORE AND AFTER FEEDING RAW MILK[119]

	Before feeding	Time of sampling after feeding		
		30 min	60 min	120 min
SCN$^-$ concentration (mM)	0·45 ±0·14 (17)	0·15±0·08 (30)	0·15±0·08 (37)	0·20±0·09 (38)
LP concentration (U ml^{-1})	0·009±0·004 (75)	1·4 ±9·7 (20)	1·14±0·88 (18)	1·09±0·68 (19)

Mean values with standard deviation are shown. Figure in brackets is sample size. Four calves were sampled over a period of 50 days. The milk feeds contained 1·9±0·85 U ml^{-1} of LP ($n=11$). At 5 h, no LP was detected.

and secretions. It is found in the mammary, salivary and thyroid glands, in the stomach and kidney, and in fluids such as synovial, cerebral, spinal, lymph and plasma. SCN^- is derived both endogenously during the detoxification reaction between thiosulphates and metabolic products of sulphur amino acids and cyanide, and exogenously from foods containing glucosides that yield SCN^- after hydrolysis. Cows on natural pastures with clover (CN^- containing) and 'non-grasses' (e.g. cruciferae containing glucosides) give milk with higher concentrations of SCN^- (0.02–0.25 mM) than cows on winter feed or lay pastures.[120,121] The health of the udder also seems to influence the SCN^- level; milk containing less than 5×10^5 leucocytes ml^{-1} contains less SCN^- than milk with higher cell numbers.[55] Obviously the increase is derived from the blood plasma by diffusion. The saliva of man and animals also contains high levels of SCN^- but is not a major source in gastric juice. Gastric juice in the abomasum of the calf contains very high concentrations of SCN^- independently of the level in the ingested milk (up to 0.45 mM before feeding) (Table 4). It is now assumed that the same parietal cells in the gastric mucosa which secrete HCl also secrete SCN^-. While HCl secretion is delayed in the calf (to prevent digestion of immunoglobulins) the gastric juice of the human baby at birth has a pH of 1.5–2, so it ought to secrete SCN^- as well. The gastric juice of infants also contains SCN^-, but more data are required before it can be established how much of it is derived from saliva and secreted in the stomach, and how variable the contribution from the milk is (J. Carlsson and L. Gothefors, pers. comm.). The gastric juices of mice, rats, guinea pigs and hamsters have been known for a long time to concentrate SCN^- from the blood plasma.[122]

4.1.3. H_2O_2

Some catalase-negative microorganisms, e.g. lactic acid bacteria, metabolically generate H_2O_2 under aerobic conditions. In the presence of lactoperoxidase and SCN^-, as in raw milk, sensitive streptococci therefore become self-inhibitory; some streptococci are 'resistant' to the lactoperoxidase system (see later). Catalase-positive organisms, e.g. coliforms, dismutate any H_2O_2 that is formed and can therefore be inhibited or killed only if an exogenous source of H_2O_2 is provided.[123–125] H_2O_2 can be supplied chemically, generated enzymatically (e.g. by glucose oxidase, xanthine oxidase, etc.) or provided by catalase-negative organisms (e.g. lactic acid bacteria) in ecological systems as in the intestinal tract[119] or oral cavity.[126,127,129,130]

4.2. Oxidation Products of Thiocyanate

Although H_2O_2 *per se* is a strong oxidizing agent at high concentrations, it reacts only very slowly with biological material; hence 10–20 mM H_2O_2 is required to kill *E. coli*. However, 0·01–0·02 mM H_2O_2 can form a complex with lactoperoxidase which stabilizes the oxidizing power of H_2O_2, catalysing the oxidation of SCN^-, Br^- and I^-. The end products of the oxidation of thiocyanate (CO_2, NH_4^+, SO_4^{2-}) are inert but the intermediate oxidation product(s) are inhibitory.[97,128] One of the intermediate products has now been definitely identified and chemically synthesized, viz. hypothiocyanite ($OSCN^-$), the anion of hypothiocyanous acid (HOSCN) with which it is in acid–base equilibrium; a pK_a value of 5·3 was calculated for HOSCN.[126] Further oxidation by excess H_2O_2 yields higher anions of cyanosulphurous acid (HO_2SCN) and cyanosulphuric acid (HO_3SCN) (Table 5).[116,130–132]

TABLE 5

OXIDATION PRODUCTS OF THIOCYANATE BY THE LACTOPEROXIDASE SYSTEM[97,116,131,132,147]

$H_2O_2 + SCN^-$	$\xrightarrow{LP}$	$OSCN^- + H_2O$
$H_2O_2 + OSCN^-$	$\xrightarrow{LP}$	$O_2SCN^- + H_2O$
$H_2O_2 + O_2SCN^-$	$\xrightarrow{LP}$	$O_3SCN^- + H_2O$

End oxidation products: SO_4^{2-}, NH_4^+, CO_2

4.3. Mode of Action

Under aerobic conditions $OSCN^-$ appears to have a chaotropic effect on the inner membrane of bacteria because, within minutes of exposure to it (or to the complete lactoperoxidase system), K^+ and amino acids are leaked into the medium;[133] the uptake of carbohydrates, amino acids, etc. is inhibited, and subsequently the synthesis of proteins, DNA and RNA.[33,134–136] Thus the early observations of inhibition of oxygen uptake, enzyme inhibition, acid formation and growth of lactic streptococci are explained.[97,104] While lactic streptococci are only temporarily inhibited, Gram-negative organisms such as coliforms (Fig. 4), pseudomonads, salmonellae, shigellae microplasmae, vibrios and multiple antibiotic-resistant strains are killed (Fig. 5), and eventually lysed (Fig. 6).[125] The difference in response between Gram-positive and Gram-negative organisms is not yet fully understood but appears to be related to the structure and composition of the cell wall of Gram-positive and the outer membrane of Gram-negative organisms, respectively. Amino

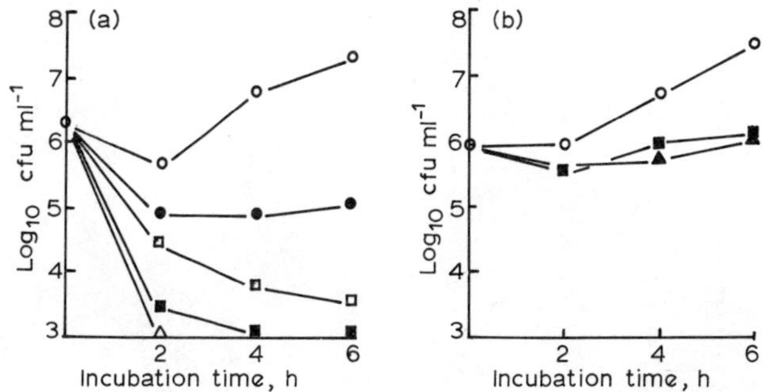

FIG. 4. Effect of various concentrations of OSCN⁻ on the viability of *E. coli* (a) and *S. lactis* (b) is synthetic medium, pH 5·5. OSCN⁻ concentrations: ○, 0; ●, 5 μM; □, 10 μM; ▨, 20 μM; △, 25 μM; ▲, 35 μM (from Ref. 133).

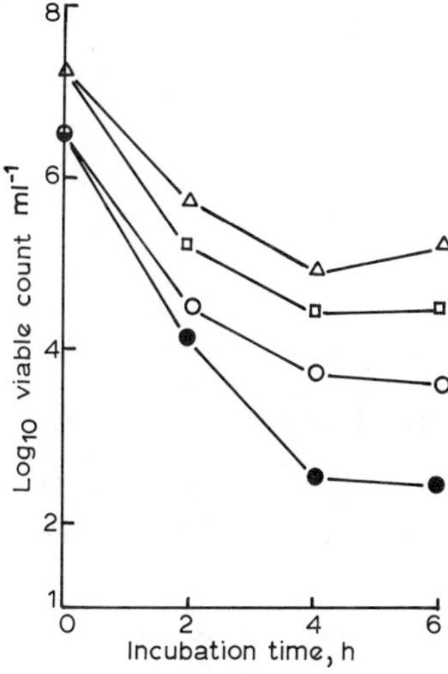

FIG. 5. Bactericidal effect of the lactoperoxidase system against *E. coli* 0101 (serum resistant) at a concentration of 0·15 mM SCN⁻ (○) and 0·30 mM SCN⁻ (●); against *P. aeruginosa* at 0·15 mM SCN⁻ (△); and against *S. typhimurium* at 0·15 mM SCN⁻ (□) (from Ref. 125).

acid and K⁺ leakage studies indicate that the inner membrane of *S. lactis* is affected less than that of *E. coli* because less leakage occurs (Fig. 7). It appears that the streptococcal cell wall is a greater barrier to the penetration of OSCN⁻ than the outer membrane of *E. coli*.[133] Indeed, a

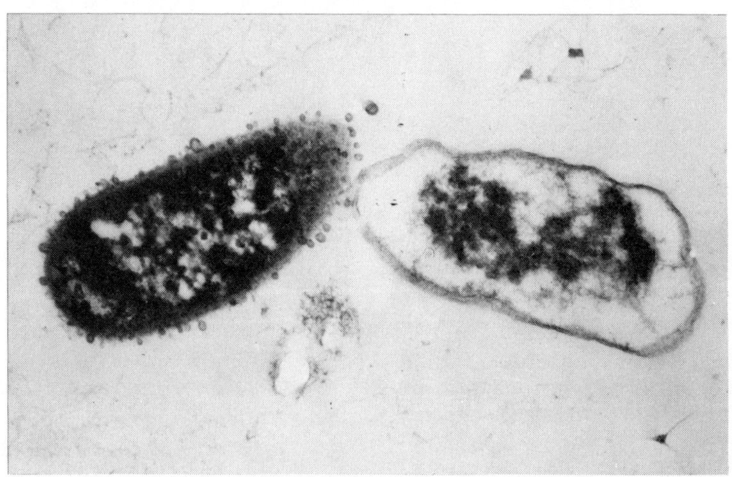

FIG. 6. Damage and lysis of *E. coli* 9703 (0111) after exposure to be lactoperoxidase system (note 'blibbing' on unlysed cell), ×30 000. (Courtesy of B. E. Brooker, NIRD.)

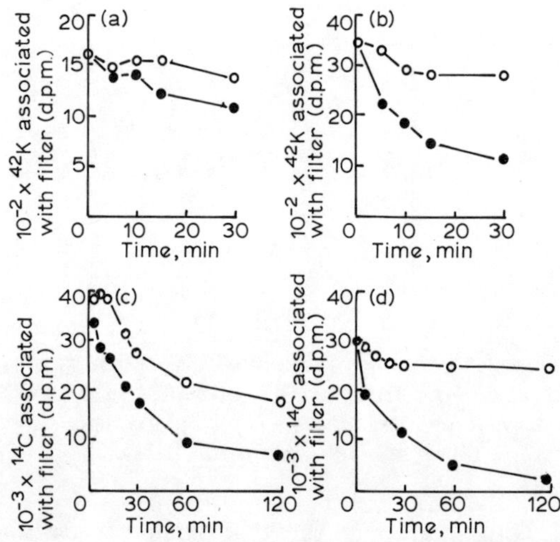

FIG. 7. Effect of OSCN⁻ on leakage of ⁴²K⁺ from *S. lactis* (a) and *E. coli* (b) and on leakage of ¹⁴C-labelled amino acids from *S. lactis* (c) and *E. coli* (d): ○, no OSCN⁻; ●, 25 μM OSCN⁻ (from Ref. 133).

variation in susceptibility amongst Gram negative organisms had been observed previously; at a given concentration of oxidizable SCN⁻, the sensitivity decreased in the order: *E. coli* 0111 (serum susceptible, i.e. killed by the complement-activated bacteriocidal activity of specific antibodies); *E. coli* 0101 (serum resistant; compare Figs 4 and 5); *Salmonella typhimurium*; *Ps. aeruginosa*. Increasing the SCN⁻ concentration and/or lowering the pH or cell density increases susceptibility.[125] (On the resistance of a strain of *E. coli*, see later.[137]) The 'barrier' hypothesis seems to be confirmed by more detailed studies on *S. typhimurium*;[138] organisms in the log phase were more resistant to killing than those in the stationary phase. Even more significantly, rough mutants were more susceptible than the smooth parent strain, the former possessing decreasing amounts of the polysaccharide core component of lipopolysaccharide and hence may provide a lesser barrier to permeability of OSCN⁻ than the complete antigen. Incidentally, susceptibility of bacteria to phagocytosis and serum sensitivity depend on the same properties, being either virulent (smooth) or avirulent (rough) strains.

The possible involvement of sulphydryl groups in the antibacterial effect was first suggested against *S. agalactiae*.[139] The lactoperoxidase-catalysed incorporation of SCN⁻ into sulphydryl groups of bovine serum led to the realization that OSCN⁻ was rapidly consumed in the oxidation of bacterial sulphydryls to sulphenyl thiocyanate and sulphenic acid derivatives.[132,140,141] Hence it is not surprising that some of the key glycolytic enzymes of streptococci were found to be inhibited.[104,142,143] The susceptibility of an organism appears to depend also on the total sulphydryl content. The resistance of *Streptococcus mutans* was increased by preincubation with some sulphydryl compounds.[144] Nevertheless, reactions other than with sulphydryl groups cannot be excluded, since it was found that the non-sulphydryl-dependent enzyme D-lactate dehydrogenase, bound to the inner membrane or in purified form, is also inhibited.[145] Organisms exposed to the lactoperoxidase system fail to take up lactate because they cannot generate proton motive forces across the inner membrane. Unfortunately these findings do not explain the rapid leakage of K⁺ and amino acids. Besides the 'relative resistance', depending on the cell envelope and sulphydryl level of the bacteria, there exist also completely resistant organisms. Certain streptococci were shown to be killed (*S. pyogenes*) or inhibited in raw milk (*S. agalactiae*) while others multiplied (*S. disgalactiae*) as well as in heated milk.[5] These observations were extended[4] to the then known

304 BRUNO REITER

Lancefield's sero-groups of streptococci and it was shown that different strains of the same sero-group could be either susceptible or resistant;[9] in the case of Group N streptococci, the different susceptibilities of starter strains to inhibition in raw milk[9] led initially to the identification of SCN⁻ as *the* oxidizable substrate because other oxidizable substrates occurring in milk, such as iodide and indican, inhibited both susceptible and resistant organisms.[104,105]

The early observations[104] of resistance amongst streptococci (Fig. 8) proved eventually to be particularly important in ecological systems such as the intestinal and oral flora. As far as starters are concerned, the obvious solution to the problem was to test the starter strains for inhibition in raw milk (free of bacteriophage) and avoid their use.

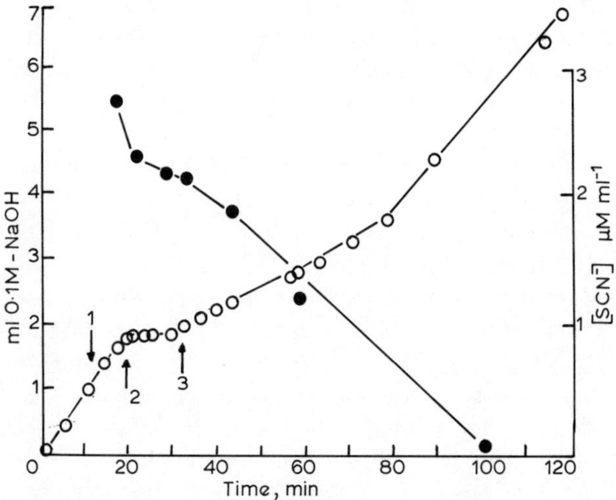

FIG. 8. Inhibition of *Streptococcus cremoris* (1972) by the lactoperoxidase system and its reversal by a cell-free extract of a resistant strain of *S. cremoris* (803) (○, NaOH added; ●, SCN⁻ added). Additions to a growing culture of *S. cremoris* 792: 1, 5 μM SCN⁻; 2, 70 units of lactoperoxidase; 3, 1 ml of cell-free extract of resistant *S. cremoris* 803 (from Ref. 104).

Resistant strains were shown to possess a 'reversal' factor (enzyme) which reverses the inhibition of glycolysis; it catalyses the oxidation of $NADH_2$ in the presence of an intermediate oxidation product (now known to be $OSCN^-$). This factor was partly purified from resistant strains and was absent in susceptible strains.[97,128] The resistance factor has now been confirmed to be present in some oral streptococci (*S. mutans*) and renamed $NADH:OSCN^-$-oxidoreductase, which converts $OSCN^-$ into the inert thiocyanate (SCN^-).[143] There is some circumstantial evidence that $OSCN^-$ only inhibits *E. coli* while the higher oxyacids were suggested to be bacteriocidal.[146] The existence of very short-lived higher oxyacids was first proposed as antistreptococcal agents by Hogg and Jago.[131] Pruitt *et al.*,[147] in their kinetic and polarographic studies on the oxidation of SCN^-, came to the conclusion that the higher oxyacids are produced at increased H_2O_2 concentrations. These findings explain why SCN^- appeared to be abruptly further oxidized at SCN^- to H_2O_2 ratios above 1:1.[97,104] Carlsson *et al.*[148] confirmed that $OSCN^-$ at increased H_2O_2 concentrations becomes bacteriocidal and cytotoxic probably because of the higher oxyacids produced.

4.4. Biological Significance and *In Vivo* Activity

Research on the intestinal flora of the calf established that a high proportion of the lactobacilli colonizing the oesophagus, abomasum and upper duodenum produce H_2O_2.[119,149] That this natural source of H_2O_2 could activate the lactoperoxidase/thiocyanate system was proven as follows. A culture of *E. coli* was fed to cannulated calves which were either given raw milk or milk heated to destroy the lactoperoxidase. Samples withdrawn at intervals from the abomasum showed that *E. coli* were killed when raw milk was fed (reduction by $\sim 95\%$) but no reduction occurred when heated milk was fed. Addition of a solid source of H_2O_2 (magnesium peroxide, MgO_2) reduced the inoculum by 99·99% (Fig. 9).[150] So far, it has not been possible to detect any oxidation products of SCN^- in the abomasal fluid because it was not possible to clarify it sufficiently. Samples from the duodenum should be more suitable because raw milk fed to calves is rapidly clotted in the abomasum and clear straw-coloured whey flows through the pylorus within minutes of feeding.

The natural occurrence and activity of the lactoperoxidase system has now been established also in a different way. Research on dental caries showed that saliva contains not only SCN^- but also the oxidation product, hypothiocyanite $(OSCN^-)$, catalysed by salivary peroxidase and H_2O_2 generated by resistant oral streptococci. When freshly collected saliva was incubated, the $OSCN^-$ level increased appreciably, H_2O_2 being the limiting component. This was the first time that $OSCN^-$ was detected in a secretion under physiological conditions, thus proving that the lactoperoxidase system operates naturally *in vivo*. Concurrently,

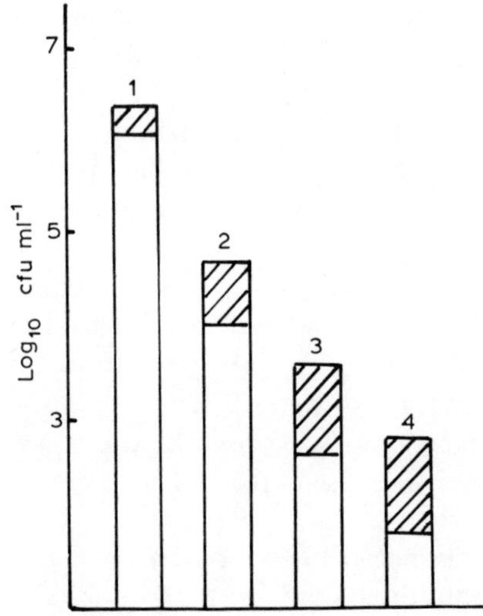

FIG. 9. Recovery of *E. coli* from abomasal fluid 30 min (shaded area) or 60 min (unshaded area) after feeding a cannulated calf with milk containing approximately 10^7 cfu ml^{-1} *E. coli* 9703. 1, Heated milk (no lactoperoxidase); 2, raw milk; 3, raw milk plus glucose/glucose oxidase; 4, raw milk plus MgO_2 (from Ref. 149).

acid production in dental plaques is inhibited, suggesting that the system reduces caries (e.g. Refs 124 and 144).

Considering that the lactoperoxidase system is known to inhibit many of the bacterial species that cause bovine mastitis, it is surprising how few *in vivo* experiments have been performed to date. The non-lactating udder has long been recognized as a latent source of bacteria which affect the udder after calving (e.g. Ref. 151). Intensive research led subsequently to the development of the so-called NIRD method of mastitis control which consists of hygiene measures, routine antibiotic treatment of the udder after drying-off (see Ref. 152) and milking machine control.[153] The 'NIRD method' of mastitis control produces a marked decrease in intramammary infections caused by staphylococci and streptococci with the exception of *S. uberis*, and *E. coli*, which are now considered the major mastitis-causing organisms (e.g. Ref. 154). However, although treatment with antibiotics proved to be successful, and is being widely used, there is always doubt about the indiscriminate use of antibiotics and the emergence and spread of multiple antibiotic-resistant strains. It was and remains therefore of theoretical and practical interest to study the natural defence mechanisms, both to understand them and hopefully to increase their effectiveness.

Attempts[32,155] to evaluate the *in vivo* role of phagocytosis, complements and antibodies in mastitic infections proved to be inconclusive but yielded unexpected results (Table 6).[155] A limited number of glands were

TABLE 6

RATE OF INFECTION AFTER INFUSION OF STAPHYLOCOCCI OR STREPTOCOCCI INTO QUARTERS IMMEDIATELY AFTER LAST MILKING OR INTO QUARTERS DRY FOR 14 DAYS[155]

| | *Number of quarters infected after infusion with* | | | |
| | *Staphylococci* | | *Streptococci* | |
No. of days quarters dry	*20 cfu/qr*	*2000 cfu/qr*	*20 cfu/qr*	*2000 cfu/qr*
0	1/6	6/6[a]	0/6	0/6
14	1/6	4/6[b]	1/6	4/6

Six cows infused with staphylococci, 6 cows infused with streptococci. In each cow, 2 drying-off quarters and 2 dry quarters infused with low (20 cfu/qr) and high (2000 cfu/qr) inocula, respectively.
[a] Infection delayed in 1 quarter.
[b] Infection delayed in 2 quarters.

infused with 2000 cfu of either *S. uberis* or *Staph. aureus* on the day of drying-off and 14 days afterwards. *S. uberis* failed to infect any glands (0/6) when infused at day 0 (immediately after drying-off) but infected 4 of 6 glands 14 days after drying-off (2 delayed infections). *S. aureus* infected every gland at day 0, and 4 of 6 glands at day 14 (2 delayed). Although the small number of glands investigated did not allow firm conclusions, it was suggested that at the beginning of the dry period the lactoperoxidase system may inhibit *S. uberis* but not *Staph. aureus*. This interpretation, however, depends on the presence of O_2 which would allow the formation of H_2O_2. Previous attempts to measure the O_2 pressure of milk in the bovine teat sinus indicated a 15% saturation with O_2; these results were obtained by cannulating milk through a layer of paraffin and they need to be confirmed by direct measurements *in vivo* by introducing an O_2 electrode through the teat canal. It is, however, of some interest that 15% O_2 saturation was found to be sufficient to inhibit *S. uberis* in milk (unpublished). There is an analogy in phagocytosis: intracellular killing is reduced under anaerobic conditions and the leucocytes of children with chronic granolomatous disease can kill catalase-negative bacteria *in vitro* (e.g. H_2O_2-producing streptococci); catalase-positive bacteria (e.g. staphylococci) survive because the leucocytes are metabolically unable to generate sufficient H_2O_2. The *in vivo* information about the infections suffered by such children confirms the *in vitro* results (e.g. Ref. 156). Determination of the O_2 pressure *in vivo* could also have an important bearing on the efficiency of the leucocytes in preventing infections both in the lactating and in the non-lactating mammary gland. At the same time it would be of interest to investigate whether the intracellular killing depends, at least partially, on the level of SCN^- as previously suggested.[21,32,128]

More recently the experiments involving the infection of non-lactating udders with *S. uberis* were repeated, confirming the earlier results:[157] 30–600 cfu of *S. uberis* were infused into glands at drying-off (day 0) and after 7, 14 and 21 days. The rate of infection was: day 0, 0/9; day 7, 1/8; day 12, 2/6; day 21, 8/10. After 21 days and up to 1 day pre-partum, 100% of the glands became infected (total 18). Those results basically confirm the findings of Reiter *et al.*[155] and Roguinsky[158] who suggested that the lactoperoxidase system is most active against *S. uberis* at the beginning of the dry period. This suggestion appears to be substantiated by the findings of Brown and Mickelson[159] who attributed the declining lactoperoxidase activity against *Streptococcus agalactiae* in the non-lactating udder to the increasing concentrations of cystine and cysteine in the secretion. The authors regarded the former as a growth stimulant, counteracting the system, while the latter would reduce the oxidized $OSCN^-$ to SCN^-.[160–163]

The most acceptable way to determine the possible role of the lactoperoxidase system would be to reverse the effect by a strong reducing agent such as sodium sulphite or mercaptoethanol, or better yet, to assay whether $OSCN^-$ is formed in the lactating and/or non-lactating bovine udder *in vivo*.

4.5. Practical Applications

From 1976 onwards, animal trials were undertaken to feed newborn calves with whole raw milk containing the complete activated lactoperoxidase system, a source of H_2O_2 (glucose oxidase/glucose) and increased SCN^- concentrations. Trials performed under various conditions showed that scouring (diarrhoea) was reduced and the live weight gains significantly increased especially when the incidence of scouring in the controls (no artificially activated lactoperoxidase system) was high

(Fig. 10). When scouring was low (as in the herd of the National Institute for Research in Dairying, Shinfield, UK), live-weight gains were still higher, but less so, for the calves fed milk with the lactoperoxidase system compared with the control animals (Table 7).[164]

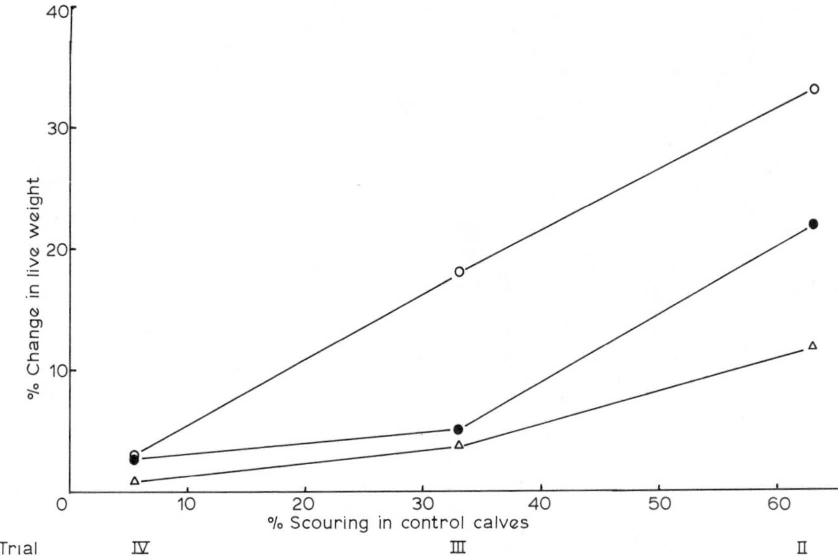

FIG. 10. Relationship between the effect of the lactoperoxidase system on live weight change (%) and the level of scouring (%) in calves given whole milk only in three separate trials (II, III, IV) over the periods 0–2 weeks (○), 0–5 weeks (●) and 0–8 weeks (△) (from Ref. 164).

These calf trials were performed by feeding milk and not calf starters; this means that even the natural, host-specific milk can be improved by artificially activating the lactoperoxidase system. Most calf starters are based on skim milk powders, and it was shown that lactoperoxidase can be preserved and activated by a suitable source of H_2O_2. Successful trials have been made with such milk powders in the rearing of calves.[165,166]

Another use of the lactoperoxidase system was first suggested to the Food and Agricultural Organization, Rome, in 1967 (cited in Ref. 167). It was argued that the supply of low concentrations of H_2O_2 (0·1 mM) or the continuous production of H_2O_2 by enzymes (xanthine oxidase, native to milk) and hypoxanthine, xanthine, or a meat extract might be

TABLE 7

OVERALL MEAN LIVE-WEIGHT CHANGES FOR CALVES FED WHOLE MILK WITH OR
WITHOUT THE LACTOPEROXIDASE SYSTEM[164]

	Whole milk + LPS	*Whole milk*	*Difference ±S.E.[a]*
No. of calves	92	93	−1 ±
Birth weight (kg)	36	37	
Live-weight change (g day^{-1})			
0–1 week	306	269	37 ±35·0
0–2 weeks	340	294	46[b] ±21·7
0–3 weeks	368	322	45[c] ±16·8
0–5 weeks	371	341	31[b] ±12·4
0–8 weeks	477	454	23 ±13·2

[a]S.E. based on the variation between calves within treatments and trials.
Levels of significance: [b]$p<0.05$; [c]$0.001<p<0.01$.

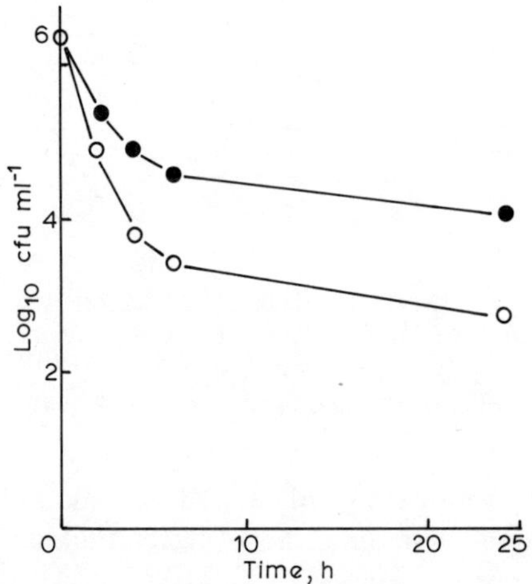

FIG. 11. Effect of the lactoperoxidase system on *Ps. fluorescens* in milk at 4°C in the presence of 0·17 mM SCN$^-$ (●) or 0·26 mM SCN$^-$ (○). Glucose (0·3%) and glucose oxidase 0·1 U ml^{-1}) were added to aseptically drawn milk (from Ref. 169).

used to preserve milk collected under difficult conditions and subjected to long journeys at high ambient temperatures. At that time this approach was considered to be impractical. However, Reiter *et al.*,[123] in co-operation with Alfa Laval, Sweden, showed at first that psychrotrophs were killed in cooled milk by the activated lactoperoxidase system (see also Refs 124, 125, 168) (Fig. 11). To investigate the effect on cheese making, milk was experimentally inoculated with *Ps. fluorescens*, incubated for 3 days at 4°C, and pasteurized before being made into cheese; the cheese became rancid and unacceptable.[169] However, when another batch of milk, inoculated to the same level, was treated with the lactoperoxidase system, the cheese developed normal flavour, without rancidity. Although pasteurization killed the psychrotrophs, their enzymes, which are extremely heat-resistant, survived in the control cheeses and spoiled the flavour (Table 8).

TABLE 8

EFFECT OF LACTOPEROXIDASE SYSTEM ON MULTIPLICATION OF *Ps. fluorescens* AND CHEESE QUALITY[169]

Treatment of milk	*No. of Ps. fluorescens (cfu ml^{-1} × 10) in milk stored at 5°C for (days)*				*Cheese quality at 4 months*	
	0	*1*	*2*	*3*	*FFA^a (µmol 10 g^{-1})*	*Flavour*
None	15	29	150	1400	248	Rancid
LP system	21	1·1	0·2	0·1	50	Normal

^{a}Free fatty acids.

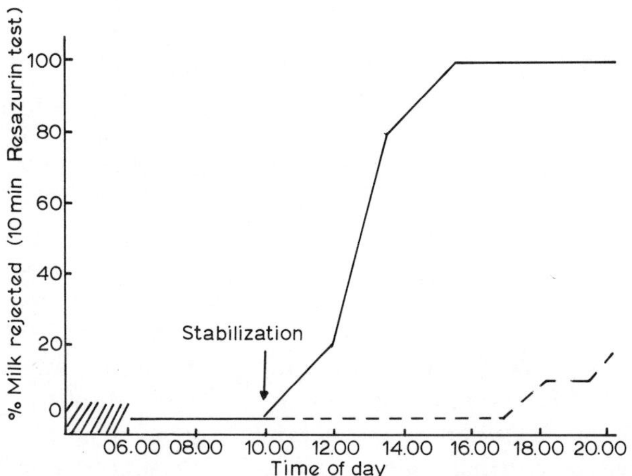

FIG. 12. Effect of stabilization of milk by activation of the lactoperoxidase system on the keeping quality of raw milk delivered to the Narahenpita Collection Centre, Sri Lanka. Ambient temperature: 29–30°C, shaded area indicates milking period, ——, control samples; – – –, stabilized samples (from Ref. 171).

In large-scale trials (Kenya, Mexico, Sri Lanka, Pakistan), the lacto-peroxidase system was activated to preserve raw milk collected under difficult conditions in developing countries. In spite of ambient tempera-tures of up to 37°C, the keeping quality of the milk was increased appreciably by the activation of the lactoperoxidase system measured by the 10 min Resazurin test.[170] The preservative effect was not due to the reduction of the Gram-negative flora but, as expected,[167] to the sup-pression of acid development by contaminating lactic acid bac-teria[150,171] (Fig. 12 and Table 9).*

TABLE 9

DEVELOPMENT OF ACIDITY IN SAMPLES OF MILK TAKEN AT THE GIRIULLA COLLECTION CENTRE, SRI LANKA[150]

	Acidity (% lactic acid) at (time)				
	10 a.m.	12 noon	2 p.m.	4 p.m.	6 p.m.
Stabilized milk (LP)	0·15	0·15	0·16	0·16	0·17
Untreated controls	0·16	0·17	0·19	0·22	0·26

Cows were milked at 4–6 a.m. and samples were stabilized by activation of the lactoperoxidase (LP) system at 7.30 a.m. Ambient temperature exceeded 30°C at around 9 a.m. and reached 32°C in the afternoon.

5. SYNERGISM

The best-known example of synergism is the complement-mediated bacteriocidal effect of specific antibodies, mainly of the IgM class. The multiple lesions produced in the outer membrane of Gram-negative organisms facilitate the access of lysozyme to the peptidoglycan layer, thus rapidly lysing the organism. Secretory IgA (sIgA) by itself is not complement-binding but may bind it in the presence of lysozyme,[172,173] although this has been disputed.[174] Secretory IgA has also been shown to enhance the antimicrobial effect of the lactoperoxidase system against *S. mutans*.[175] The main function of sIgA is now recognized to be the prevention of the attachment of pathogens (e.g. *E. coli*, *Vibrio cholera*) to the epithelium of the intestines *in vivo*. The sIgA must be specific for the

*The use of the activated lactoperoxidase system is protected by British Patents 1468405 (Method of improving the keeping qualities of milk and other liquids) and 1546747 (Antibacterial pharmaceutical and feedstuff compositions).

antigenic adhesins (plasmid-controlled) which determine the capability of an organism to attach in a particular host. Exposure to the lacto-peroxidase system has an effect similar to specific sIgA *in vitro*; it prevents the specific attachment to brush borders isolated from either pig or calf[33,36] (Figs 13–15, Table 10). This effect needs to be confirmed *in vivo* because it may be an important defence mechanism, replacing or augmenting the action of sIgA. Moreover, if motility in the mucin layer contributes to the process of attachment, the system also immobilizes bacteria because it inhibits energy metabolism.

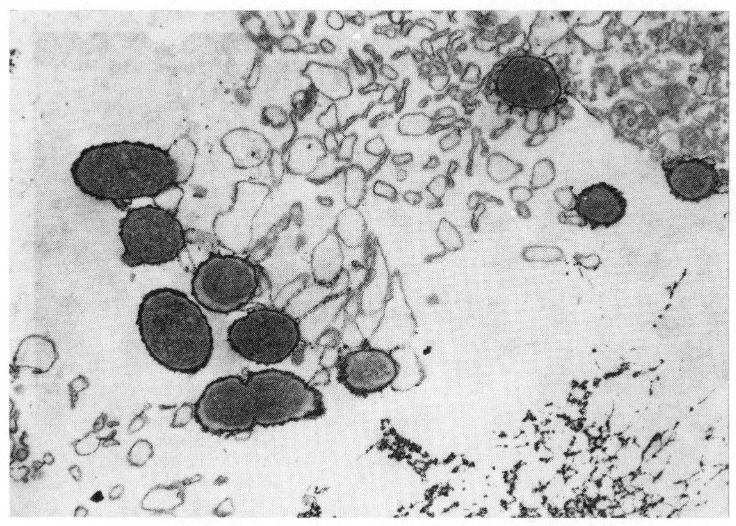

FIG. 13. Attachment of *E. coli* 0149:K_{91},K_{88}ac to porcine brush border cells. Treated with colloidal iron hydroxide at pH 1·8 revealing anionic surface antigen, ×18 000. (Courtesy of B. E. Brooker, NIRD.)

Under physiological conditions, the most likely limiting factors in the lactoperoxidase system are SCN^- and H_2O_2; the former is derived from the diet, feed, saliva or secretion in the stomach, and the latter from various sources. Catalase-negative organisms can produce enough H_2O_2 to become self-inhibitory in the presence of lactoperoxidase and SCN^- or, if resistant to the system, provide H_2O_2 to inhibit or kill other bacteria as in ecological systems such as in the intestinal tract or oral cavity. Actively phagocyting leucocytes may also be a source of H_2O_2. Eukaryotic cells (e.g. in the salivary gland) are also capable of producing

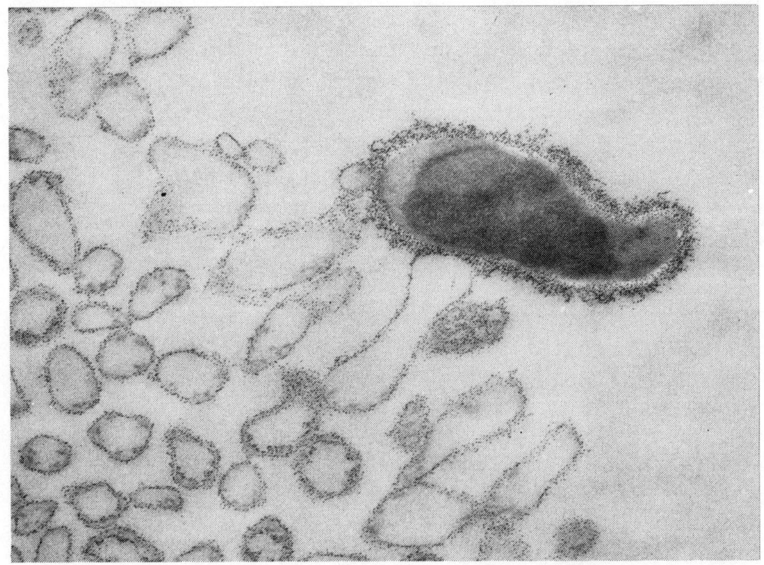

FIG. 14. Attachment of *E. coli* 0149:K_{91},K_{88}ac to porcine brush border cells treated with cationic ferritin to reveal cationic surface antigen, $\times 25\,000$. Note the minimal even staining of the microvilli on the brush border cells and the large bacterial capsule. (Courtesy of B. E. Brooker, NIRD, also Refs 33 and 105.)

sufficient H_2O_2 to oxidize SCN^- because $OSCN^-$ was detected in sterile samples of stimulated human perotid saliva collected directly from Stenson's duct.[176] In this context, the question arises of whether achlorhydric patients (who are unable to secrete normal levels of hydrochloric acid) are also unable to secrete SCN^-; moreover, such patients are known to have an abnormal intestinal flora and hence may lack H_2O_2-producing lactic acid bacteria.

Several workers reported that specific antibodies promote the bacteriostatic activity of lactoferrin,[89,95,177,178] while others disputed it.[179,180] Interestingly, it appears now that the antibodies are not directed towards the somatic antigens, as thought previously, but against the low molecular weight iron chelators (siderophores) and outer membrane proteins synthesized by the organisms to overcome iron deficiency.[69,69a,69b,178,181] Siderophores and outer membrane protein receptors are now recognized to belong to the virulence factors of Gram negative pathogens. They have been now detected and isolated from *E. coli* 'grown' in the peritoneum of mice.[69b] It is therefore not surprising

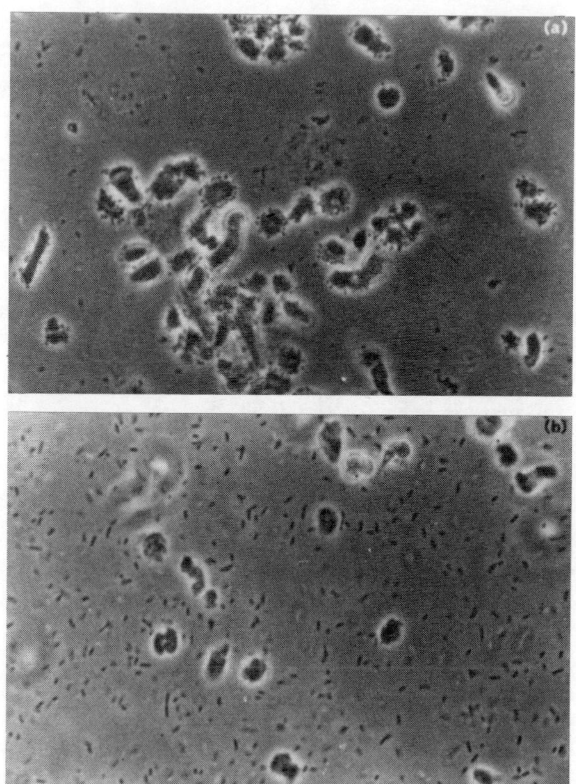

FIG. 15. Attachment of *E. coli* 0149:K_{91},K_{88}ac to isolated porcine brush border cells. (a) attachment of *E. coli* (control); (b) reduced attachment of *E. coli* after exposure to lactoperoxidase system (from Refs 33 and 105).

TABLE 10

ATTACHMENT OF *E. coli* 0149:K_{91},K_{88}ac TO PORCINE
BRUSH BORDER CELLS

Control	Treatment with lactoperoxidase system
14, 17, 18, 16	0, 0, 4, 3
17, 20, 12, 8	3, 7, 6, 3
16, 17, 14, 16	2, 4, 0, 5
15, 18, 20, 12	1, 0, 2, 2
11, 14, 14, 13	4, 2, 6, 3
$x = 14 \cdot 6$	$x = 2 \cdot 95$
$s = \pm 2 \cdot 9$	$s = \pm 2 \cdot 00$

that antibodies against siderophores have been demonstrated in human serum[181] and human milk (J. H. Brock, Glasgow University, pers. comm.). Antibodies against iron-regulated outer membrane proteins were detected in adult sera but at best only in trace amounts in sera of infants between 2 months and 3 years.[182–184] Human milk, however, contains appreciable amounts of such antibodies (E. Griffith, National Institute Biological Standards and Control, London, pers. comm.). There are no data available for bovine sera or milk, although such natural antibodies may be important in relation to mastitis and the protection of newborn animals against Gram-negative pathogens. The contradictions may be resolved by ascertaining in future whether the organisms used in experiments synthesize siderophores and their receptors or not.

A direct interaction between lysozyme and lactoferrin was observed with *Micrococcus luteus*:[185] the protoplasts produced by the lytic activity of lysozyme were strongly agglutinated by lactoferrin; it is suggested that the lysozyme–lactoferrin complex may increase this effect. The capacity of lactoferrin to form complexes may even effect the attachment of pathogens to intestinal epithelia. It was suggested[58] that the fucose residues of lactoferrin-glycans may complex with fucose-rich glycopeptides of intestinal mucus and thus inhibit fucose-sensitive adhesion of some pathogens.

Lactoferrin was reported[186] to increase haemolysis of red blood cells by activating, or at least stimulating, C-complement. Rivier *et al.*[187] reported that transferrin (blood plasma iron chelator) increases the susceptibility of *E. coli* to the complement-mediated bacteriocidal activity of specific antibodies, possibly because iron-deprived organisms become energy-deficient and are therefore less able to repair damage (by complement-antibody) to their cell envelope.

6. COMMENTS AND PROSPECTS

Bovine milk is an important source of human food but its antimicrobial activity has so far not been exploited. Our knowledge about the occurrence, distribution and mode of action of the protective proteins offers new opportunities for its utilization.

As detailed in this chapter, the colostrum and milk of all species contain a multifactorial antibacterial system, albeit at different concentrations and in different proportions. (The antiviral properties have so far been concerned with specific antibodies although non-antibody factors are beginning to be investigated; see e.g. Ref. 188.) These proteins protect the newborn and thus help to bridge the immunological gap until the newborn synthesizes its own defence systems. The same protective systems appear to contribute to the defence of the mammary gland against infections but the evidence for it is scant as yet.

There is little doubt that natural feeding provides the best protection for the newborn. Artificial feeding depends for its success on the nutritive composition and degree of hygiene obtainable but remains unsatisfactory as long as the protective proteins are not either preserved during manufacture, taking account of their different heat sensitivity,[189,190] or added as purified proteins afterwards. Feeding dried bovine milk causes difficulties even in the rearing of calves, partly because the heat treatment destroys or reduces the clotting ability of the reconstituted milk and partly because the lactoperoxidase is destroyed (see experiments with heated milk). Dried bovine milk is even more unsuitable for other animals such as piglets, rabbits and guinea piglets which cannot be successfully reared on such products. The human infant seems to be remarkably resilient; its digestive system deals with cow's milk rather well. However, the standard of hygiene needs to be adequate to avoid infections, as events in developing countries have recently highlighted.

At least it is now recognized that the milk of each species is nutritionally best suited to its own offspring whose requirements depend on the maturity at birth, digestive system, rate of growth and the environment into which it is born (the calf of a whale or a porpoise *does* need ~45% fat in the milk). It took a very long time for paediatricians and nutritionists to appreciate the difference in the composition of different milks but nowadays babies are given, nearly exclusively, 'humanized' cow's milk. It is to be hoped that the significance of the heat-sensitive protective proteins will eventually be appreciated as well.

6.1. Immunoglobulins

Although these proteins are not the subject of this chapter, they interrelate with the other protective proteins and at least need to be summarized.

Long before secretory IgA was discovered, Petersen and his collaborators[191,192] (see also Refs 33, 34, 36) fought for his concept of the mammary gland as an 'exocrine endothelial gland' capable of producing antibodies against bacteria and viruses. (The latter was confirmed by intramammary infusion of bacteriophage as a model for viruses.[193]) It was also demonstrated[191,192] that the process of suckling by calves infected with salmonellae stimulated the production of specific antibodies in the udder. This 'on demand' antibody production was termed 'dia-thelic' (i.e. incoming through the duct). In 1975, Salajka *et al.*[194] con-firmed the phenomenon with *E. coli* infected piglets.

Petersen aimed to produce 'immune' milk by vaccinating cows with human pathogens; this concept has now been taken up and led to the manufacture of 'milk immunoglobulin concentrate' (e.g. Refs 195–198) which contains antibodies against the main serotypes of strains of *E. coli* (and rotavirus) affecting infants. Good results have been reported in the treatment of pre-term, low-weight and hospitalized infants in two trials.[196,199] It is suggested that the bovine class of immunoglobulins, IgG, representing 73–81% of the preparation, withstands proteolytic diges-tion similarly to secretory IgA.[200,201] In this context, it is of interest that bovine colostrum contains naturally specific antibodies, bacteriocidal in the presence of complements, against human enteropathogenic strains of *E. coli*, indicating a considerable cross-reaction between human and bovine strains.[33,34] Bovine and human rotavirus strains are also known to have common antigens, and hence it is not surprising that human rotavirus antibodies occur in bovine colostra.[197]

6.2. Lactoperoxidase

The major non-antibody protective protein in bovine milk is lacto-peroxidase which was first exploited to select resistant starter strains, to preserve milk and then to improve the growth of newborn calves. The importance of using, whenever possible, 'host-specific' milk (or at least milk adjusted in its composition) was experienced during a series of experiments with newborn piglets. Cow's milk frequently formed a bolus in the stomach which could become extended to three times its normal size. Failure to empty the stomach regularly influenced the results greatly and the benefit of the lactoperoxidase system remained equivocal. In a later exploratory experiment, diluted colostrum proved to be nut-ritionally more acceptable and appeared to give clear-cut, albeit very limited, results. Eight piglets were infected with *E. coli* 0149, K_{88}, at birth (no colostrum) and hand-fed with diluted bovine colostrum with and without the activated lactoperoxidase system. All the piglets (4) receiving the lactoperoxidase system remained well and unaffected while 3/4 of the controls developed severe diarrhoea within 48–60 h and 1 piglet died (Reiter, Knutsson, Ewos, A. B., unpublished). Out of 10 infected piglets that remained with the dam, 6 died and 4 remained unaffected.

Before the treatment of milk with the lactoperoxidase system (or indeed the direct treatment of infants and adults by the activated system) can be accepted by the FAO and WHO, it must be proven that the system occurs naturally in man and animals and has no undesirable side-effects.[202]

Besides proving that the lactoperoxidase system occurs naturally, a number of investigations have indicated that the system has no toxic or damaging effects. Erythrocytes, which are considered to be useful models for investigating the damaging effects of agents on mammalian membranes, were not lysed following exposure to the lactoperoxidase system, which is in complete contrast to the effect of the myeloperoxidase–H_2O_2–Cl^- system which not only lyses erythrocytes but also inhibits spermatozoa (as well as being virucidal) (reviewed by Clark[204]). Bovine spermatozoa exposed to the lactoperoxidase system showed decreased mobility and penetration into cervical mucus,[205] which points only to interference with the energy metabolism.

The concentration of SCN^- required for the preservation of milk is so low (12 ppm) that it falls within the limits observed in the milk of cows on natural pastures. Alternatively, kale or rape seed, which contain glucosides that yield SCN^-, or clover which contains CN^- (detoxified to SCN^-), could 'naturally' increase the SCN^- level. Activation of the system requires little H_2O_2 (~ 8 ppm), which is in contrast to the 300–800 ppm recommended[203] by the FAO for the preservation of milk produced under difficult conditions, such as lack of cooling facilities, high ambient temperatures and distant transport. The addition of H_2O_2 could not be regarded as 'adulterating' the milk. If it can be shown that $OSCN^-$ occurs in milk, this would indicate that H_2O_2 occurs naturally in milk oxidizing SCN^- (cf. saliva); the addition of SCN^- would only correct the shortage due to current feeding regimes. Furthermore, H_2O_2 is rapidly used up in the oxidation of SCN^-, and $OSCN^-$ is both short-lived and sensitive to pasteurization. This is in contrast to the high concentrations of H_2O_2 which, according to FAO recommendations, have to be removed by the addition of catalase.

More stringent criteria will have to be applied before the lactoperoxidase system can be approved for the treatment of diarrhoea in infants and adults. It will be necessary to assay salivary peroxidase and SCN^- in the gastric juice and ascertain whether the lactic acid flora in the stomach and duodenum is resistant to the lactoperoxidase system and produces H_2O_2.

As detailed above, the lactoperoxidase system uniquely affects the inner or cytoplasmic membrane and membrane-bound enzymes.

Respiration is immediately inhibited in bacteria but is not affected, for instance, in whole liver cells; only the respiration of isolated mitochondria is inhibited. The lack of haemolysis and inhibition of respiration of these two cell types indicates that $OSCN^-$ does not penetrate the mammalian cell membranes (Finch, Baum and Reiter, to be published). The effect on the isolated mitochondria is not surprising because they are commonly considered to be 'internalized' bacterial membranes.

It appears therefore that the lactoperoxidase system is not toxic; indeed it may be even protective against the cytotoxic effect of H_2O_2 generated *in vitro* and *in vivo*. *E. coli* and some oral streptococcal species[148a,148b] can be protected against lethal concentrations of H_2O_2 (1 mM) by the presence of lactoperoxidase thiocyanate, but only under anaerobic conditions. Organisms incubated under anaerobic conditions proved to be viable and apparently had the capacity to repair any lesions in the inner membrane caused during the aerobic exposure to lactoperoxidase and SCN^- or to the oxidation products (see Fig. 7).[133] However, they remain inhibited.

During clinical investigations on a toothpaste containing H_2O_2-producing enzymes for the activation of salivary peroxidase, it was observed that patients who had suffered from recurrent aphthous lesions (oral ulcers) were cured in the majority of cases and generally remained free from lesions.[117,206] This apparent healing and protective effect provoked recent research on various cultured cells *in vitro*. Hänström *et al.*[207] and Tenovuo and Larjava[208] showed, independently, that human gingival fibroblasts could be protected against the toxic effect of H_2O_2; the former workers demonstrated the same effect against HeLa and Chinese hamster ovary cells which retained their capacity to proliferate. However, their rate of lactic acid production in the presence of glucose was reduced, as in bacteria, but only at much higher concentrations of $OSCN^-$, although mammalian glyceraldehyde-3-phosphate dehydrogenase was inhibited like the bacterial enzyme. $NADH:OSCN^-$ oxido-reductase, which occurs in resistant bacteria,[97,148] was not detected in mammalian cells. It appears that the lower sensitivity of mammalian cells compared with that of bacteria might be due to an impermeability of the mammalian cell membrane to $OSCN^-$ (see results on haemolysis).

Not only does H_2O_2 damage the mammalian cell membrane[209] but it also causes single-strand breaks in DNA and is mutagenic.[210] White *et al.*[211] demonstrated that the lactoperoxidase system and $OSCN^-$ does not oxidize calf thymus DNA (no change in UV spectra) and has no mutagenic effect on mutagen-sensitive strains of *Salmonella typhimurium* and *Saccharomyces cerevisiae*. $OSCN^-$ synthesized chemically also fails to kill streptococci[116] or *E. coli*,[146] inhibiting only their metabolism and growth. It was therefore suggested that not $OSCN^-$ but higher oxidation products of SCN^- are responsible for the killing of *E. coli*.[146] Such higher oxidation products are produced only when H_2O_2 exceeds SCN^- in the presence of lactoperoxidase or when $OSCN^-$ is exposed to H_2O_2 in the absence of lactoperoxidase. It is now suggested that such further intermediary oxidation product (HO_2SCN^-, cyanosulphurous acid) is bactericidal for bacteria and toxic for mammalian cells. The end oxidation products have now been confirmed to be inert.[11,97,116,143,147,148]

All these observations bear out the axiom that oxygen metabolism leads to toxic end-products such as oxygen free radicals and H_2O_2 but that enzymes such as superoxide dismutase (which occurs also in milk[212]), catalase and peroxidase protect eukaryotic cells.

There is little doubt that the explanation of the effect of the lactoperoxidase system in calves appears to be more complex than presented previously.[119] We did take into account the effect of low pH in the abomasum, the secretion of SCN^- and H_2O_2 production by colonizing lactic acid bacteria. This is why raw milk through its level of lactoperoxidase was effective under control conditions without supplying a source of H_2O_2 and SCN^-, and the addition of these components only increased the bactericidal effectiveness of the system.[119,150] *E. coli* was always assayed aerobically and we do not know how well the organisms would have survived anaerobically.[149] *In vivo* this depends on the O_2 tension in the stomach and the upper duodenum or on how soon the conditions become truly anaerobic. The recovery under anaerobic conditions may be prevented by other inhibitors such as lactoferrin and lysozyme (particularly in breast-fed infants). We should also like to know whether the lactoperoxidase system prevents the attachment to the intestinal tissue *in vivo*. Although these questions and doubts need to be answered, it has been shown recently that scouring in calves due to the ingestion of *E. coli* carrying the K99 adhesin was convincingly prevented by feeding both the lactoperoxidase system *and* lactoferrin (J. P. Prieels and J. P. Paraudin, Oleofina SA, Brussels, pers. comm.). Foremost,[165,166] of course, extensive field trials—over 1000 calves—and the successful commercial use of calf starters containing the activated lactoperoxidase system (M. Knutsson, Enos AB, Sweden, pers. comm.) do show that the lactoperoxidase system increases the live weight gains of calves. Whatever the explanation of the working of the system *in vivo*, attempts are now being made to explain the system in the rearing of piglets and eventually to extend it to infants who cannot be breast-fed.

Reiter[35] and Korhonen[107] discussed the question of whether the antibacterial factors in milk, and particularly the lactoperoxidase system, protect the mammary gland or the newborn against infection. Both authors deplored the lack of *in vivo* evidence for the role of the lactoperoxidase system in mastitis. However, recent confirmation[157-163] of earlier work[32,155] showing that the lactoperoxidase system may protect the drying-off bovine udder against infection by *S. uberis* (but not *Staph. aureus*) should stimulate further work. It is clear that positive information on the O_2 pressure in the lactating and non-lactating udder (H_2O_2 production by *S. uberis*) and identification of the intramammary oxidation product of SCN^-, $OSCN^-$, are required. The influx of leucocytes, mainly macrophages, during drying-off may generate an additional source of H_2O_2 during the ingestion of fat and casein. Also, an attempt ought to be made to demonstrate whether the high numbers of leucocytes actively phagocyte and kill bacteria in the dry udder.

Early work[32,97] suggested that the myeloperoxidase of polymorphonuclear leucocytes may oxidize SCN^- to antibacterial oxidation products of SCN^-. This concept was completely overshadowed in the medical literature by the initial erroneous concept that myeloperoxidase utilizes I^- and later Cl^- also (reviewed by Clark[204]). I^- was discarded[104] from the beginning because its concentration is far too low in secretions such as milk, and Cl^- was of no interest because lactoperoxidase, unlike myeloperoxidase, does not oxidize Cl^-. Recent findings may revive the concept that leucocytes utilize SCN^-. In the past it was accepted that human milk contains lactoperoxidase,[118] and it had been shown that *E. coli* are killed only after addition of SCN^- (and a source of H_2O_2).[36] Indeed, in 1976 raw human milk was successfully preserved at $4°C$ (Fig. 16) (J. Hewitt, Public Health Laboratory Service, Dulwich Hospital, London, pers. comm.). Since then, however, it appears that the peroxidative activity of human milk is due partly to myeloperoxidase (derived from leucocytes) and partly to salivary lactoperoxidase. K. M. Pruitt (pers. comm.) confirmed that the 'milk peroxidase' requires SCN^- (and H_2O_2). Considering that milk contains Cl^-, it appears that the enzyme oxidizes SCN^- in preference to Cl^-; furthermore, the human milk peroxidase–SCN^- system proved to be dramatically more bacteriocidal than the bovine milk lactoperoxidase–SCN^- system because this strain was relatively resistant to the latter (see under mode of action, Section 4.3). These unexpected results appear to warrant a reassessment of the contribution of the human milk peroxidase–Cl^- system to the intracellular killing of bacteria by polymorphonucleated leucocytes.[204] Cl^- is oxidized to OCl^- which is highly toxic; amongst other effects, it lyses erythrocytes and would be expected

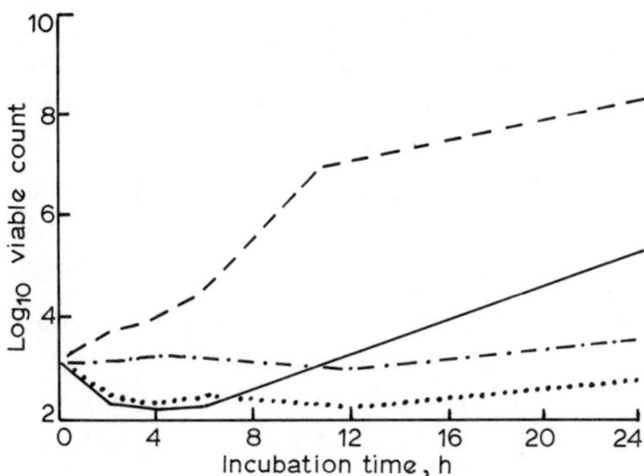

FIG. 16. Effect of the lactoperoxidase system on the bacterial quality of human milk collected for a milk bank, – – –, Untreated control at 37°C; ——, complete system at 37°C; –·–·–, untreated control at 4°C; ··, complete system at 4°C. (Personal communication from Dr J. Hewitt, Public Health Laboratory Service, Dulwich Hospital, London.)

to damage the membranes of the phagocytic vacuola and fusion of lysosomes involved in intracellular killing. It therefore needs to be tested whether the highly permeable anion SCN^- diffuses from the blood into the phagocytic cells and thus becomes oxidized as previously suggested.[32,36,97,105] (The myeloperoxidase–Cl^- system is also supposed to produce free radicals[204] while there is no direct proof for the lactoperoxidase–SCN^- system.[105]

6.3. Lactoferrin

Up to now it was not feasible to consider the fortification of artificial feeds based on bovine milk with milk lysozyme and lactoferrin, which are the major non-immunoglobulin protective proteins in human milk. It now appears that a technically feasible purification method for lactoperoxidase and lactoferrin has been developed (J. P. Perraudin and J. P. Prieels, Oleofino SA., Brussels, pers. comm.), and it is to be hoped that this process can be extended to the purification of lysozyme. This advance in dairy technology increases the significance of the results obtained on feeding a formula feed fortified with conalbumin to infants with diarrhoea. Considering that adults with chronic diarrhoea (achlorhydric) who did not respond to antibiotic treatments appear to benefit from conalbumin (U. Cornelli, Ricordati SPA, Milan, pers. comm.), it is expected that the effect of lactoferrin (and/or some of the other protective proteins) could become useful for both infants and adults.

6.4. Lysozyme

There is enough evidence available to show that lysozyme is resistant to digestion, but there is no direct evidence that intestinal bacteria are lysed. Ideally, germ-free animals, mono-contaminated with susceptible bacteria, ought to be fed with lysozyme (preferably milk lysozyme) to assess its *in vivo* activity instead of the attempts to deduce its activity by analysing the faecal flora. Other interesting aspects are its synergistic activity with the other systems and its possible adjuvant (immunomodulator) effect.

Recent research appears to show that endotoxins (lipopolysaccharides) of Gram-negative organisms regulate some immuno responses, at least in mice and piglets. Endotoxins seem to have an important role in inducing precursor suppressor cells in the gut-associated lymphoid tissue (GALT) and in mediating one form of oral tolerance to small dietary changes in the newborn (e.g. foreign proteins, ovalbumin, fed orally to mice and piglets[214–218]). It appears therefore that the natural encounter of antigens with GALT in the presence of endotoxin (as B-cell mitogen) may have profound effects on the host and mediate one form of oral tolerance in the newborn.

It follows that the Gram-negative intestinal flora has a role to play. The natural defence systems do not eradicate the Gram negative flora like broad-spectrum antibiotics but establish a balanced intestinal flora.

6.5. Milk Fat Globule Membranes

Although this subject has not been treated in the main text, it warrants attention because earlier work has now been confirmed which ought to increase interest in non-immunoglobulin glycopeptides, and lipids, in colostrum and milk.[219] The milk fat globule (FG) is surrounded by a true cell membrane (unit membrane). This can be shown by electron microscopy and is supported by the serological cross-reaction (agglutination) between bovine, but not human, red blood cells (RBC) and fat globule membranes (FGM); FG are also agglutinated by the cold agglutinins present in blood and milk.[220] The agglutination of RBC by *E. coli* possessing K_{88} or K_{99} antigens was shown to be inhibited by FG or FGM. The enzyme produced by *Vibrio cholerae*, which destroys the RBC receptors, also destroyed the receptor on the FG and FGM for coliforms.[221] The agglutination of RBC's, however, is nonspecific in the sense that RBC's of several species can be used for agglutination. The ability of *E. coli* to attach directly to FG or FGM was therefore investigated on the supposition that the FGM, as a true cell membrane, may possess the same specific receptors as intestinal brush borders.

Indeed, it was found that enteropathogenic bovine strains of *E. coli* possessing K_{99} antigen attach only to bovine FGM, porcine enteropathogenic strains possessing K_{88} antigen attach only to porcine FGM, while human enteropathogenic strains possessing CFA_1 and CFA_2 colonization factors attach to human FGM but there was also a low attachment rate to porcine FGM (Fig. 17, Table 11) (for review on

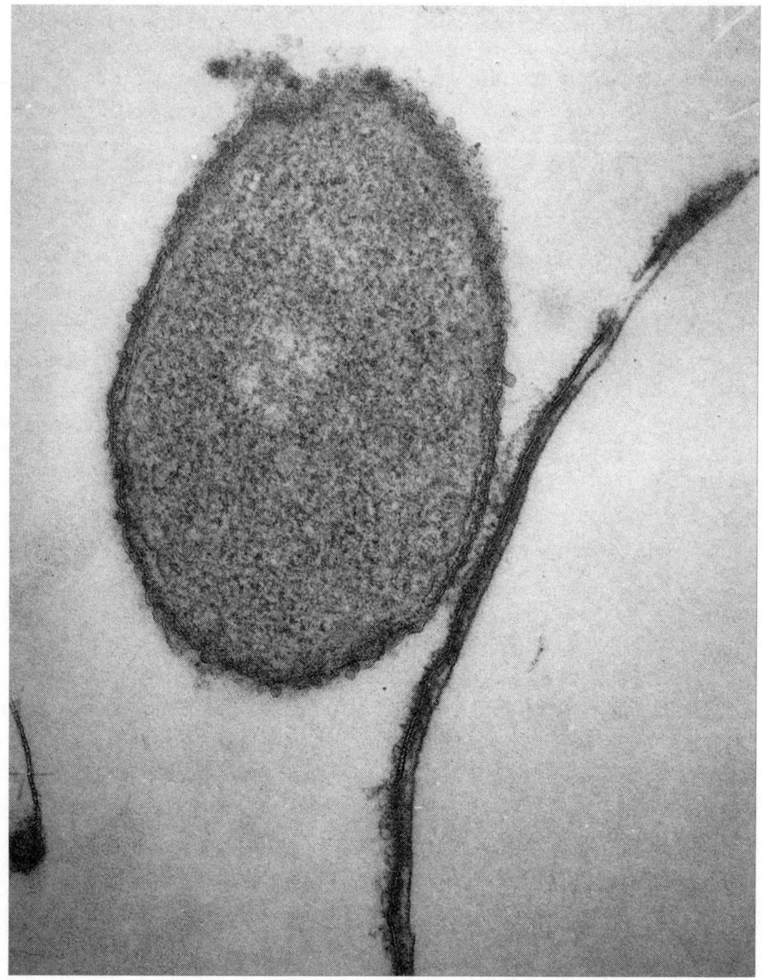

FIG. 17. Attachment of *E. coli* to fat globule membrane. (Courtesy of B. E. Brooker, NIRD, also Ref. 36.)

TABLE 11

SPECIFIC ATTACHMENT OF *E. coli* TO FAT GLOBULE MEMBRANES

Antigens	Log_{10} reduction[a]		
	Porcine	Bovine	Human
K_{88}	2·5–3·5[a]	0	1·0
K_{99}	0	2·5–3	0
CFA I and II (human)	0	0	2–3

[a] Difference between inoculum and number of bacteria attached to cream layer after centrifugation.

attachment, see Ref. 222). The bovine and porcine strains failed to attach to FGM after exposure to the lactoperoxidase system, while the human strains remained unaffected. The latter strains, which are mannose-sensitive (i.e. fail to attach to RBC's in the presence of mannose), were also prevented by mannose from attaching to FGM. Clearly, the mode of attachment between these strains differs. These experiments raise the question of whether this attachment is so strong that it hinders attachment to the intestine *in vivo*.[36]

An attempt was also made to develop a simple test for the determination of attachment of strains of *E. coli*. Wright and Tramer[8] showed that some strains of *Streptococcus cremoris* are agglutinated in milk and, on standing, are carried with the fat globules to the cream layer. This phenomenon explained why some starters are inhibited when grown in raw milk. For a long time a similar test has been used to detect brucellosis: the 'ring test' for *Brucella abortus* agglutinins whereby the organisms also rise to the cream layer if the milk contains specific antibodies (agglutinins). A photomicrograph published by Kenyon *et al.*[223] clearly shows the clustering of fat globules by *Brucella* organisms attributed to IgA and IgG. IgA described by the authors is probably secretory IgA, and the attachment of the major part of sIgA to fat globules was confirmed only recently.[224] A similar test based on agglutination of fat globule membranes has also been described recently.[225]

E. coli specifically clusters fat globules, which can be observed under the microscope. Alternatively, 10^7–10^8 cfu ml^{-1} of the organisms can be inoculated into whole milk, which after incubation for 30 min at $37°C$ is centrifuged to separate the fat. The bacterial count in the skim milk indicates the degree of attachment to the fat (Table 10); reductions to

10^2–10^3 cfu ml^{-1} were common (unpublished). The brush border membranes of susceptible and resistant phenotypes of pigs show different glycolipid patterns; their galactosyl residues are important.[226,227] Carbohydrate compositions for erythrocyte receptors from *E. coli* possessing K_{99} (calf enteropathogen) and CFA/I adhesins (human) were suggested by Faris *et al.*[228]

Considering the well-known similarities between the structural and chemical nature of erythrocytes and FGM, it is reasonable to speculate that some of their constituents (glycolipids and glycoproteins) possessing inhibitory activities can be isolated from various milk fractions. Indeed, various milk fractions, non-immunoglobulin in nature, were shown to have interesting biological activities: inhibition of heat-labile enterotoxins of *E. coli* and *Vibrio cholerae*;[229–231] receptor-like glyco compounds in human milk which inhibit *V. cholerae* adherence (haemagglutination[231,232]); and a rotavirus inhibitor.[188]

It appears therefore that the FGM can be used as a model for the attachment to the intestinal brush border cells, and a comparison between the composition of the membranes—erythrocytes, FGM and brush border—ought to be fruitful. If it emerges that the FGM plays an active part in the prevention of attachment to the intestinal brush border, the dairy industry could produce an unlimited supply of membranes isolated from butter-milk.

It is well known that infants can suffer allergic reactions to cow's milk. Several of the milk components have been implicated, but it is astonishing that the fat globule membrane containing foreign proteins has not been considered.

In conclusion, it may be of interest to cite a paragraph written by Freter[233] in his contribution to the symposium on the Association of enterotoxigenic bacteria with the mucosa of the small intestine: 'Many foodstuffs, especially milk, contain toxins as well as oligosaccharides and proteins, some of which resemble similar substances in the intestine and therefore may also resemble receptor for bacterial adhesion (attachment). It should also be possible to enrich foodstuffs by the addition of receptor materials. Many foodstuffs may therefore block bacterial association with mucosa by competing with mucosal receptor for adhesins or with bacterial receptors for toxins ... Such mechanisms may explain some of the still enigmatic relations between diet and susceptibility to diarrhoeal diseases and may be responsible for at least part of the well known protective effect of mother's milk discussed elsewhere in this symposium'. Weiser *et al.*[234] in 1969 were first to describe and acknowledge the

various non-antibody protective proteins in a textbook of immunology:
'Indeed, the attention of immunologists, which continues to be focussed
largely on antibodies, has led to the virtual neglect of a second broad
area in immunology concerned with non-antibody factors of host resistance to harmful parasites and other foreign agents, which play an
important role by augmenting antibody action, but also by affording
protection before the specific immuno response is effective'.

This statement can now be rephrased and extended. 'There is such a
strong analogy between phagocytosis and intracellular killing by leucocytes and milk—complements, immunoglobulins, lactoferrin, lysozyme, peroxidases—that ... if we accept that leucocytes are the
primary defence against invading organisms, we could regard colostrum
and milk as liquid leucocytes because so many of the antimicrobial
factors are common to both'.[33]

Hopefully the dairy industry will eventually realize some of the aims
discussed in this chapter.

REFERENCES

1. FLEMING, A., *Proc. Roy. Soc. Biol. Sci.* (London), 1922, **93**, 306.
2. FLEMING, A., *Proc. Roy. Soc. Med.*, 1932, **24** (pathology section), 1.
3. HESSE, W., *Z. Hyg. Infekt.*, 1894, **17**, 238.
4. WILSON, A. T. and ROSENBLUM, R., *J. Exp. Med.*, 1952, **95**, 25, 39.
5. JONES, F. S. and SIMMS, H. T., *J. Exp. Med.*, 1930, **51**, 327.
6. AUCLAIR, J. E. and HIRSCH, A., *J. Dairy Res.*, 1953, **20**, 45.
7. AUCLAIR, J. E., *J. Dairy Res.*, 1954, **21**, 323.
8. WRIGHT, R. C. and TRAMER, J., *J. Dairy Res.*, 1957, **24**, 174.
9. WRIGHT, R. C. and TRAMER, J., *J. Dairy Res.*, 1958, **25**, 104.
10. JAGO, G. R. and MORRISON, M., *Proc. Soc. Exp. Biol. Med.*, 1962, **111**, 585.
11. REITER, B., PICKERING, A., ORAM, J. D. and POPE, G. S., *J. Gen. Microbiol.*, 1963, **33**, xii.
12. BORDET, J. and BORDET, M., *C. R. Acad. Sci.*, 1924, **179**, 1109.
13. ROSENTHAL, L. and LIEBERMAN, H., *J. Infect. Dis.*, 1931, **18**, 226.
14. EHRLICH, P., *Z. Hyg. Infekt.*, 1902, **12**, 183 (cited by Campbell and Petersen, 1963).
15. JOLLÈS, P. and JOLLÈS, J., *Nature*, 1961, **192**, 1187.
16. MONTREUIL, J. and MULLET, S., *C. R. Acad. Sci.*, 1960, **250**, 176.
17. JOHANSSEN, B., *Acta Chem. Scand.*, 1960, **14**, 510.
18. GROVES, M. L., *J. Am. Chem. Soc.*, 1960, **82**, 3345.
19. MASSON, P. L., In: *La Lactoferrine*, S. A. Arscia (ed.), 1970, Librairie Maloine, Paris.
20. MASSON, P. L., HEREMANS, J. F., PRIGNOT, J. and WAUTERS, G., *Thorax*, 1966, **21**, 538.

21. ORAM, J. D. and REITER, B., *Rep. Nat. Inst. Res. Dairying*, 1966, 93.
22. ORAM, J. D. and REITER, B., *Biochim. Biophys. Acta*, 1968, **170**, 351.
23. POLIS, B. D. and SHMUKLER, H. W., *J. Biol. Chem.*, 1953, **201**, 475.
24. MOLDOVEANO, Z., TENOVUO, J., MESTECKY, J. and PRUITT, K. M., *Biochim. Biophys. Acta*, 1982, **718**, 103.
24a. LANGBAKK, B. and FLATMARK, T., *FEBS Lett.*, 1983, **174**, 300.
25. TENOVUO, J., In: *The Lactoperoxidase System: Chemistry and Biological Significance*, K. M. Pruitt and J. Tenovuo (eds), 1985, Marcel Dekker, New York, pp. 102–22.
26. REITER, B., In: *International Workshop on Recent Results in the Composition and Physiological Proportions of Human Milk*, 1985, Elsevier, Amsterdam, in press.
27. BRAUN, G. H., *Klin. Paediatr.*, 1976, **188**, 297.
28. GOLDMAN, A. S. and SMITH, C. W., *J. Paediatr.*, 1973, **83**, 1082.
29. HANSON, L. A. and WINBERG, J. W., *Arch. Dis. Childhood*, 1972, **47**, 845.
30. MCCLELLAND, D. B. L., GRATH, J. and SAMSON, R. R., *Acta Paed. Scand. Suppl.*, 1978, **271**, 1.
31. REITER, B., In: *Inhibition and Inactivation of Vegetative Microbes*, F. A. Skinner and W. H. Hugo (eds), 1976, Academic Press, London, pp. 31–60.
32. REITER, B. and ORAM, J. D., *Nature*, 1967, **216**, 328.
33. REITER, B., *Ann. Rech. Vet.*, 1978, **9**, 205.
34. REITER, B., *J. Dairy Res.*, 1978, **45**, 131.
35. REITER, B., In: *Resistance Factors and Genetic Aspects of Mastitis Control*, L. Bassalik–Chabielska (ed.), 1980, Ossolineum, Jablonna, Poland, pp. 351–392.
36. REITER, B., In: *Immunological Aspects of Infection in the Fetus and Newborn*, H. P. Lambert and C. B. S. Wood (eds), 1981, Academic Press, London, pp. 155–195.
37. REITER, B., *Int. J. Tiss. Reac.*, 1983, **5**, 87.
38. REITER, B., In: *Human Milk Processing, Fractionation and the Nutrition of the Very Low Birthweight Infant*, D. Baum and T. Williams (eds), 1984, Raven Press, New York, p. 29.
39. OSSERMAN, E. F., CANFIELD, R. F. and BEYCHO, K. S., *Lysozyme*, 1974, Academic Press, New York.
40. REITER, B. and ORAM, J. D., *Nature*, 1962, **193**, 651.
41. VAKIL, J. R., CHANDAN, R. C., PARRY, R. M. and SHAHANI, K. M., *J. Dairy Sci.*, 1969, **52**, 1192.
42. METCALF, R. H. and DEIBEL, R. H., *Infect. Immun.*, 1972, **6**, 178.
43. METCALF, R. and DEIBEL, R. H., *J. Bacteriol.*, 1973, **113**, 278.
44. GOODMAN, H., POLLOCK, J. J., KATOMA, L. T., IACONO, V. J., CHO, M.-L. and THOMAS, E., *J. Bacteriol.*, 1981, **146**, 764.
45. TORTOSA, M., CHO, M.-L., WILKENS, J. T., IACONO, V. J. and POLLOCK, J. J., *Infect. Immun.*, 1981, **32**, 1261.
46. NAKAMURA, O., *Z. Immun. Forsch. Allergy Forsch.*, 1923, **38**, 425.
47. PETERSON, R. G. and HARTNELL, E., *J. Infect. Dis.*, 1955, **96**, 75.
48. MCCLELLAND, D. B. L. and VAN FURTH, R., *Immunology*, 1975, **28**, 1099.
49. ISAACSON, P., *Gut*, 1982, **23**, 578.

50. FALCHUK, K. R., PERROTTO, J. L. and ISSELBACHER, K. J., *New Engl. J. Med.*, 1975, **292**, 395.
51. DONALDSON, D. M., ELLSWORTH, B. E. and MATHESON, H., *J. Immunol.*, 1974, **92**, 896.
52. SELSTED, M. E. and MARTINEZ, R. J., *Infect. Immun.*, 1978, **220**, 782.
53. HANNEBERG, B. and FINNE, P., *Acta Paediatr. Scand.*, 1974, **63**, 588.
54. KRAWCZUK, J., SAWICKI, Z. and KRAWCZYNSKI, J., *J. Clin. Chem. Clin. Biochem.*, 1978, **16**, 343.
55. KORHONEN, H., PhD Thesis, Helsinki, 1973.
56. PADGETT, G. A. and HIRSCH, J. E., *Aust. J. Exp. Biol. Med. Sci.*, 1967, **45**, 569.
57. SUKEWA, K., KIKUCHI, T. and HONDA, H., *J. Jap. Soc. Food Nutr.*, 1967, **20**, 45.
58. SPIK, G., JORIEUX, S., MAZURIER, J., NAVARIO, J., ROMOND, C. and MONTREUIL, J., In: *Human Milk Processing, Fractionation and the Nutrition of the Very Low Birthweight Infant*, D. Baum and T. Williams (eds), 1984, Raven Press, New York, pp. 133–44.
59. JOLLÈS, P., *Biomedicine*, 1976, **25**, 275.
60. LODINOVÁ, R. and JOUJA, V., *Acta Paediatr. Scand.*, 1977, **66**, 709.
61. NAMBA, Y., HIDAKA, Y., TAKI, K. and MORIMOTO, T., *Infect. Immun.*, 1981, **31**, 580.
62. CARINI, S. and LODI, R., *Industria Latte*, 1982, **17**, 35.
63: WASSERFALL, F. and PROKOPEK, D., *Milchwissenschaft*, 1978, **33**, 228.
64. WASSERFALL, F. and TEUBER, M., *Appl. Environ. Microb.*, 1979, **38**, 197.
65. LODI, R., OGGIONI, F., VEZZONI, A. M. and CARINI, S., *Industria Latte*, 1983, **19**, 41.
66. SIONHON, A., DOUGLAS, J. R., DRASER, B. S. and SOOTHILL, J. F., *Arch. Dis. Child.*, 1982, **57**, 54.
67. KLOCKERS, M. and ROBERTS, P., *Acta Haemat.*, 1976, **55**, 289.
68. BULLEN, J. J., ROGERS, H. J. and GRIFFITH, E., *Curr. Top. Microbiol. Immunol.*, 1978, **80**, 1.
69. WEINBERG, E. D., *Microbiol. Rev.*, 1978, **42**, 45.
69a. WEINBERG, E. D., *Physiol. Rev.*, 1984, **64**, 65.
69b. GRIFFITH, E., *Philos. Trans. Roy. Soc. London Ser. B*, 1983, **303**, 85.
69c. WAGEGG, W. and BRAUN, V., *J. Bacteriol.*, 1981, **145**, 156.
70. BROCK, J. H., *Arch. Dis. Child.*, 1980, **55**, 417.
71. BROCK, J. H., In: *Topics in Molecular and Structural Biology: Metal Protein*, P. M. Hoorison (ed.), 1984, Macmillan, London, pp. 1–53.
72. SPIK, G. and MONTREUIL, J., *Bull. Europ. Physiopath. Resp.*, 1983, **19**, 123.
73. MASSON, P. L. and HEREMANS, J. T., *Eur. J. Biochem.*, 1977, **6**, 579.
74. REITER, B., BROCK, J. H. and STEEL, E. D., *Immunology*, 1975, **28**, 83.
75. BISHOP, J. G., SCHANBACHER, F. L., FERGUSON, I. C. and SMITH, K. L., *Infect. Immun.*, 1976, **14**, 911.
75a. GRIFFITH, E. and HUMPHREY, J., *Infect. Immun.*, 1977, **15**, 396.
76. AISEN, P. and LEIBMAN, A., *Biochem. Biophys. Res. Commun.*, 1968, **30**, 407.
77. ROGERS, H. J. and SYNGE, C., *Immunology*, 1978, **34**, 19.
78. REITER, B. and ORAM, J. D., *J. Dairy Res.*, 1968, **35**, 67.
79. ARNOLD, R. R., COLE, M. P. and McGHEE, J. R., *Science*, 1977, **197**, 263.

80. ARNOLD, R. R., BREWER, M. and GAUTHIER, J. J., *Infect. Immun.*, 1980, **28**, 893.

81. ARNOLD, R. R., RUSSEL, J. E., CHAMPION, W. J. and GAUTHIER, J. J., *Infect. Immun.*, 1981, **32**, 655.

82. ARNOLD, R. R., RUSSEL, J. E., CHAMPION, W. J., BREWER, M. and GAUTHIER, J. J. *Infect. Immun.*, 1982, **35**, 792.

83. HARMON, R. J. and SCHONBACHER, F. L., FERGUSON, L. E. and SMITH, K. L., *Infect. Immun.*, 1976, **13**, 533.

84. SENFT, B., KLOBASA, F., MEYER, F. and PFLCIDERER, V. E., *Züchtkel.*, 1976, **48**, 278.

85. SENFT, B., MEYER, F. and ERHARDT, G., In: *Resistance Factors and Genetic Aspects of Mastitis Control*, L. Bassalik-Chabielska (ed.), 1981, Ossolineum, Jablonna, Poland, pp. 441–457.

86. REITER, B. and BRAMLEY, A. J., In: *Seminar on Mastitis Control*, F. H. Dodd, T. K. Griffin and R. G. Kingwill (eds), 1975, Document 85, IDF, Reading, pp. 210–215.

87. BROCK, J. H., ARZABA, F., LAMPREAVE, F. and PIÑEIRO, A., *Biochim. Biophys. Acta*, 1976, **446**, 214.

88. BROCK, J. H., PIÑEIRO, A. and LAMPREAVE, F., *FEBS Lett.*, 1978, **17**, 23.

89. SPIK, G., CHERON, A., MONTREUIL, J. and DOLBY, J., *Immunology*, 1978, **35**, 663.

90. SPIK, G., BRUNET, B., MAZURIER-DEHAINE, C., FONTAINE, G. and MONTREUIL, J., *Acta Paediatr. Scand.*, 1982, **71**, 979.

91. COX, T. M., MAZURIER, J., SPIK, G., MONTREUIL, J. and PETERS, T. J., *Biochim. Biophys. Acta*, 1979, **588**, 120.

91a. FRANSSON, G. B., THOREN-THOLLING, K., JONES, B., HAMBREOS, L. and LÖNNERDAL, B., *Nutr. Res.*, 1983, **3**, 373.

91b. FRANSSON, G. B., KEEN, C. L. and LÖNNERDAL, B., *J. Pediatr. Gastroenterol. Nutr.*, 1983, **2**, 693.

92 FRANSSON, G. B. and LÖNNERDAL, B., *J. Paediatr.*, 1980, **96**, 380.

93. HUEBERS, H., HUEBERS, E. and FINCH, C. A., In: *The Biochemistry and Physiology of Iron*, P. Saltman and J. Hegenaur (ed.), 1982, Elsevier Biomedical, New York, pp. 311–319.

94. PEAKER, M. and LINZELL, J. L., *Nature*, 1975, **253**, 464.

95. BULLEN, J. J., ROGERS, H. J. and LEIGH, L., *Brit. Med. J.*, 1972, **446**, 69.

96. STEPHENS, S., HARKNESS, R. A. and COCKLE, S. M., *Br. J. Exp. Path.*, 1979, **60**, 252.

97. ORAM, J. D. and REITER, B., *Biochem. J.*, 1966, **100**, 373.

98. PHELPS, C. F. and ANTONINI, E., *Biochem. J.*, 1975, **147**, 385.

99. ANTONINI, E., ORSI, N. and VALENTI, F., *G. Mall. Infett. Parasit.*, 1977, **2**, 481.

100. VALENTI, P., DESTASIO, A., SEGANTI, L., MASTROMARINO, P., SINIBALDI, L. and ORSI, N., *J. Clin. Microbiol.*, 1980, **11**, 445.

101. VALENTI, P., ANTONINI, E. and ORSI, N., *Int. J. Tiss. Reac.*, 1983, **5**, 124.

102. CORDA, R., BIDDAN, P., CORRIAS, A. and PUXEDDU, E., *J. Tiss. Reac.*, 1983, **5**, 117.

103. PORTMAN, A. and AUCLAIR, J. E., *Lait*, 1959, **39**, 147.

104. REITER, B., PICKERING, A. and ORAM, J. D., In: *International Symposium:*

Food Microbiology, 4th edn, N. Molin (ed.), 1964, Almqvist and Wicksell, Uppsala, Sweden, pp. 297–305.

105. REITER, B., In: *Oxygen Free Radicals and Tissue Damage* (CIBA Foundation Symposium No. 65, new series), 1979, Excerpta Medica, Amsterdam, Oxford, New York, pp. 285–294.

106. KNEIFEL, W., *Oster. Milchn.*, 1981, **15**, 1.

107. KORHONEN, H., In: *Resistance Factors and Genetic Aspects of Mastitis Control*, L. Bassalik-Chabielska (ed.), 1981, Ossolineum, Jablonna, Poland, pp. 421–440.

108a. REITER, B. and PRUITT, K. M., In: *The Lactoperoxidase System*, R. M. Pruitt and J. Tenovuo (eds), 1985, Marcel Dekker, New York, pp. 123–42.

108b. PRUITT, K. M. and REITER, B., In: *The Lactoperoxidase System*, R. M. Pruitt and J. Tenovuo (eds), 1985, Marcel Dekker, New York, pp. 143–78.

109. KIERMEIER, F. and KAYSER, C., *Z. Lebensm. Unters. Forsch.*, 1960, **11**, 481.

110. KERN, R., WILDBRETT, G. and KIERMEIER, F., *Z. Naturforsch.*, 1962, **186**, 1082.

111. RUSSEL, M. W. and REITER, B., *J. Reticuloendothelial Soc.*, 1975, **18**, 1.

112. RUSSEL, M. W., BROOKER, B. E. and REITER, B., *Res. Vet. Sci.*, 1976, **20**, 30.

113. RUSSEL, M. W., BROOKER, B. E. and REITER, B., *J. Comp. Path.*, 1977, **87**, 43.

114. SCHINDLER, J. S. and BARDSLEY, W. G., *Biochem. Biophys. Res. Commun.*, 1975, **67**, 1307.

115. STELMASZYNSKA, T. and ZGLICZYNSKI, J. M., *Eur. J. Biochem.*, 1971, **19**, 56.

116. HOOGENDOORN, H., PIESSENS, J. P., SCHOLTES, W. and STODDARD, L. A., *Caries Res.*, 1977, **11**, 77.

117. HOOGENDOORN, H., In: *The Lactoperoxidase System: Chemistry and Biological Significance*, K. N. Pruitt and J. Tenovuo (eds), 1985, Marcel Dekker, New York, pp. 217–29.

118. GOTHEFORS, L. and MARKLUND, S., *Infect. Immun.*, 1975, **11**, 1201.

119. REITER, B., MARSHALL, V. M. E. and PHILLIPS, S. M., *Res. Vet. Sci.*, 1980, **28**, 116.

120. BOULANGÉ, M., *C. R. Soc. Biol.*, 1959, **153**, 2019.

121. LAWRENCE, A. J., *Proc. XVIII Int. Dairy Congr.*, 1970, **D**, 99.

122. LOGOTHETOPULOS, J. H. and MYANT, N. B., *J. Physiol.*, 1956, **133**, 213.

123. REITER, B., BJÖRCK, L., MARSHALL, V. M. E., LONGMAN, A. G. and COUSINS, C. M., *Ann. Rep. Nat. Inst. Res. Dairying*, 1973/74, 98.

124. BJÖRCK, L., ROSÉN, C. G., MARSHALL, V. M. E. and REITER, B., *Appl. Microbiol.*, 1975, **30**, 199.

125. REITER, B., MARSHALL, V. M. E., BJÖRCK, L. and ROSÉN, C. G., *Infect. Immun.*, 1976, **13**, 800.

126. THOMAS, E. L., BATES, K. P. and JEFFERSON, N. M., *J. Dent. Res.*, 1981, **60**, 785.

127. THOMAS, E. L., In: *The Lactoperoxidase System: Chemistry and Biological Significance*, K. M. Pruitt and J. Tenovuo (eds), 1984, Marcel Dekker, New York.

128. ORAM, J. D. and REITER, B., *Biochem. J.*, 1966, **100**, 382.

129. PRUITT, K. M. and TENOVUO, J., *Biochim. Biophys. Acta*, 1982, **704**, 204.

130. PRUITT, K. M., TENOVUO, J., FLEMING, R. H. and ADAMSON, M., *Caries Res.*, 1982, **16**, 315.

131. HOGG, D. H. and JAGO, G. R., *Biochem. J.*, 1970, **117**, 779, 791.

132. AUNE, T. M. and THOMAS, E. L., *Eur. J. Biochem.*, 1977, **80**, 209.

133. MARSHALL, V. M. E. and REITER, B., *J. Gen. Microbiol.*, 1980, **120**, 513.

134. MARSHALL, V. M. E. and REITER, B., *Proc. Soc. Gen. Microbiol.*, 1976, **3**, 109.

135. MARSHALL, V. M. E., PhD Thesis, Reading University, UK, 1978.

136. MICKELSON, M. N., *J. Bacteriol.*, 1977, **132**, 541.

137. THOMAS, E. L. and AUNE, T., *Antimicrob. Agents Chemother.*, 1978, **13**, 1000.

138. PURDY, M. A., TENOVUO, J., PRUITT, K. M. and WHITE, W. E., *Infect. Immun.*, 1983, **39**, 1187.

139. MICKELSON, M. N., *J. Gen. Microbiol.*, 1966, **43**, 31.

140. MICKELSON, M. N., *Appl. Env. Microb.*, 1979, **38**, 821.

141. THOMAS, E. L. and AUNE, T. M., *Infect. Immun.*, 1978, **20**, 456.

142. ADAMSON, M. and PRUITT, R. M., *Biochim. Biophys. Acta*, 1982, **658**, 238.

143. CARLSSON, J., IWAMI, Y. and YAMADA, T., *Infect. Immun.*, 1983, **40**, 70.

144. THOMAS, E. L., PERA, A., SMITH, K. W. and CHIANG, A. R., *Infect. Immun.*, 1983, **39**, 767.

145. LAW, B. A. and JOHN, P., *FEMS Microbiol. Lett.*, 1981, **10**, 67.

146. BJÖRCK, L. and CLAESSON, O., *J. Dairy Sci.*, 1980, **63**, 919.

147. PRUITT, K. M., TENOVUO, J., ANDREWS, R. H. and McKENE, T., *Biochemistry*, 1982, **21**, 562.

148. CARLSSON, J., EDLUND, M. B. W. and HÄNSTRÖM, L., *Infect. Immun.*, 1984, **44**, 581.

148a. CARLSSON, J., *Infect. Immun.*, 1980, **29**, 1190.

148b. ADAMSON, M. and CARLSSON, J., *Infect. Immun.*, 1982, **35**, 20.

149. MARSHALL, U. M., PHILLIPS, S. M. and TURVEY, A., *Res. Vet. Sci.*, 1982, **32**, 259.

150. REITER, B. and HÄRNULV, B. G., *Dairy Ind. Int.*, 1982, **47**(5), 13.

151. NEAVE, F. K., DODD, F. H. and HENRIGUES, E., *J. Dairy Res.*, 1950, **17**, 37.

152. DODD, F. H. and JACKSON, E. R. (eds), *The Control of Bovine Mastitis*, 1971, NIRD, Shinfield, Berks, UK; Unwin Brothers Ltd.

153. THIEL, C. C. and DODD, F. H. (eds), *Machine Milking*, 1979, Technical Bulletin 1, NIRD, Shinfield, Berks, UK.

154. COUSINS, C. L., HIGGS, F. M. and JACKSON, E. R., *J. Dairy Res.*, **47**, 11.

155. REITER, B., SHARPE, M. E. and HIGGS, T. M., *Res. Vet. Sci.*, 1970, **11**, 18.

156. KAPLAN, E. L., LAXDAL, T. and QUIE, P. G., *Pediatrics*, 1968, **41**, 591.

157. McDONALD, J. S. and ANDERSON, A. J., *Am. J. Vet. Res.*, 1981, **42**, 465.

158. ROGUINSKY, M., *Ann. Rech. Vet.*, 1977, **8**, 153.

159 BROWN, R. W. and MICKELSON, M. N., *Am. J. Vet. Res.*, 1979, **40**, 250.

160. BROWN, R. W., *J. Dairy Sci.*, 1967, **50**, 1572.

161. BROWN, R. W., *J. Dairy Sci.*, 1974, **57**, 797.

162. BROWN, R. W. and BAETZ, A. L., *Am. J. Vet. Res.*, 1976, **37**, 75.

163. MICKELSON, M. N., *Appl. Env. Microb.*, 1976, **32**, 238.

164. REITER, B., FULFORD, R. J., MARSHALL, V. M. E., YARROW, N., DUCKER, M. J. and KNUTSSON, M., *Anim. Prod.*, 1981, **32**, 297.

165. WATERHOUSE, A. and MULLAN, W. M. A., *Anim. Prod.*, 1980, **30**, 458.

166. WATERHOUSE, A. and MULLAN, W. M. A., *Ir. J. Fd. Sci. Technol.*, 1980, **4**, 69.

167. TENTONI, R., PASTORE, M. and OTTOGALLI, G., *Ann. Microbiol. Enzymol.*, 1968, **18**, 85.

168. BJÖRCK, L., *J. Dairy Res.*, 1978, **45**, 109.

169. REITER, B. and MARSHALL, V. M. E., In: *Cold Tolerant Microbes in Spoilage and Environment*, A. D. Russel and R. Fuller (eds), 1979, Appl. Bact. Technical Series No. 13, Academic Press, London, pp. 153–164.

170. BJÖRCK, L., CLAESSON, O. and SCHULTESS, W., *Milchwissenschaft*, 1979, **34**, 726.

171. HÄRNULV, B. G. and KANDASAMY, C., *Milchwissenschaft*, 1982, **37**, 454.

172. ADINOLFI, M., GLYN, A. H., LINDSAAND, M. P. and MILNE, C. M., *Immunology*, 1966, **10**, 517.

173. HILL, I. R. and PORTER, P., *Immunology*, 1974, **26**, 1239.

174. HEDDLE, R. J., KNOP, J., STEELE, E. J. and ROWLEY, D., *Immunology*, 1975, **28**, 1061.

175. TENOVUO, J., MOLDOVEANU, Z., MESTECKY, J., PRUITT, K. M. and MANSSON-RAHEMTULLA, B., *J. Immunol.*, 1982, **128**, 726.

176. PRUITT, K. M., MANSON-RAHEMTULLA, B. and TENOVUO, J., *Arch. Oral Biol.*, 1983, **28**, 517.

177. STEPHENS, S., DOLBY, J. B., MONTREUIL, J. and SPIK, G., *Immunology*, 1980, **41**, 597.

178. BROCK, J. H., PICKERING, M. G., McDOWELL, M. C. and DEAKON, A. G., *Infect. Immun.*, 1983, **40**, 453.

179. SAMSON, R. R., MIRTLE, C. and McCLELLAND, D. B. L., *Acta Paediatr. Scand.*, 1980, **59**, 517.

180. LAW, B. A. and REITER, B., *J. Dairy Res.*, 1977, **44**, 595.

181. MOORE, D. G., YAMCEY, T. J., LANGFORD, C. E. and EARHART, C. F., *Infect. Immun.*, 1980, **27**, 418.

182. GRIFFITH, E., STEVENSON, P. and JOYCE, P., *FEMS Microbiol. Lett.*, 1983, **16**, 95.

183. GRIFFITH, E., STEVENSON, P., THORPE, R. and CHART, H., *Infect. Immun.*, 1985, **47**, 808.

184. CHART, H. and GRIFFITH, E., *Infect. Immun.*, in press.

185. PERRAUDIN, J. P. and PRIEELS, J. P., *Biochim. Biophys. Acta*, 1982, **170**, 351.

186. MORGAN, O. S., BANKAY, J. and QUASH, G. A., *W. I. Med. J.*, 1975, **24**, 46.

187. RIVIER, D., PAGE, N. and ISLIKER, H., *Ann. Immunol.* (Inst. Pasteur), 1983, **134C**, 25.

188. ØTNAESS, A.-B. and ØRSTAVIK, I., *Infect. Immun.*, 1981, **33**, 459.

189. FORD, J. E., LAW, B. A., MARSHALL, V. M. E. and REITER, B., *J. Pediat.*, 1977, **90**, 29.

190. RAPTOPOULOU-GIGI, M., MARVICK, K. and McCLELLAND, D. B. L., *Brit. Med. J.*, 1977, **1**, 12.

191. CAMPBELL, B., SARVA, M. and PETERSEN, W. E., *Science*, 1957, **125**, 932.

192. CAMPBELL, B. and PETERSEN, W. E., *Dairy Sci. Abstr.*, 1963, **25**, 345.

193. DI BIASE, C. and REITER, B., *Ann. Rep. Nat. Inst. Res. Dairying*, 1962, 71.

194. SALAJKA, K., CERWOHOVS, E. J. and SARMANOVA, F., *Document Vet. Sci.*, 1975, **20**, 30.

195. HILPERT, H. and LINK-AMSTER, H., In: *Acute Diarrhoea: Its Nutritional Consequences in Children*, J. Bellanti (ed.), 1983, Raven Press, New York, pp. 123–128.

196. MIETENS, C., KLEINHORST, H., HILPERT, H., GERBER, H., LINK-AMSTER, H. and PAHUD, J. J., *Eur. J. Pediatr.*, 1979, **132**, 239.

197. MIETENS, C., HILPERT, H. and WERCHAU, H., In: *Acute Diarrhoea: Its Nutritional Consequences in Children*, J. Bellanti (ed.), 1983, Raven Press, New York, pp. 111–115.

198. HILPERT, H., In: *Human Milk Processing, Fractionation and the Nutrition of the Very Low Birthweight Infant*, D. Baum and T. Williams (eds), 1984, Raven Press, New York, pp. 17–28.

199. BALLABRIGA, A., FARRIAUX, J. P., HILPERT, H., GERBER, H. and ARCALIS, L., *Acta Paediatr. Belg.*, 1976, **29**, 126.

200. MICHAEL, J. G., RINGENBOCK, R. and HOTTENSTEIN, S., *J. Infect. Dis.*, 1971, **124**, 445.

201. ZINKERNAGEL, R. and COLOMBINI, A., *Med. Microbiol. Immunol.*, 1975, **162**, 1.

202. REITER, B. and HÄRNULV, G., *J. Food Protection*, 1984, **47**, 724.

203. FOOD AND AGRICULTURAL ORGANIZATION, Rome, Document 57/11/8655, 1957.

204. CLARK, R. A., In: *Advances in Inflammation Research*, G. Weissman (ed.), Vol. 5, 1983, Raven Press, New York, pp. 107–146.

205. REITER, B. and GIBBONS, R. J., *Ann. Rep. Nat. Inst. Res. Dairying*, 1964, 87.

206. HOOGENDOORN, H. and SCHOLTES, W., *Tijdschr. Tandheelk*, 1979, **86**, 36.

207. HÄNSTRÖM, L., JOHANSSON, A. and CARLSSON, J., *Med. Biol.*, 1983, **61**, 268.

208. TENOVUO, J. and LARJAVA, H., *Arch. Oral Biol.*, 1984, **29**, 445.

209. WEISS, S. J., YOUNG, J., LO BUGLIO, A. F. and SLIVKA, A., *J. Clin. Invest.*, 1981, **68**, 714.

210. BRADLEY, M. O. and ERICKSON, L. C., *Biochim. Biophys. Acta*, 1981, **654**, 135.

211. WHITE, W. E., PRUITT, K. M. and MANSSON-RAHEMTULLA, B., *Antimicrobiol. Agents Chemother.*, 1983, **23**, 267.

212. HILL, R. D., *Aust. J. Dairy Technol.*, 1975, **30**, 26.

213. MOLDOVENNO, G., TENOVUO, J., MESTECKY, J. and PRUITT, K. M., *Biochim. Biophys. Acta*, 1982, **718**, 103.

214. NEWBY, T. J., STOKES, C. R. and BOURNE, F. J., *Clin. Exp. Immunol.*, 1980, **39**, 349.

215. NEWBY, T. J., STOKES, C. R. and BOURNE, F. J., *Clin. Exp. Immunol.*, 1983, **52**, 117.

216. STOKES, T. J., NEWBY, T. J. and BOURNE, F. J., *Clin. Exp. Immunol.*, 1983, **52**, 1.

217. MICHALEK, S. M., KYONO, H., WANNEMUEHLER, M. J., MOSTELLEA, J. M. and MCGHEE, J. R., *J. Immunol.*, 1981, **128**, 1992.

218. WANNEMUEHLER, M. J., KYONO, H., BOBB, J. L., MICHALEK, S. M. and MCGHEE, J. R., *J. Immunol.*, 1982, **129**, 959.

219. GIBBONS, R. J. and VAN HOUTE, J., *Ann. Rev. Microbiol.*, 1975, **29**, 19.
220. REITER, B., *Ann. Rep. Nat. Inst. Res. Dairying*, 1967, 89.
221. REITER, B. and BROWN, P., *Proc. Soc. Gen. Microbiol.*, 1976, **3**, 109.
222. GAASTRA, W. and DE GRAP, F. K., *Microbiol. Rev.*, 1982, **46**, 129.
223. KENYON, A. J., JENNESS, R. and ANDERSEN, R. K., *J. Dairy Sci.*, 1966, **49**, 1144.
224. HONKANEN-BUZALSKI, T. and SANDHOLM, M., *Comp. Immun. Microbiol. Infect. Dis.*, 1981, **4**, 329.
225. ATROSHI, F., ALAVIUHKOLA, T., SCHILDT, R. and SANDHOLM, M., *Comp. Immun. Microbiol. Infect. Dis.*, 1983, **6**, 235.
226. KEARNS, M. J. and GIBBONS, R. A., *FEMS Microbiol. Lett.*, 1979, **6**, 165.
227. SELWOOD, R., *Biochim. Biophys. Acta*, 1980, **632**, 326.
228. FARIS, A., LINDAHL, M. and WADSTRÖM, T., *FEMS Microbiol. Lett.*, 1980, **7**, 265.
229. ØTNAESS, A.-B. and HALVORSEN, S., *Acta Path. Scand.* (C), 1980, **88**, 249.
230. ØTNAESS, A.-B. and SVENNERHOLM, A.-M., *Infect. Immun.*, 1982, **35**, 738.
231. HOLMGREN, J., SVENNERHOLM, A.-M. and ÅHRÉN, C., *Infect. Immun.*, 1981, **33**, 136.
232. HOLMGREN, J. and SVENNERHOLM, A.-M., *Infect. Immun.*, 1983, **39**, 147.
233. FRETER, R., In: *Cholera and Related Diarrhoea: Molecular Aspects of a Global Health Problem* (Proc. Nobel Symp. 43), 1978, Kargar, Basel, pp. 68–84.
234. WEISER, R. S., MYRVIK, Q. N. and PEARSALL, N. N., *Fundamentals of Immunology*, 1969, Lee and Febiger, Philadelphia.

Chapter 11

VITAMINS IN BOVINE AND HUMAN MILKS

F. M. CREMIN and PAUL POWER

Department of Nutrition, University College, Cork, Republic of Ireland

1. INTRODUCTION

Vitamins are organic chemicals required in trace amounts for growth and maintenance of health. This definition distinguishes them from other organic and inorganic nutrients in the diet, including essential amino acids which are required in relatively large quantities, and trace elements which are required in trace amounts. 'Vitamin' is an operational term referring to a given chemical for a given animal species; these compounds can be synthesized by certain species and are therefore not vitamins for those species. Vitamin C is a vitamin for primates and guinea pigs only, as they are deficient in the enzyme gulonolactone oxidase which is essential for vitamin C synthesis from dietary D-glucose or D-galactose. The vitamins have no chemical resemblance to each other, but because of a similar role in metabolism they are considered together. They are generally divided into two major groups: fat-soluble and water-soluble. Vitamins A, D, E and K are generally found in the lipids fraction of foods and are termed the fat-soluble vitamins. The water-soluble vitamins include vitamin C and the vitamins of the B complex: thiamine, riboflavin, niacin, pantothenic acid, pyridoxine, vitamin B_{12}, folic acid and biotin. While choline, inositol and *p*-aminobenzoic acid are sometimes classed as vitamins, they will not be considered as such here because they do not meet many of the criteria established at the outset for a vitamin.

In metabolism, the water-soluble vitamins and vitamin K function predominantly as coenzymes and thereby exert metabolic effects that are

TABLE 1

RECOMMENDED DAILY ALLOWANCES (RDA), CONCENTRATIONS OF VITAMINS IN HUMAN AND BOVINE MILKS, AND THE CONTRIBUTION OF MILK AND MILK PRODUCTS TO SELECTED WESTERN DIETS

Vitamin	RDA[a]	Vitamin concentration (typical) per litre		Contribution (%) of dairy products to the vitamin content of Western diets	
		Bovine milk	Human milk	Liquid milk: UK (1980–1982)[c]	Milk products: industrialized countries[e,f]
A (µg RE)	1000	400	600	20·5	12–45[e], 13[f]
D (IU)	400	40	15–40	4·4[d]	5–20[e]
E (µg)	10 000	1000	5000	3·3[d]	10[e]
K (µg)	2000[b]	50	20	0·8[d]	—
B₁ (µg)	1400	450	150	10·8	6–20[e], 8·6[f]
B₂ (µg)	1600	1750	370	36·7	35–70[e], 38·6[f]
Niacin (µg)	—	900	1800	—	—
Niacin equivalents (mg)	18	8·2	2·9	15	—
B₆ (µg)	2200	500	100	9·2	10–20[e], 10·3[f]
Pantothenic acid (µg)	8000[b]	3500	2500	14·9[d]	20–30[e]
Biotin (µg)	200[b]	35	7	3·3	—
Folic acid (µg)	400	55	40	5	—
B₁₂ (µg)	3	4·5	0·4	46·2	20·1[f]
C (mg)	60	20	45	8	4–13[e]

[a] RDA values for 23–50 year old adult males.[8]

[b] These vitamins have not been assigned an RDA value, and the contribution of dairy products to their intake have been calculated on the basis of recommended daily intakes.[16]

[c] Calculated as % of RDA from the data of Scott et al.,[14] assuming a per capita daily intake of 0·33 litre of whole milk.[88]

[d] Calculated from concentrations typically occurring in milk.

[e] Contribution expressed as a percentage of total vitamin intake.[16]

[f] Contribution of dairy foods, other than butter, to total vitamin intake.[108]

out of all proportion to their trace requirement in the diet (Table 1). The metabolic function of the fat-soluble vitamins, A, D and E, where elucidated, is other than as a coenzyme. Thus, vitamin A is central to the visual process as a constituent of the visual pigment rhodopsin. In contrast, in metabolism, vitamin D meets all the criteria of a hormone, while the antioxidant effect of vitamin E is dominant.

Milk and dairy products contribute significantly to the recommended daily dietary allowance (RDA) for many of the vitamins (Table 1) in Western societies. The vitamin content of milk is altered by many factors including nutrition, genetics, stage of lactation, season and food processing. Consequently, many variables may alter the quantity of vitamins consumed by an individual from a given volume of milk. This has two important consequences: reduced vitamin intake and potential vitamin imbalance which may result in adverse effects on the metabolism of other vitamins and nutrients. Examples of the latter include the essential roles of niacin, riboflavin and vitamin C in the metabolism of vitamin A, pyridoxine and folic acid, respectively.

This chapter is structured to include a general review of the chemical structure(s), physical properties, physiological functions, deficiency symptoms, human requirements and food sources of each vitamin. For further information on these topics, the reader is referred to some general texts.[1-5] The chapter also includes a more detailed discussion of factors affecting the concentration of individual vitamins in human and bovine milks.

2. VITAMIN A (RETINOL, AXEROPHTHOL)

2.1. Chemical Structure and Physical Properties

Retinol (vitamin A_1) and 3-dehydroretinol (vitamin A_2) are alcohols with the structures shown in Fig. 1; they occur in nature mainly as fatty acid esters. Their absorption is greatly enhanced by emulsifiers, e.g. bile salts. Vitamin A exists naturally in several *cis/trans* isomeric forms that result from configurational differences of the double bonds in the side chain (Fig. 2). The molecular structure of vitamin A indicates that sixteen isomers are theoretically possible, of which six are known. However, only two isomers are of real importance, namely all-*trans*-vitamin A (Fig. 1) which is the major naturally occurring form of vitamin A and is the form with highest biological activity, and neo-vitamin A, the 13-*cis* isomer, which has a biological activity of $\sim 75\%$ of the all-*trans* isomer. Natural

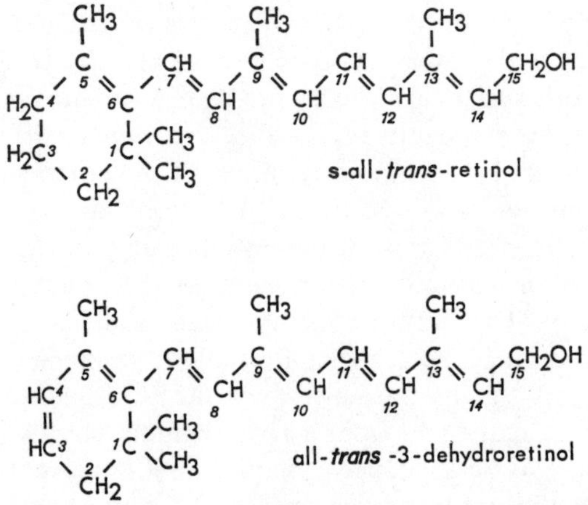

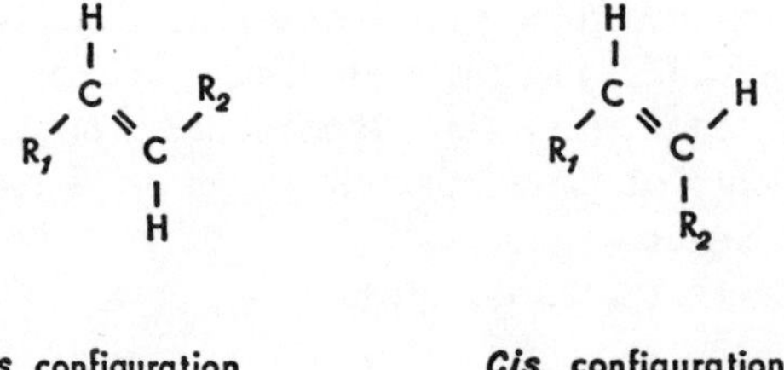

FIG. 1. Chemical structure of retinol (vitamin A_1) and 3-dehydroretinol (vitamin A_2).

FIG. 2. *Cis/trans* isomerism as in vitamin A and carotenes.

preparations of the vitamin contain ~33% neo-vitamin A, whereas good synthetic products generally contain considerably less. Vitamin A acetate and vitamin A palmitate are the principal commercial forms. Vitamin A esters are hydrolysed by a pancreatic hydrolase before absorption from the small intestine. The liberated vitamin A alcohol is re-esterified intracellularly with long-chain fatty acids, especially palmitic acid, during its absorption through the intestinal wall. The esters are marketed as oily solutions, stabilized powders or aqueous emulsions. Both are insoluble in water although water-dispersible forms of vitamin A acetate are commercially available and are commonly used.

Vitamin A esters show a characteristic absorption spectrum, the position of the absorption maximum depending on the solvent. In isopro-

panol and cyclohexane its absorption maximum occurs at 326 nm and
328 nm, respectively. Vitamin A alcohol and its esters are rapidly destroyed by light, oxygen and acids. For stability they must be stored in
nitrogen gas in hermetically sealed, opaque containers. Vitamin A may
be determined in foods and feedstuffs, after extraction, by a number of
methods including fluorescence spectrophotometry, the Carr–Price reaction or conversion of vitamin A into anhydrovitamin A by acid.

2.2. Physiological Function and Deficiency Symptoms

In the body, β-carotene (Fig. 3) can be converted into vitamin A by
cleavage at the midpoint of the polyene chain connecting the two β-
ionone rings. The biosynthesis of vitamin A from β-carotene is most
likely mediated by a dioxygenase-catalysed reaction in which molecular
oxygen reacts with the two central carbon atoms of β-carotene followed

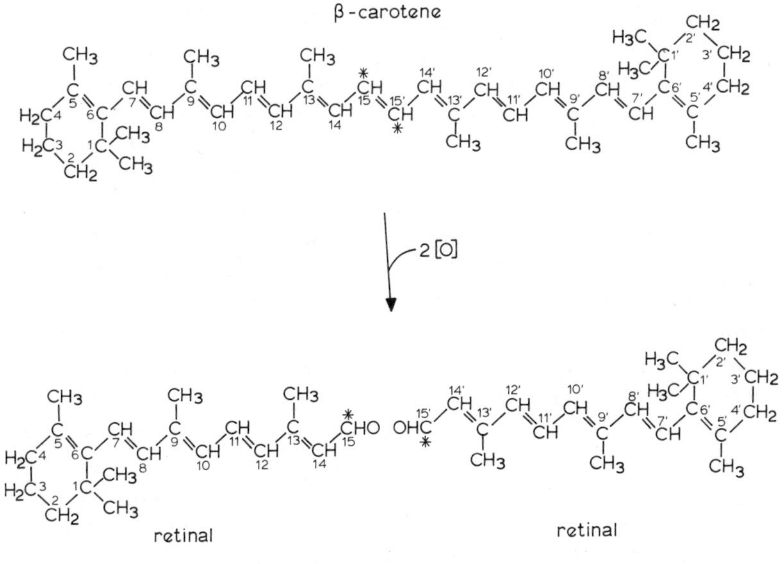

FIG. 3. Oxidation of β-carotene to 2 molecules of retinal in the body.

by cleavage of the central double bond of β-carotene to yield two
molecules of vitamin A aldehyde (retinal). Vitamin A alcohol (retinol) is
then produced by reduction of the aldehyde in an NADH-dependent
reaction catalysed by retinene reductase (Fig. 4). This conversion occurs
in the intestinal wall and liver of many species, but liver is believed to be

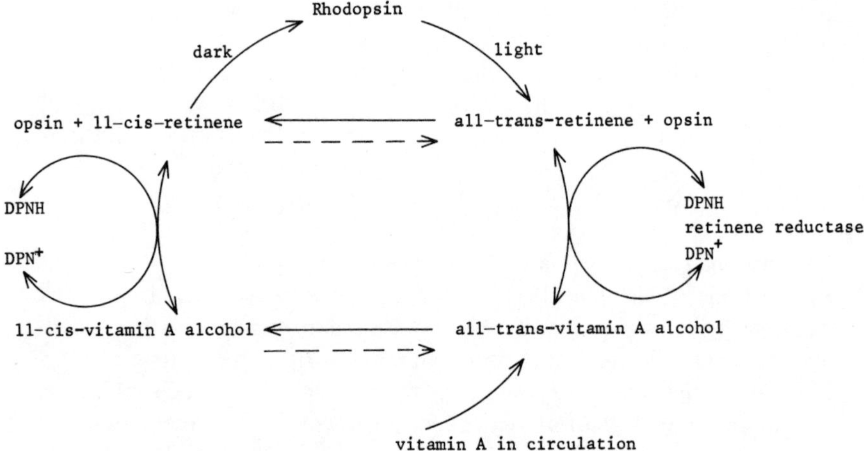

FIG. 4. Biochemical function of vitamin A in rod vision.

the only organ capable of this conversion in humans. Of the various isomers with provitamin A activity, only β-carotene yields two molecules of vitamin A alcohol on oxidative cleavage and subsequent reduction. In humans, the efficiency of absorption of provitamin A is estimated at about one-third of that ingested. Efficiency of conversion of provitamin A into vitamin A activity in liver also varies. The overall utilization of β-carotene is taken as one-sixth that of retinol. Other carotenoids with vitamin A activity, e.g. α-carotene and cryptoxanthin, are only half as active as β-carotene; therefore their efficiency as sources of vitamin A is one-twelfth that of retinol. The carotenoids are highly coloured, insoluble in water, and possess stability properties similar to those of vitamin A. The absorption maxima of β-carotene, apocarotenal and apocarotenoic ester in cyclohexane are 456 and 484 nm, 461 and 488 nm, and 449 and 475 nm, respectively. The β-carotene content of foods and feedstuffs is determined spectrophotometrically after clean-up by application of suitable absorption and partition chromatography techniques.

Vitamin A (all-*trans*-retinol) has an important role in the visual process (Fig. 4). The alcohol is oxidized enzymatically in an oxidation–reduction equilibrium system to the aldehyde (all-*trans*), formerly known as retinene but now called retinaldehyde or retinal. The all-*trans*-retinaldehyde is isomerized to the 11-*cis* form which combines with the protein opsin to form the visual pigment rhodopsin which is the photoreceptor for vision at low light intensities. As shown in Fig. 4, when light falls on the retina (the outer segments of the rods) the pigment complex is broken down through the formation of all-*trans*-retinaldehyde which is not bound by opsin. Ion transport and membrane potentials are affected by this decomposition which causes the transmission of an impulse up the optic nerve, resulting in vision.

The all-*trans*-retinaldehyde is isomerized back to 11-*cis*-retinaldehyde, which recombines with opsin, regenerating the visual pigment and thus continually renewing the light-sensitivity of the retina. Because of this essential function of vitamin A in vision, deficiency leads to impairment of the regeneration of visual pigment which manifests itself in man as night blindness which is one of the first symptoms of vitamin A deficiency; others include functional disturbances of the skin or mucosa ('epithelial protective' function), disturbed bone growth and impaired food utilization which results in slow growth and in weight loss. Absence of the epithelial protective function results in degeneration and keratinization of the ovaries, vagina, testes and cornea. The injury to the cornea leads to the most widely recognized symptom of vitamin A deficiency, xerophthalmia. Individual symptoms may or may not be highly pronounced, and can vary from one individual to another, from one animal species to another; a particular symptom may be outstanding or masked by others, depending on the specific nutritional conditions and other circumstances. In contrast to the well understood mechanism of the function of vitamin A in visual processes, its function at the molecular level of cell physiology and its metabolism in body tissues are still obscure. Vitamin A appears to have effects on subcellular systems such as the structure of cell membranes and cell particles, as well as on certain enzymes responsible for cell metabolism and on the biosynthesis and interactions of steroid hormones.

Recent research suggests that β-carotene has a special function in the reproductive cycle of bovine species, one not performed by Vitamin A. The ovaries contain high concentrations of β-carotene during the luteal phase, and it has been postulated that β-carotene is involved in the synthesis of progesterone. Moreover, in cows on β-carotene-free diets, 'silent' heat is frequently observed as well as delayed ovulation, follicular cysts, delayed and reduced formation of the corpora lutea and decreased progesterone concentrations in the serum. These conditions are corrected by the addition of β-carotene to the ration.

2.3. Human Requirement and Food Sources
The highest naturally occurring concentrations of vitamin A are found in

cod and shark liver oils. Mammalian liver, meat offals, egg yolk, milk and milk products are important sources of vitamin A in the human diet. A wide variety of plant foods contain β-carotene as a major pigment, including fruits (rose-hips, paprika, pumpkins, apricots and oranges) and vegetables (carrots, spinach, kale, cabbage, lettuce); it also occurs in kidney, liver, spleen, milk, butter, cream and cheese. The human daily requirement for vitamin A is a minimum of 20–30 IU per kg body weight. However, this is only sufficient to prevent the appearance of deficiency symptoms; about two to three times this quantity is required to assure an optimal supply. It may generally be assumed that the optimal requirement of an adult is about 5000 IU per day.

2.4. Vitamin A Activity in Milk

Vitamin A activity is expressed as retinol equivalents (RE) or international units (IU). In milk it is derived almost exclusively from retinol and β-carotene by the equation, $\mu g\,RE = \mu g$ retinol $+ (\mu g$ β-carotene)$/2$,[6] although FAD/WHO[7] and NRC[8] recommend that a divisor of 6 be used for all foods. β-Carotene accounts for $\sim 90\%$ of the provitamin A activity in milk. Approximately 97% of the retinol in bovine milk is esterified,[9,10] being produced in the mammary gland from plasma retinol. Usually, human milk has more vitamin A activity than cow's milk, typical values being 600 and 400 RE per litre, respectively. Breed and nutrition, especially the latter, effect considerable variation in the vitamin A activity of cow's milk. The importance of breed has been demonstrated by Krukovsky and Whiting,[11] Wagner[12] and Thompson

TABLE 2
VITAMIN A ACTIVITY IN MILK OF DIFFERENT BREEDS OF COWS[a]

	Channel Island breeds		Non-Channel Island breeds	
	Summer	*Winter*	*Summer*	*Winter*
Retinol (μg litre^{-1})	649	265	619	412
β-Carotene (μg litre^{-1})	1143	266	315	105
Vitamin A activity (μg RE)[b]	1221	398	777	465
Retinol/β-carotene ratio	0·6	11·0	2·0	4·0
Contribution (%) of β-carotene to vitamin A activity	46·8	33·4	20·3	11·4

[a] Calculated from data of Scott *et al.*[14]
[b] Assuming 2 μg β-carotene $= 1$ μg RE in milk.

et al.,[13] and more recently by Scott *et al.*[14] who measured the concentration of retinol and β-carotene in the milks of Channel Island (CI) and non-Channel Island (NCI) breeds of cattle. The vitamin A activities of the milks, calculated from these data by the present authors, are included in Table 2. The β-carotene content of the milk of CI cattle is ~2–3 times that of NCI breeds, as is the relative contribution of the β-carotene to the total retinol equivalent. The contribution of β-carotene to total vitamin A activity decreases by 28% and 45% in winter milks of CI and NCI breeds, respectively. The ratio of retinol to β-carotene is significantly different between both groups of cattle, possibly because CI cattle are unable to convert β-carotene into vitamin A as efficiently as other breeds.[15]

Seasonal variations in the vitamin A activity of cow's milk (Fig. 5) are a reflection of the carotene content of the diet, which typically is high in summer and low in winter. Similar variations are observed in human milk.[17] Carotene in fodder is easily destroyed, especially by wilting of freshly cut material, whereby the carotene content may be reduced by more than 50%. Field drying can result in losses of 98% or more (Table 3), especially in hostile weather.[5,18] In green fodder, lowest (10–15%) carotene losses are associated with conservation by artificial drying. Consequently, the concentration of vitamin A in milk is very much a consequence of the quality of fodder fed to cows (Table 4).

The vitamin A and carotene contents of milk do not exhibit a linear relationship with the β-carotene content of the diet.[20] Thus, when the carotene intake from the ration is increased from 130 to 200 mg per day, it is several times more effective in raising the vitamin A and carotene contents of milk fat than an additional increase to 300 mg per day. In addition to breed, many constituents of the ration may adversely affect the utilization of dietary β-carotene including raw soya beans, vitamin E and phosphorus deficiencies, ingestion of clover species containing cyanate, nitrate in plants, overuse of nitrogenous fertilizers and the carrier of the β-carotene, e.g. 1 g of β-carotene in the form of microcrystalline beadlets is more potent than 4 g of β-carotene in pasture grass.[21] Thus, attempts to increase the vitamin A content of milk by feeding carotene will not always be successful; it is better achieved by dietary supplements of retinol.

The vitamin A activity of cow's colostrum is 30 times higher than that of mid-lactation milk[22] but it decreases rapidly, such that 1 week after calving the vitamin A and carotene contents of milk are within the normal range. Feeding of a diet high in carotene during the last weeks of

 F. M. CREMIN AND PAUL POWER

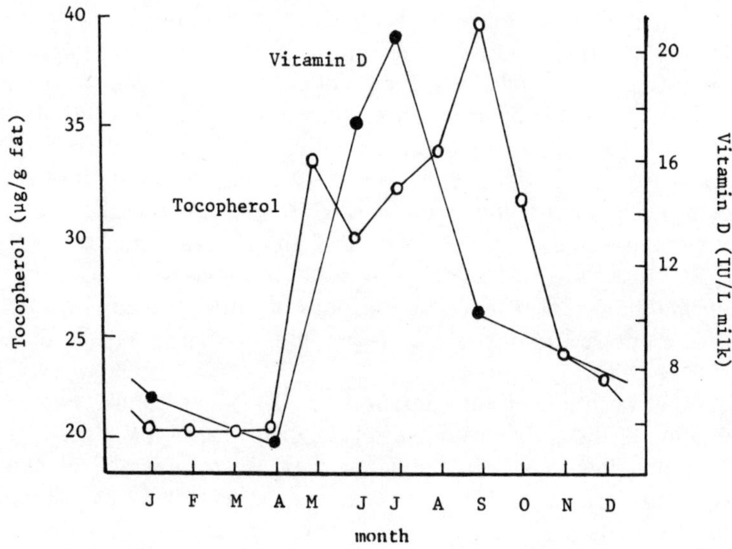

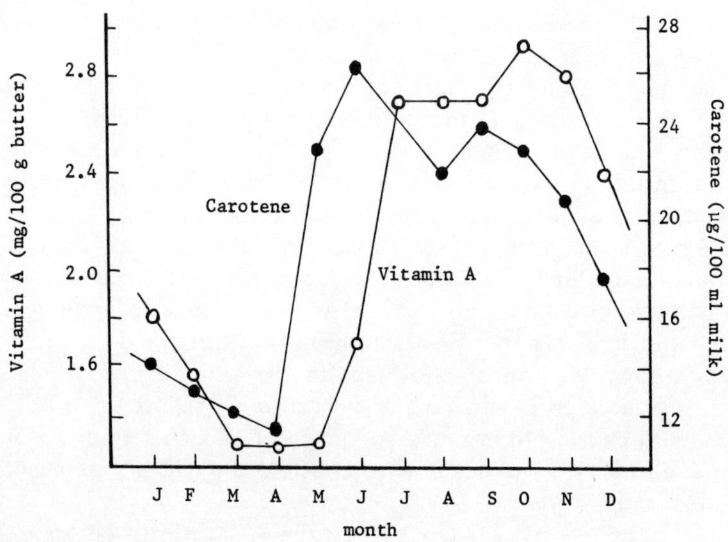

FIG. 5. Seasonal variations in the concentration of carotene and vitamins A, D and E. (Reproduced from Ref. 16 with permission.)

TABLE 3
LOSS OF CAROTENE DURING DRYING AND STORAGE OF HAY[5]

	Carotene (mg per kg of dry matter)
Freshly cut grass	213 (100%)
Fresh hay	29 (14%)
Hay stored for 28 weeks	4 (2%)

TABLE 4
VITAMIN A CONTENT OF MILK FROM COWS FED DIFFERENT FODDERS[19]

Fodder	Vitamin A (IU) per gram of butterfat
Hay	13·0
Wilted silage	24·9
Silage preserved with preformed acid	32·7

pregnancy effects an increase in the vitamin A activity of cow's colostrum.[23-25] Carotene loading increases the β-carotene[26] but not the vitamin A[27] concentration in human milk. The concentration of vitamin A in human colostrum is influenced by diet during pregnancy and is reported to be $\sim$10-fold[28] or $\sim$2-fold[26,29,30] higher than that in mature milk. These discrepancies may be attributable to variations in the time after parturition when the samples of colostrum were taken and analysed; for example, the vitamin A content of cow's milk exhibits an enormous decrease over the initial five milkings following parturition.[22] The vitamin A content of mature human milk can be elevated $\sim$5-fold by acute loading of the mother, by manipulation of her diet or by dietary vitamin A supplementation.[31]

3. VITAMIN D (CHOLECALCIFEROL, ERGOCALCIFEROL)

3.1. Chemical Structure and Physical Properties

Vitamin D occurs in several chemical forms; the two most nutritionally important are vitamin D_2 (ergocalciferol) and vitamin D_3 (cholecalciferol). These compounds are formed by ultraviolet irradiation of their provitamins, ergosterol and 7-dehydrocholesterol, respectively (Fig. 6).

Commercial forms of vitamin D include the crystalline compounds, oily solutions and water-dispersible stabilized powders. Vitamins D_2 and

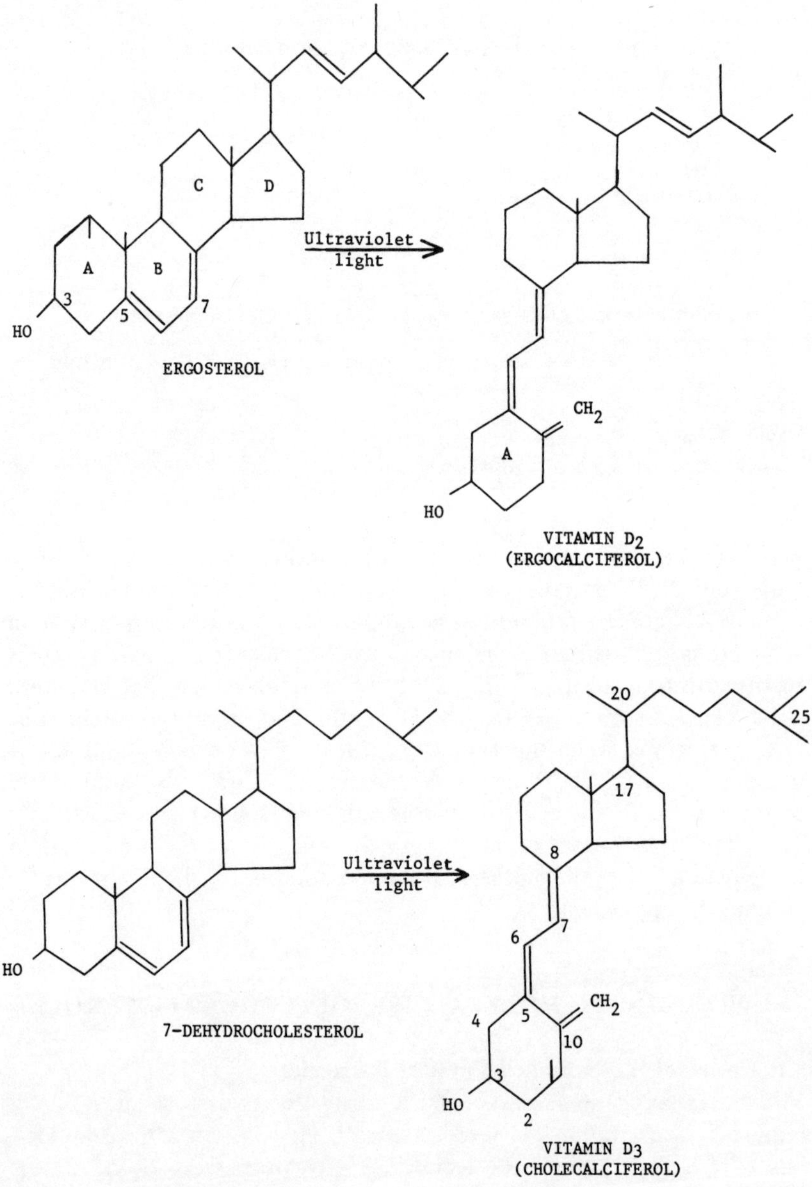

FIG. 6. Transformation of ergosterol and 7-dehydrocholesterol into vitamin D_2 and vitamin D_3, respectively, by irradiation with ultraviolet light.

D_3 exhibit absorption maxima at 245 nm in ethanolic solution. They are destroyed relatively rapidly by light, oxygen and acids, and should be stored as for vitamin A and β-carotene. The crystalline compounds are relatively stable to heat, but they tend to isomerize in oily solution. Vitamin D in foods is stable to storage, processing and cooking. Vitamin D occurs in foods and feedstuffs at very low concentrations, and consequently its activity in foods is determined by biological assay, or more recently by high performance liquid chromatography.

3.2. Physiological Function and Deficiency Symptoms
Vitamin D is essential for calcification processes in the body, including bone and teeth formation. It facilitates this process by (i) increasing absorption of dietary calcium and possibly phosphorus from the intestinal lumen, and (ii) increasing the re-absorption of calcium and phosphorus in the renal tubules. Vitamin D helps to ensure that sufficient calcium and phosphorus are available in the blood for the normal calcification of bones. A deficiency of vitamin D therefore results in poor bone calcification and soft bones which are liable to bend under body load.

In the body, vitamin D is converted into physiologically active metabolites which carry out the above functions. The principal functional form appears to be 1,25-dihydroxycholecalciferol. It is formed by two enzymatic hydroxylation steps involving the formation of 25-hydroxycholecalciferol in the liver and subsequent hydroxylation to 1,25-dihydroxycholecalciferol in the kidneys. The latter step helps to explain the observation of vitamin D resistant rickets in patients with chronic renal disease.

Vitamin D deficiency, especially when combined with an inadequate and/or imbalanced dietary intake of calcium and phosphorus, results in rickets in infants and children and osteomalacia in adults. Rickets is characterized by reduced deposition of calcium in growing bone, whereas osteomalacia is characterized by a loss of calcium from fully developed bone.

3.3. Human Requirement and Food Sources
In man, periods of growth are characterized by a high requirement for vitamin D; a daily intake of 400–500 IU ensures optimal supply. Adults involved in sufficient outdoor activity should obtain adequate vitamin D from natural sources and from the conversion of provitamin D_3 (7-dehydrocholesterol) in skin into vitamin D_3. Those working at night or

underground and the elderly may require additional vitamin D. Vitamin D requirements are increased to 500–1000 IU per day in pregnancy and lactation.

Vitamin D_2 is found in small quantities in fish-liver oils and some sponges. Vitamin D_3 is more widely distributed in nature, and is found in relatively large amounts in fish-liver oils and fatty tissues, and smaller amounts are present in hen's eggs, milk, butter, cream and cheese. In many countries milk is fortified with vitamin D making it a major dietary source.

3.4. Vitamin D Activity in Milk

Vitamin D activity in milk stems from its content of cholecalciferol (D_3) and ergocalciferol (D_2) and their metabolites, especially the 25-OH D_3 metabolite. Vitamin D in human and bovine milks has been measured primarily by bioassay, but recently several studies report the use of chemical analysis, including high performance liquid chromatography and a competitive radio-binding bioassay.[32–36] A summary of results of vitamin D analysis in human and bovine milks obtained by a number of investigators using chemical analysis and/or bioassay is shown in Table 5. Vitamin D_3 compounds contribute most to the antirachitic properties of human milk, especially the 25-OH vitamin D_3 metabolite, which

TABLE 5

CONCENTRATION OF VITAMIN D IN HUMAN AND COW'S MILK

	Human milk		Cow's milk	
	Chemical analysis (ng litre^{-1})	*Bioassay* (IU litre^{-1})	*Chemical analysis* (ng litre^{-1})	*Bioassay* (IU litre^{-1})
Vitamin D activity	—	$40^a, 15^d$	—	$40^e, 38^d$
D_3	$338^a, 124^b$	—	$281^e, 340^f$	—
D_2	$41^a, 56^b$	—	—	—
25-OH D_2	$163^a, 297^b, 270^c$	—	145^d	—
25-OH D_3	$<20^a, 48^b$	—	—	—
24,25-$(OH)_2 D_2/D_3$	$<21^a, 25^c$	—	15^e	—
1,25-$(OH)_2 D_2/D_3$	$<0\cdot6^a, 2^c$	—	—	—

[a] Data obtained on 3 women consuming 400–800 IU vitamin D per day.[32]
[b] Subjects consumed a vitamin D_3 supplement of 10 μg per day for 3 months.[35]
[c] Data obtained on 18–36 women, 3–21 days post-partum.[36]
[d] Human data obtained on 2 women consuming normal dietary intakes of vitamin D. Cow's milk samples were obtained from retail sources.[37]
[e] Data obtained from Reeve *et al.*[33]
[f] Data obtained on bulk HTST milk by Scott *et al.*[14]

contributes $\sim 70\%$ of total activity. Vitamin D_3 compounds account for $\sim 85\%$ of the antirachitic properties of cow's milk; the 25-OH vitamin D_3 metabolite accounts for 62% of the total activity. Conversion of the vitamin D_3 content (µg) of both milks to antirachitic activity (IU) using the conversion factor $1\,IU = 0.025\,µg\,D_3$, and allowing for the 5-fold greater antirachitic potency of 25-OH vitamin D_3,[38] yields values comparable to those obtained by bioassay involving stimulation of calcium absorption in the vitamin D deficient rat. Weisman et al.[36] reported that, in human milk, $\sim 15\,IU\,litre^{-1}$ of vitamin D was due to 25-OH vitamin D_3/D_2, based on these compounds having only 1·5 times more biological, i.e. antirachitic, activity than vitamin D_3. If a conversion factor of 5 had been used by these authors, the results would have concurred with those of Reeve et al.,[32] i.e. 40–$50\,IU\,litre^{-1}$. The vitamin D activity observed in a rat tibia bioassay of human milk in Europe is lower ($15\,IU\,litre^{-1}$ [37]) than that reported by Reeve et al.[32] This probably reflects a lower maternal dietary intake of vitamin D in European countries, due to consumption of unfortified milk and/or differences in bioassay criteria.

Like vitamin A, the concentration of vitamin D_3 in milk shows a seasonal variation (Fig. 5) but, unlike vitamin A, it is unrelated to diet since vitamin D_3 in mammals is derived from skin by conversion of 7-dehydrocholesterol by sunlight. The short wavelength rays (280–315 nm) of sunlight effect this conversion, and the seasonal influence is related to amount and type of exposure to these rays. However, Scott et al.[14] found no significant difference in vitamin D_3 (only) concentrations between summer and winter bulk HTST milk in England. Vitamin D_2 (ergocalciferol) also occurs in milk, but in smaller concentrations than vitamin D_3. Unlike vitamin D_3, it is derived by ultraviolet irradiation of ergosterol in the animal's feed. Consequently, if the vitamin D_2 content of milk is to be increased, ergosterol, or feeds containing ergosterol, must be irradiated before consumption.[39–43] Therefore ergosterol, unlike 7-dehydrocholesterol, cannot be converted by the animal into compound(s) possessing vitamin D activity. Wilting increases the vitamin D_2 but decreases the β-carotene and vitamin E contents of forage crops. However, the above experiments with vitamin D_2 and others[44] indicate that only a fraction (2–3%) of dietary vitamin D_2 is normally secreted from blood to milk, probably because of its rapid degradation in the body to metabolites with no antirachitic activity.[40,45,46]

Oral administration of 8000–16 000 IU (0·2–0·4 mg) of vitamin D_3 per cow per day for 4–12 months, an amount similar to current recommended dietary allowances,[47] did not increase the vitamin D potency of milk.[46,48] Reeve *et al.*[33] demonstrated that a 14-fold increase in dietary vitamin D resulted in milk having twice the vitamin D concentration of normal milk. Massive supplementation of two cows with 10^6 IU (25 mg) each of vitamins D_2 and D_3 resulted in the secretion in milk of similar maximum concentrations of both vitamins, i.e. 250 IU litre^{-1} of D_2 and D_3 in one animal and 450 IU litre^{-1} of D_2 and D_3 in the other[49] at 2–3 days. The concentration of D_2 and D_3 increased and then decreased rapidly in the milk of both animals. In humans, Hollis[35] demonstrated that maternal dietary vitamin D_2 supplementation (60 µg D_2 per day for 2 weeks) or exposure of the mother to ultraviolet irradiation (1·5 minimal erythema dose) resulted in a marked increase in the concentrations of vitamin D_2 and vitamin D_3, i.e. ∼25–50-fold, but only slightly increased their 25-OH metabolites (1·3–3-fold) in the milk. The vitamin D activity of bovine colostrum is several times greater than that of mature milk,[50] and a diet rich in vitamin D activity prior to calving increases its concentration in colostrum.[51] A water-soluble form of vitamin D (vitamin D sulphate) was identified in bovine and human milks at a concentration of 4 and 10 µg litre^{-1} respectively,[52–55] but this has recently been shown to possess no antirachitic activity.[37,56–58]

4. VITAMIN E (TOCOPHEROLS, TOCOTRIENOLS)

4.1. Chemical Structure and Physical Properties

Tocopherols and tocotrienols (Fig. 7) contribute to the vitamin E activity of foods. Tocopherols are methylated 6-chromanol (dihydrobenzo-γ-pyran) compounds containing a saturated isoprenoid 16-carbon side chain. In contrast, tocotrienols contain three double bonds in the attached side chain. The α-, β-, γ- and δ-tocopherols differ from each other only in the position and number of the methyl substituents on the benzo moiety of the chromanol nucleus. Because of inadequate information on the occurrence of many of these compounds in foods, only α-tocopherol can be considered when calculating the vitamin E content of foods. Tocopherols and tocotrienols other than α-tocopherol have low biological activity, varying between 1% and 50% of that of α-tocopherol. Vitamin E is commercially available as α-tocopherol acetate in the form of an oil or stabilized water-dispersible powder. Tocopherol (*dl*-) and its acetate exhibit absorption maxima at 292 nm and 284–285 nm, respectively, in ethanol. The acetate form is relatively stable to atmospheric oxygen, but is hydrolysed to free tocopherol by moisture in the presence

A

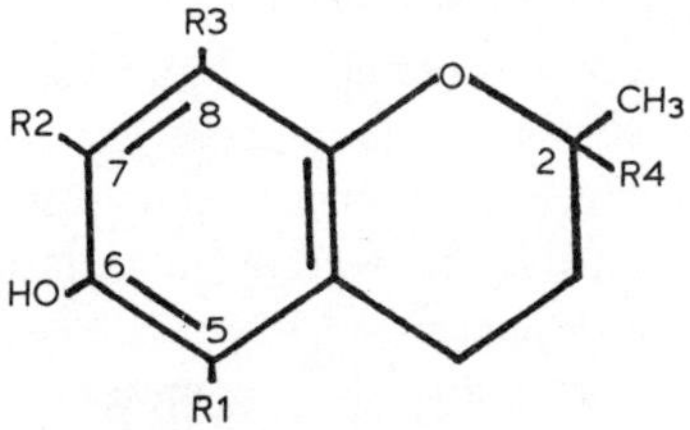

TOCOLS
$R4 = CH_2(CH_2CH_2\overset{\displaystyle CH_3}{C}HCH_2)_3H$

TOCOTRIENOLS
$R4 = CH_2(CH_2CH=\overset{\displaystyle CH_3}{C}CH_2)_3H$

B

Tocol	Tocotrienol	Methyl positions
α- (alpha)	ζ- (zeta)	5, 7, 8
β- (beta)	ε- (epsilon)	5, 8
γ- (gamma)	η- (eta)	7, 8
δ- (delta)	8-methyltocotrienol	8

FIG. 7. Structure of the cores (A) and positions of attachment of methyl groups (B) of naturally occurring compounds with vitamin E activity.

of alkalis or strong acids. The free tocopherol is rapidly oxidized in air with darkening of colour. Where the instrumentation is available, high performance liquid chromatography is the method of choice for tocopherol analysis in foods and feedstuffs.

Vitamin E is present in practically all body tissues; uterus, testes, adrenals and pituitary have particularly high vitamin E concentrations compared with other organs, which agrees well with the specific physiological functions of the vitamin in these organs. In the liver, vitamin E is localized mainly in the metabolically active cellular particles, the mitochondria and microsomes.

4.2. Physiological Function and Deficiency Symptoms

Vitamin E has a variety of important physiological effects in the body through its antioxidative properties. These include (i) stabilization of unsaturated fatty acids by inhibiting the formation of toxic lipoperoxides, (ii) inhibition of vitamin A oxidation and destruction, (iii) prevention of damage to blood vessels and capillary bed permeability, and (iv) stabilization of red blood cell membranes. Animal experiments indicate that vitamin E exerts a variety of other effects by mechanisms that are not yet understood, including maintenance of testicular function, prevention of foetal resorption, muscular degeneration and liver necrosis. Recent studies indicate that vitamin E is involved in the metabolism of nucleic acids and polyunsaturated fatty acids, and that it has a regulatory action on the pituitary–midbrain system. It also promotes the production of thyrotrophic and adrenocorticotrophic hormones and the gonadotrophins, and the hormone content of the pituitary falls in vitamin E deficiency. In these cases too, the mechanism of action of vitamin E and the possible connections between the metabolic disturbances and the outward clinical symptoms have not yet been determined. This may account for the fact that in man little or no clinical vitamin E deficiency symptoms have been established with certainty.

Investigations of experimental vitamin E deficiency in laboratory animals have shown that a wide range of very varied symptoms, with little in common, may occur, depending on external and internal factors such as composition of diet or animal species. For example, the effect of a chronically deficient supply of vitamin E is accentuated if, in addition, there is a deficiency of protein or of selenium which result in the occurrence of anaemia and liver necrosis, respectively. Consumption of diets containing large amounts of polyunsaturated fatty acids, e.g. fish liver oils, also accentuates the effects of chronic vitamin E deficiency.

4.3. Human Requirement and Food Sources

The daily requirement of man for vitamin E is estimated to be 30 IU; in pregnancy and for the elderly about twice as much is necessary to achieve an optimal supply, and the daily requirement of an infant is 5–10 IU. The vitamin E requirement of the body increases with increased dietary administration of polyunsaturated fatty acids. Among the richest sources of vitamin E are cereal germs and most oilseeds, and it is also found in leafy vegetables (lettuce, spinach, cabbage, leek), in animal organs (pituitary, adrenals, pancreas, spleen) and in milk, butter and abdominal fat.

4.4. Vitamin E Activity in Milk

Human and bovine milks typically contain an average of 5 and 1 mg litre^{-1} of vitamin E, respectively, but diet causes large variations in these concentrations. Vitamin E occurs in cow's milk exclusively as α-tocopherol,[59] but in human milk it occurs as α-tocopherol ($\sim 75\%$) and γ-tocopherol (~ 15–20%) together with very small concentrations of β- and δ-tocopherols.[60] The vitamin E potency of β-, γ- and δ- relative to α-tocopherol (100%) are 56, 16 and 0.5%, respectively, as measured by rat foetal re-absorption assay.[61] The ratio of vitamin E (mg) to polyunsaturated fatty acids (g) is 1·1 and 0·84 in human and bovine milks, respectively,[17] and maternal ingestion of oils rich in polyunsaturated fatty acids increases their concentrations and that of vitamin E in human milk.[62]

The concentration of vitamin E in bovine milk exhibits a seasonal pattern (Fig. 5) reflecting variations in dietary vitamin E intake. Brown[63] reported that the concentration of vitamin E in different grasses reaches a maximum (270–350 mg kg^{-1}) in April and a minimum (9–19 mg kg^{-1}) in September. The timing of this seasonal effect on forage composition may vary from one country to another, but the net effect is a reduction in the vitamin E concentration in milk of grazing cows. As with β-carotene, hay is a very poor source of vitamin E (7–14 mg kg^{-1}) compared to fresh grass (60–100 mg kg^{-1});[64] silage contains considerably more vitamin E (305 mg kg^{-1}) than hay (15 mg kg^{-1}).[63] More than 95% of the tocopherols in grass, clover and lucerne is α-tocopherol whereas the corresponding value in barley and maize is $\sim 10\%$.[63] Consequently the forage fed to cows during winter will greatly influence the α-tocopherol content of milk.

Fish oils[65,66] and certain leguminous plants[67,68] when included in the diet of cows decrease the vitamin E content of milk by antagonizing and/or decreasing its utilization. Rations with the same vitamin E content may therefore produce milk with different vitamin E potencies. The latter may explain the considerable variation reported[16] in the vitamin E content of cow's milk. Bovine and human colostra have ~ 4–5

TABLE 6

EFFECT OF DIETARY VITAMIN E SUPPLEMENTATION ON COLOSTRAL VITAMIN E CONCENTRATION[10]

Daily tocopherol supplement (g)	No. of cows	Tocopherol (μg per gram of fat) in various colostral milks on day:						
		1	2	3	4	5/6	7/8	15/16
0	5	98	97	73	51	41	28	20
4	1	426	390	266	281	170	124	34
10	3	485	483	376	252	182	98	39

times more vitamin E than their mature milks. An illustrated in Table 6, its concentration in bovine milk is determined primarily by the vitamin E content of the diet consumed during pregnancy;[10] administration of 4 g per day of vitamin E more than quadrupled and doubled the vitamin E content of colostrum on day 1 and day 15/16, respectively.

The vitamin E concentration in milk is of public health importance as premature infants have low body stores and are consequently susceptible to clinical deficiency symptoms, including haemolytic anaemia.

5. VITAMIN K

5.1. Chemical Structure and Physical Properties

The essential structural feature of all chemicals possessing vitamin K activity is that of 2-methyl-1,4-naphthoquinone (Fig. 8) which is designated vitamin K_3 or menadione. The naturally occurring form of vitamin K in plants possesses a phytyl radical at position 3 and is designated vitamin K_1 or mephyton (Fig. 9). The form of vitamin K that

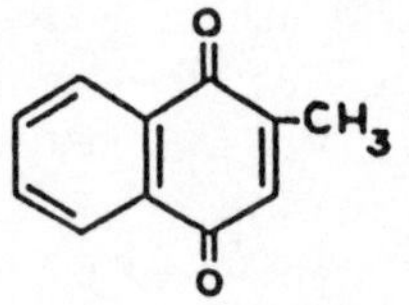

FIG. 8. Structural formula of vitamin K_3; menadione (2-methyl-1,4-naphthoquinone).

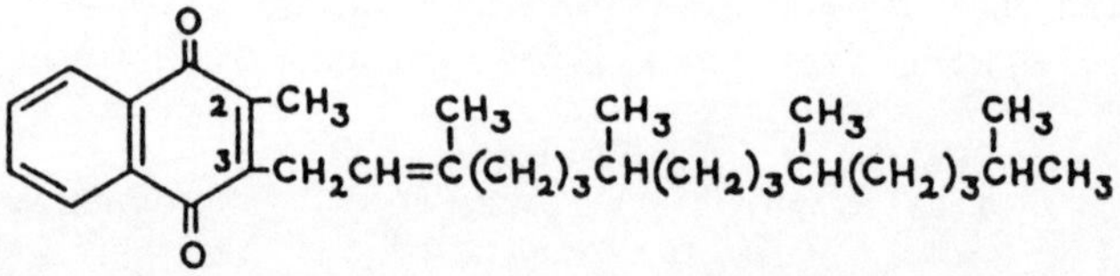

FIG. 9. Structural formula of vitamin K_1; phytonadione (phylloquinone; mephyton [2-methyl-3-phytyl-1,4-naphthoquinone]).

occurs in animals and as metabolites of intestinal bacteria is designated vitamin K_2 or menaquinone or farnoquinone (Fig. 10). Vitamin K_3 and several synthetic derivatives based on it possess vitamin K activity; it is the form of vitamin K used most often in clinical practice. Vitamin K is a

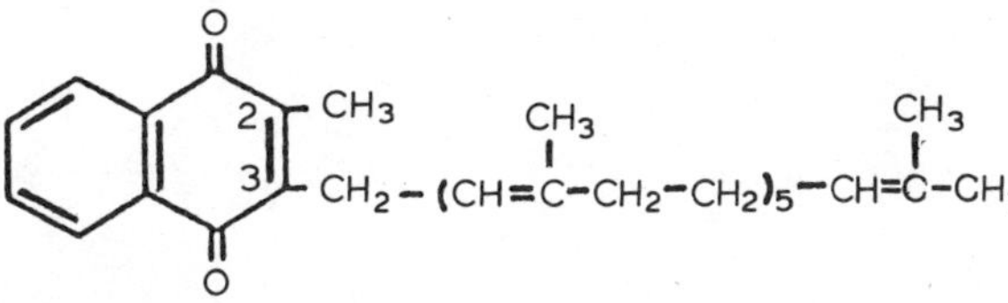

FIG. 10. Structural formula of vitamin K_2 (2-methyl-3-difarnesyl-1,4-naphtho-quinone; farnoquinone) (side chain). It may also occur with 7 or 9 isoprene units $-CH=C(CH_3)-CH_2-CH_2-$.

fat-soluble vitamin but vitamin K_3 forms a similarly active water-soluble sodium bisulphite salt (water-soluble K_3). Similarly, the sodium salt of naphthohydroquinone diphosphate is water-soluble and has vitamin K activity. Vitamin K_1, the most common naturally occurring form of vitamin K, is insoluble in water, sparingly soluble in ethanol and readily soluble in ether, chloroform, fats and oils. It exhibits ultraviolet absorption maxima at 243, 249, 261 and 270 nm. Vitamin K_1 is slowly degraded by atmospheric oxygen but is fairly rapidly destroyed by light. It is relatively stable to heat but is decomposed by alkalis. It is quantified in foods and feedstuffs by biological assay; the test measures, preferably curatively, the readiness of the blood of chicks (or rats) to coagulate, before and after ingestion of graded intakes of foods containing vitamin K of unknown concentration and appropriate pure vitamin K standards.

5.2. Physiological Function and Deficiency Symptoms

Vitamin K is most often associated with the synthesis of the plasma clotting factor, prothrombin, in the liver. It is also required for the synthesis of the plasma clotting factors VII, IX and X. Vitamin K deficiency or administration of vitamin K antagonists results in a reduced activity of vitamin K dependent plasma clotting factors, resulting in delayed clotting and haemorrhagic problems. Clotting is initiated by the conversion of soluble fibrinogen into insoluble fibrin by thrombin derived from prothrombin by the action of plasma and tissue thromboplastin. A variety of clotting factors initiate this complex process, as illustrated in the highly simplified scheme shown in Fig. 11. Advanced liver disease, as in carcinoma and cirrhosis, may be associated with a prothrombin deficiency which is not responsive to vitamin K. The effect of vitamin K on prothrombin synthesis is mediated in cells through a vitamin K dependent carboxylase system. This enzyme catalyses the carboxylation of selected glutamic acid residues on the peptide chain to

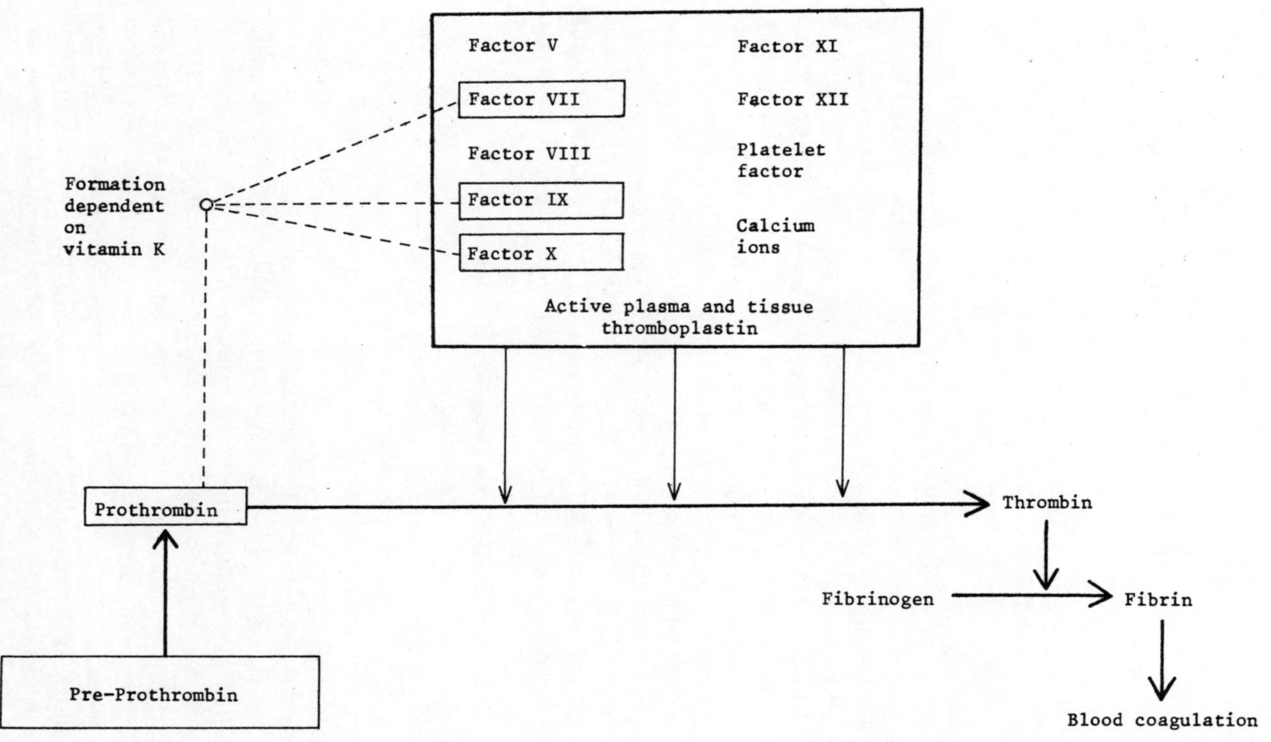

FIG. 11. Role of vitamin K in blood coagulation. (Modified from Ref. 5.)

γ-carboxyglutamic acid residues. This pre-prothrombin is then glycosylated to form prothrombin which is secreted into plasma. The vitamin K dependent carboxylase system occurs in liver, bone, kidney, placenta and spleen. The γ-carboxyglutamic acid residues resulting from its action on glutamic acid residues act as Ca^{2+} binding sites, which may be associated with the mineralization and organization of mineralized tissues. It is suggested that salt deposition, crystallographic polymorphism of the mineral phase, and turnover of bone mineral may be, in part, regulated by proteins containing carboxyglutamate. The latter represents a significant extension of the role of fat-soluble vitamin K in metabolism.

In the body, menadione (vitamin K_3) is probably converted into a vitamin K_2 type of compound containing twenty carbon atoms in the side-chain. The linkage of menadione with an isoprenoid side-chain, as in naturally occurring vitamin K, appears to be necessary for the biological activity of menadione.

Vitamin K deficiency can result not only from inadequate dietary supply but also from absorption disorders. Deficiency symptoms can also appear in animals if adequate synthesis by the intestinal flora cannot take place, for example, on administration of antibiotics or sulphonamides or through the effect of anticoagulants, such as dicoumarol, which may be present in spoiled sweet-clover and in certain rodenticides (dicoumarin or indanedione derivatives).

Vitamin K deficiency leads uniformly in man and in all investigated animals to a fall in the prothrombin content of the blood, eliciting a haemorrhagic tendency and haemorrhages in the most varied tissues and organs (the subcutaneous tissue, muscles, brain, gastrointestinal tract, abdominal cavity, urogenital organs, etc.), which can give rise especially to complications in neonatal and premature birth cases in humans. All other symptoms must be regarded as consequences of this phenomenon.

5.3. Human Requirement and Food Sources

Since the intestinal flora is insufficiently developed in the first days after birth to provide the vitamin K requirement of the infant, and because of the low vitamin K content of the mother's milk, the prothrombin level is very low in the newborn. A daily dose of 1–2 mg vitamin K for newborn infants, or 2–5 mg daily to the mother before confinement, is therefore recommended. The daily requirement of adults is estimated at about 1 mg. Useful sources of vitamin K_1 include green plants, green vegetables, potatoes, fruits and liver oils. Vitamin K_2 is derived from animal materials and microbial synthesis.

5.4. Vitamin K Activity in Milk

Reported values for the concentration of vitamin K in milk, as in the case of vitamin D, differ greatly. Factors contributing to the observed variations in concentration include inadequacy of available analytical methods, a fact compounded by variations in the maternal intake of vitamin K.[69] The concentration of vitamin K is higher in bovine than in human milk, typical reported values being 50 and 20 μg litre^{-1}, respectively. The vitamin K status of breast-fed premature and normal infants may be marginal in the early neonatal period as feeding of cow's milk reduces prothrombin time in these infants. Breast-fed and fasting neonates require vitamin K supplementation to prevent haemorrhagic disease.[70,71] From the beginning of the second week of life the intestinal microflora of the neonate is considered sufficiently developed to contribute substantially to the vitamin K requirements of infants.[72] Infants are usually supplemented with vitamin K_1 (phylloquinone) rather than synthetic vitamin K derivatives, e.g. naphthoquinone, because of the latter's potential toxicity. Vitamin K_1 is obtained from green plants, whereas vitamin K_2 (menaquinone) is derived from bacterial synthesis in the intestine. Haroon *et al.*[73] reported values of 2·1 and 2·3 μg of vitamin K_1 per litre of mature human milk and colostrum, respectively ($n=20$), and 4·9 μg per litre of bovine milk, which indicates that vitamin K_2 contributes significantly to vitamin K activity in milk and/or that values reported previously[16,74] are too high. Administration of an oral supplement of 20 mg of vitamin K_1 to one mother led to an increase in the vitamin K concentration of her milk to 140 μg litre^{-1} and a two-fold increase in concentration persisted 12 days after administration of the supplement.

Although it is a fat-soluble vitamin, vitamin K possesses many of the features of the water-soluble vitamins and as such it is found in significant quantities in skim milk and is synthesized by bacteria. Vitamin K persisted in the rumen contents of a cow fed a diet devoid of vitamin K for 5 months,[75] indicating that the vitamin K concentration in milk is probably not influenced by dietary vitamin K at the concentration at which it normally occurs in foods.[21]

6. VITAMIN B_1 (THIAMINE)

6.1. Chemical Structure and Physical Properties

Vitamin B_1 consists of a molecule of 2,5-dimethyl-6-aminopyridine bonded through a methylene bridge to a molecule of 4-methyl-5-

hydroxyethylthiazole (Fig. 12). This structure is chemically unique in that the thiazole nucleus is not found elsewhere in nature and biotin is the only other vitamin known to contain a sulphur atom. Thiamine hydrochloride (chloride hydrochloride) is the most commonly used commercial form of vitamin B_1, although thiamine mononitrate is also used

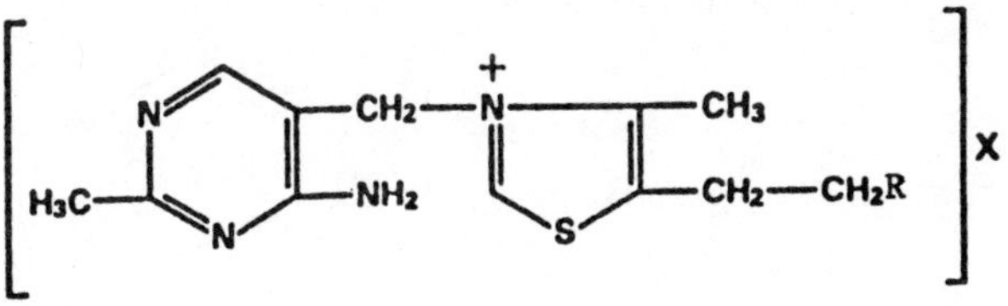

Hydrochloride: $X = Cl^-$, HCl; $R = OH$
Mononitrate: $X = NO_3^-$; $R = OH$
Diphosphate (carboxylase):

$$R = O - P - O - P - OH$$

FIG. 12. Structural formulae of thiamine and its analogues.

(Fig. 12). They are both white crystalline powders readily soluble in water; when dry they are stable and can withstand heating, but they are readily destroyed by alkali and by boiling in aqueous solutions at pH values greater than 5·5. Vitamin B_1 has a characteristic absorption spectrum in the region 200–300 nm; the maxima and respective extinctions depend markedly on the solvent used and the pH of the solution. Vitamin B_1 is most frequently measured in foods by a fluorometric assay, based on its oxidation in alkaline solution to strongly fluorescent thiochrome.

6.2. Physiological Function and Deficiency Symptoms

Vitamin B_1, as thiamine pyrophosphate (Fig. 12), participates as a cofactor for several key enzymes in carbohydrate metabolism including pyruvate dehydrogenase, α-ketoglutarate dehydrogenase, transketolase and transaldolase (Fig. 13). Pyruvate and α-ketoglutarate dehydrogenases catalyse the oxidative decarboxylation of pyruvate and α-ketoglutarate to acetyl-coenzyme A and succinyl-coenzyme A, respectively. The water-soluble vitamin pantothenic acid, which is the central constituent of coenzyme A, also participates as a cofactor in the *activation* of acetate and succinate in these reactions.

After conversion into acetyl-coenzyme A in the glycolytic pathway,

 F. M. CREMIN AND PAUL POWER

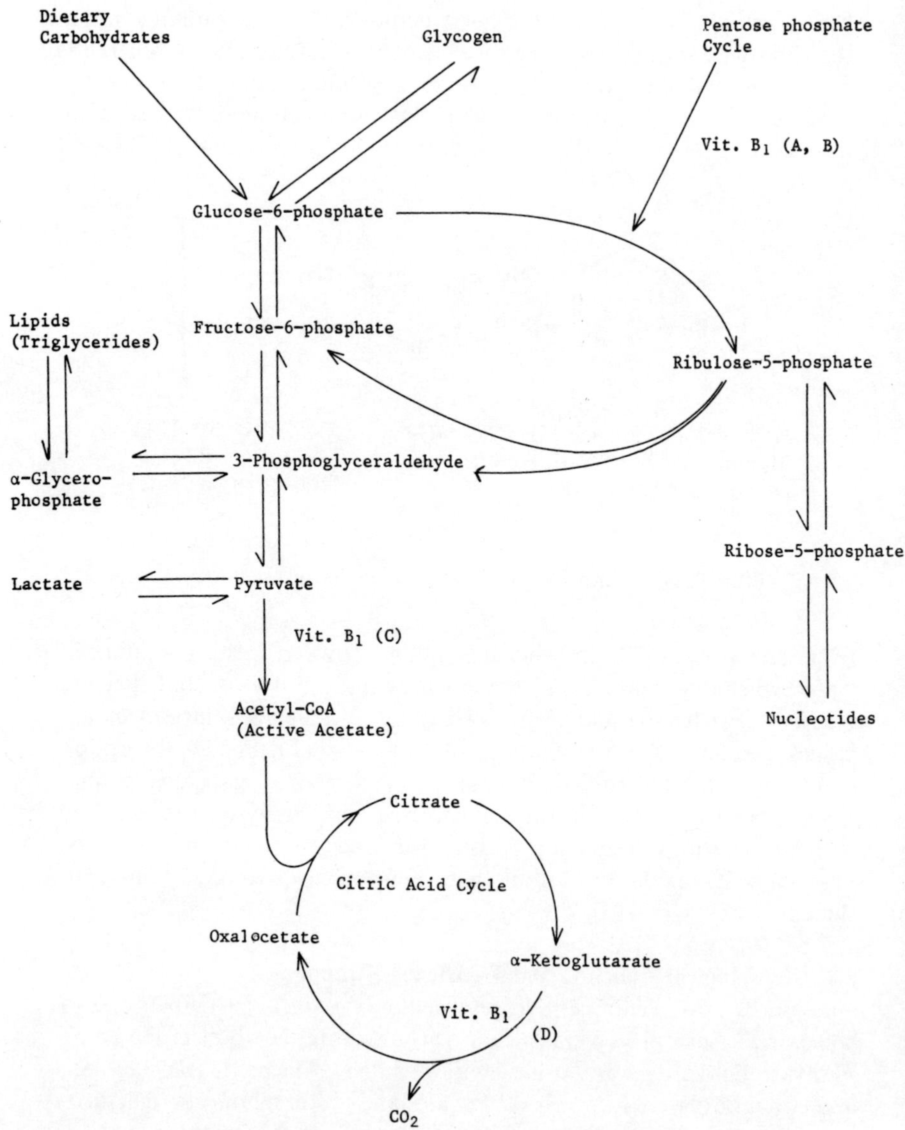

FIG. 13. Vitamin B_1 dependent enzymes in carbohydrate metabolism. A, transketolase; B, transaldolase; C, pyruvate dehydrogenase; and D, α-ketoglutarate dehydrogenase. (Adapted from Ref. 5.)

carbohydrates are sequentially oxidized to carbon dioxide and water in the citric acid cycle and respiratory chain (see niacin), with the formation of ATP (see Fig. 17). Acetyl-coenzyme A is the basic unit from which fatty acids and steroids are synthesized in the cell (see Fig. 19) so that vitamin B_1 is essential for the synthesis of fats from carbohydrate and amino acids. Acetyl-coenzyme A is also the central molecule formed in the β-oxidation of fatty acids in the cell. Pyruvate also provides the acetyl group for acetylcholine, and consequently a dietary deficiency of vitamin B_1 results in a reduction of acetyl-coenzyme A and acetylcholine concentrations in the brain (Fig. 19). Concomitant increases in pyruvate and lactate concentrations occur because of reduced cellular pyruvate dehydrogenase activity.

Acetyl-coenzyme A is oxidized in the citric acid cycle (Fig. 13). In the process, acetyl-coenzyme A combines with oxaloacetate to form citrate, which in turn undergoes a series of metabolic changes whereby oxaloacetate is regenerated and acetyl-coenzyme A is oxidized to carbon dioxide, water and energy-containing reducing equivalents, including reduced flavine adenine dinucleotide (FADH, see vitamin B_2) and reduced nicotinamide adenine dinucleotide (NADH, see niacin). The latter cofactors are derived from the vitamins riboflavin and niacin, respectively. An adequate intake of vitamin B_1 ensures the activity of α-ketoglutarate dehydrogenase which is a key enzyme in this cycle and whose functioning is essential for optimal utilization of energy from acetyl-coenzyme A and ultimately the substrates from which it is derived.

As a coenzyme for the transketolase and transaldolase enzymes of the hexose monophosphate shunt (Fig. 13), vitamin B_1 facilitates the conversion of pentose phosphates into hexose phosphates and their subsequent oxidation in the hexose energy-yielding pathways. Optimal functioning of the pentose cycle ensures a supply of pentoses for nucleotide and nucleic acid synthesis, in addition to NADPH energy-reducing equivalents which are specifically required for *de novo* fatty acid synthesis from acetyl-coenzyme A. Thus, in lipogenesis, metabolic transformation occurs whereby the hexose monophosphate shunt becomes more active, thus generating the NADPH reducing-equivalents required for fatty acid synthesis from carbohydrate (Fig. 19). A diet rich in carbohydrate increases the daily requirement for vitamin B_1.

A severe deficiency of vitamin B_1 in humans results in beri-beri which is characterized by cardiovascular disorders including oedema, shortness of breath, tightness of chest, tachycardias and sudden death and/or nervous disorders including hyper- and hypo-sensitivity, burning feet, neuritis, muscular weakness and pain and spasma extending to paralysis. The earlier stages of the disease may be characterized by neurasthenia, fatigue, disturbed emotional equilibrium, reduced appetite and disordered digestion. Sporadic outbreaks of vitamin B_1 deficiency occur in ruminants maintained on rations high in digestible carbohydrate.

6.3. Human Requirement and Food Sources

The vitamin B_1 requirement of humans is increased by diets high in carbohydrate and energy, conditions associated with increased metabolism including fever, pregnancy, heavy labour, increased excretion in the urine brought about by high urinary volumes and reduced absorption as in gastrointestinal upsets. The minimal requirement of adults is 0·5 mg/1000 cal but total intake should not fall below 1 mg per day. Vitamin B_1 is widely distributed in foods at concentrations of 10–200 µg/100 g. It occurs in the pericarp and germ of cereals, yeast, vegetables, fruits, potatoes, pork, liver, kidneys, egg yolk and milk.

6.4. Vitamin B_1 in Milk

Typical concentrations of vitamin B_1 in mature bovine and human milks are 450 and 140 µg litre^{-1}, respectively. Vitamin B_1 occurs in milk in three forms: free thiamine, phosphorylated thiamine and a proportion of both forms bound to protein. In mature cow's milk, the distribution of vitamin B_1 between the three forms is 50–70% free, 18–45% phosphorylated and 5–17% protein bound.[76] The concentration of vitamin B_1 in bovine colostrum, 850–1220 µg litre^{-1},[77] is several times higher than that in mature milk and decreases during the first months of lactation. Human milk shows the opposite lactational response, the concentration of vitamin B_1 in colostrum ($\sim$20 µg litre^{-1}) being $\sim$14% of that in mature milk.[30,31] This increase has also been observed in mothers exhibiting poor vitamin B_1 status,[78] and women nursing such infants may develop beri-beri.[79] Vitamin B_1 supplementation of deficient women results in the secretion of milk with an increased concentration of vitamin B_1.[31] From a practical viewpoint, nutritional status does not influence the concentration of vitamin B_1 in bovine milk because of the contribution of rumen bacteria to the vitamin B_1 content of milk.[80–82] However, seasonal variations have been reported in human milk, concentrations being highest in autumn.[83] Drugs, including alcohol, isoniazid, hydrolazine, penicillamine and oral contraceptive pills have all been reported to have an adverse effect on the vitamin B_1 concentration in human milk.[84] Commercial heat treatment of milks, including HTST and UHT treatments, results in a $\sim 10\%$ loss of vitamin B_1 activity,[76,85,87] This loss is not significant because of the low contribution of liquid milk to the vitamin B_1 content of western diets, e.g. $\sim 10.7\%$ in the UK.[88]

7. VITAMIN B_2 (RIBOFLAVIN)

7.1. Chemical Structure and Physical Properties

Vitamin B_2 consists of the flavin pigment lumichrome and the reduced form of the sugar ribose (Fig. 14), hence the name riboflavin. Its yellow-green fluorescence was recognized in milk in 1879, giving rise to the synonym lactoflavin. Like vitamin B_1, vitamin B_2 occurs in virtually all

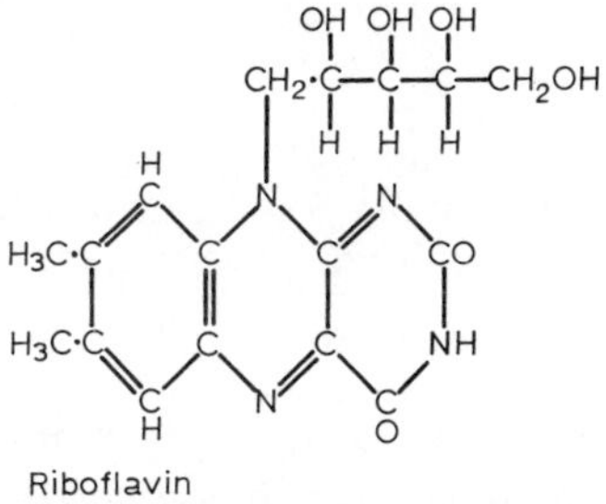

FIG. 14. Structural formula of riboflavin (vitamin B_2).

living cells. It occurs (Fig. 15) usually in its cofactor forms, flavin adenine dinucleotide (FAD) and flavin mononucleotide (FMN), free or bound to a protein. Significant amounts of vitamin B_2 occur only in milk, urine and retinal tissues. Vitamin B_2 is poorly soluble in water, unlike sodium riboflavin 5′-phosphate, which is its most important commercial form. Both forms are sensitive to destruction by light and ultraviolet radiation, but are stable to heat and atmospheric oxygen. Rapid decomposition occurs in alkaline solutions, especially if exposed to light. In 0·1 N HCl, they exhibit absorption maxima at about 223, 267, 374 and 444 nm. Vitamin B_2 analysis is most often undertaken either by direct fluorometry of the intact compound or by conversion of chloroform-insoluble riboflavin into chloroform-soluble lumiflavin by irradiation in alkaline solutions.

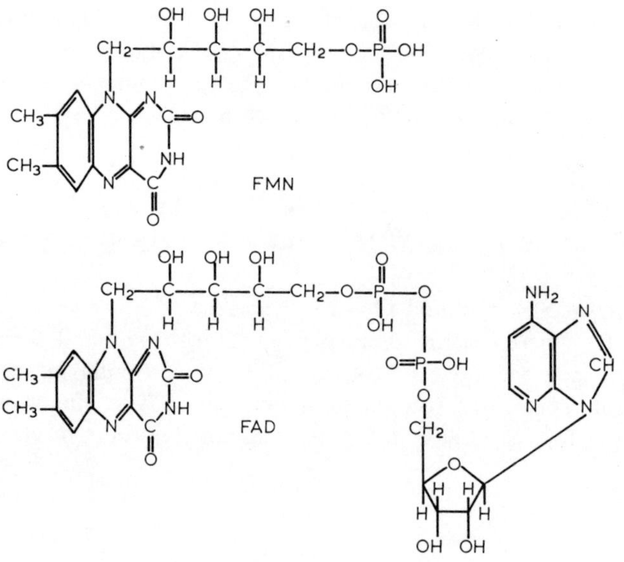

FIG. 15. Structural formulae of oxidized flavin mononucleotide (FMN) and flavin adenine dinucleotide (FAD).

7.2. Physiological Function and Deficiency Symptoms

Vitamin B_2, in the form of its phosphate ester (FMN) or of its dinucleotide (FAD) (Fig. 15), acts as a cofactor for flavin-containing enzymes concerned with hydrogen transfer. As such, they are important cofactors in the citric acid cycle and play an especially important role in the generation of ATP in the respiratory chain. Transfer of hydrogen to the coenzyme is effected by enzymes including succinate dehydrogenase, nicotine adenine dinucleotide (NADH) dehydrogenase, and nicotine adenine dinucleotide phosphate (NADPH) dehydrogenase. Subsequently, the reduced cofactor is oxidized by the cytochrome system (Fig. 17), the hydrogen electron being taken up by the Fe^{2+} of cytochrome. The hydrogen ions then form water with oxygen ions. Therefore, in the oxidation of numerous substrates, FMN and FAD play an essential role in efficient energy (ATP) generation by accepting hydrogen from the substrate-specific dehydrogenases and providing for their complete oxidation to water. Niacin, in its cofactor forms NADH and NADPH, plays an integral part in this process.

Amino acid (D and L) oxidases, xanthine oxidase, acyl-coenzyme A dehydrogenases and dehydroacyl reductases are flavin-requiring enzymes which, respectively, play a key role in the oxidation of amino acids, hypoxanthine and xanthine to uric acid, the dehydrogenation of saturated fatty acids prior to β-oxidation and synthesis of long-chain fatty acids from acetate. The latter role of vitamin B_2 (as FMN) in fatty acid metabolism illustrates the increased requirement of individuals on a high-fat diet for vitamin B_2.

As stated previously, vitamin B_2 occurs in most animal organs as FMN or FAD. However, the lens, retina and cornea of the eye contain free vitamin B_2 in relatively large amounts, but its significance is unknown. Vascularization of the cornea and clouding of the refractive media resulting in reduced sharpness of vision and rapid eye fatigue occur in vitamin B_2 deficiency. The latter is characterized by several non-specific symptoms including fatigue, inability to work, changes in the lips, the buccal mucosa and tongue, and in the skin, especially the horny skin. In man, the eyes, nose, tongue, gastrointestinal tract, anus, vulva and scrotum are vulnerable. Vitamin B_2 deficiency in pregnancy results in congenital malformations.

7.3. Human Requirement and Food Sources
Vitamin B_2 requirements in humans vary from 1 to 3 mg per day and are related to body weight, age and living conditions. Adults normally require daily intakes of 1·1–1·6 mg per day and deficiency symptoms appear at intakes less than 0·6 mg per day. The minimum daily requirement of a child is 0·4–0·5 mg. Pregnancy, lactation, growth, sickness, hyperthyroidism and increased intake of body fluids elevate the individual's requirements for vitamin B_2. Useful sources of vitamin B_2 in the diet include liver, kidney, meat, fish, eggs, yeast and vegetables, but milk is the single most important source in many Western diets (Table 1).

7.4. Vitamin B_2 in Milk
The concentrations of vitamin B_2 in mature human and bovine milks, 370 and 1750 µg litre^{-1}, respectively, are several times greater than the corresponding concentrations of vitamin B_1 in these milks. Breed has a small effect on the vitamin B_2 content of milk, e.g. Channel Island cows have ~20% more than non-Channel Island cows.[14,89] Approximately 20% of the vitamin B_2 in milk exists in the coenzyme form, i.e. flavin mononucleotide (FMN) or flavin adenine dinucleotide (FAD) bound to protein.[90–96] The vitamin B_2 concentrations in human and bovine

colostra exhibit changes similar to that of vitamin B_1 during lactation. Human colostrum has $\sim 1 \cdot 5$ times less vitamin B_2 than mature human milk,[70] whereas its concentration in bovine colostrum on day 1 of lactation is 6000–8000 µg litre^{-1},[77,97,98] and this exhibits a 4–5-fold decrease to the mature milk concentration within 7 days.[89,95]

The concentration of vitamin B_2 in bovine milk is subject to some nutritional influences. Easily digestible carbohydrates apparently stimulate rumen synthesis,[99,100] while an exclusive diet of lucerne hay of poor quality decreases rumen synthesis.[100-102] Aureomycin stimulates rumen synthesis of vitamin B_2 whereas it reduces the concentration of vitamin B_1.[103] Poppe[102] reports that 3–11% of vitamin B_2 synthesized in the rumen passes into milk, and Singh and Merilan[104] indicate that this process may be limited at the upper limit by a negative correlation between quantity of milk secreted and riboflavin concentration.

Like vitamin B_1, vitamin B_2 in liquid milk is stable to normal commercial heat treatments[76,87,105] and to frozen storage at $-40°C$ for 2 weeks.[76] However, it is extremely photosensitive, exhibiting a two-fold reduction in concentration, from 1550 to 480 µg litre^{-1}, after 6 hours exposure to sunlight. During subsequent refrigerated storage no further decrease occurs.[106] The concentration of vitamin B_2 in milk stored in glass or certain plastic containers when exposed to direct sunlight decreases by 20–40% in 1 hour. Light of wavelengths less than 610 nm, especially 490–520 nm, causes the destruction of vitamin B_2. Diffuse light can result in a 10–80% loss in a few hours. Vitamin B_2 destruction also catalyses the photochemical oxidation and destruction of ascorbic acid.[84,107] Liquid milk contributes $\sim 35\%$ of the total vitamin B_2 content of typical Western diets.[16,108]

8. NICOTINIC ACID (NIACIN)

8.1. Chemical Structure and Physical Properties

Nicotinic acid is the carboxylic acid of pyridine and is called niacin in the US to distinguish it and its amide derivative, nicotinamide, from the alkaloid nicotine (Fig. 16). Niacin is converted into niacinamide in the body and both have the same vitamin activity. Niacin and niacinamide are white crystalline powders with absorption maxima, depending on pH, at ~ 261 nm in aqueous solutions. Niacin is sparingly soluble in water and ethanol but is readily soluble in alkali. Niacinamide is readily soluble in water and slightly soluble in ethanol. Both are stable to

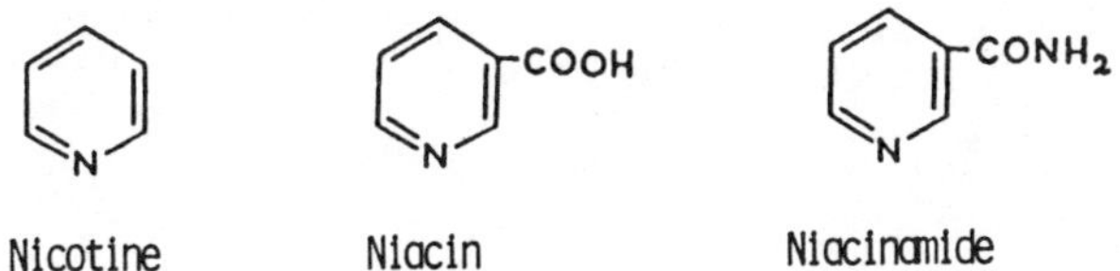

FIG. 16. Structural formulae of nicotine, niacin and niacinamide.

atmospheric oxygen, light and heat in the dry or aqueous states. Niacinamide is hydrolysed to niacin by heating in strongly acid or alkaline solutions. Fortified foods or feeds are usually assayed by chromatographic methods using the König reaction for final determination. Methods are now available for their determination by gas–liquid chromatography and high performance liquid chromatography.

8.2. Physiological Function and Deficiency Symptoms

Niacinamide acts as the functional group of the hydrogen-transferring coenzymes nicotinamide adenine dinucleotide (NAD) and nicotinamide adenine dinucleotide phosphate (NADP). Hydrogen transfer mediated by NAD and NADP plays a pivotal role in intermediary metabolism including the synthesis and degradation of fatty acids, carbohydrates and amino acids. Linked to specific apoenzymes, NAD and NADP perform an equally unique role in the complete oxidation of substrates in the citric acid cycle, accepting hydrogen during substrate combustion. Subsequently they transfer this hydrogen to the flavin enzymes of the respiratory chain (Fig. 17) and ultimately to oxygen with the formation of ATP in the process. The hydrogen-transferring coenzymes combine with apoenzymes which act quite specifically on the substrates capable of dehydrogenation, and channel these substrate hydrogens into the reactions of the respiratory chain. In the respiratory chain, there is a controlled release of substrate energy from NADH with the simultaneous formation of ATP, the 'on site' energy compound in the body. Niacin is also formed from tryptophan in the body through a vitamin B_6 dependent step, but this pathway is normally not sufficient to provide for the niacin requirements of the body. In the process, 60 mg of tryptophan is equivalent to 1 mg of niacin, hence the concept of niacin equivalents which includes an allowance for that derived from tryptophan in the diet. Niacin is also an essential cofactor for retinene reductase which regenerates retinal from retinol (see Section 2) in the visual cycle. These roles of niacin serve to highlight its great importance in human nutrition.

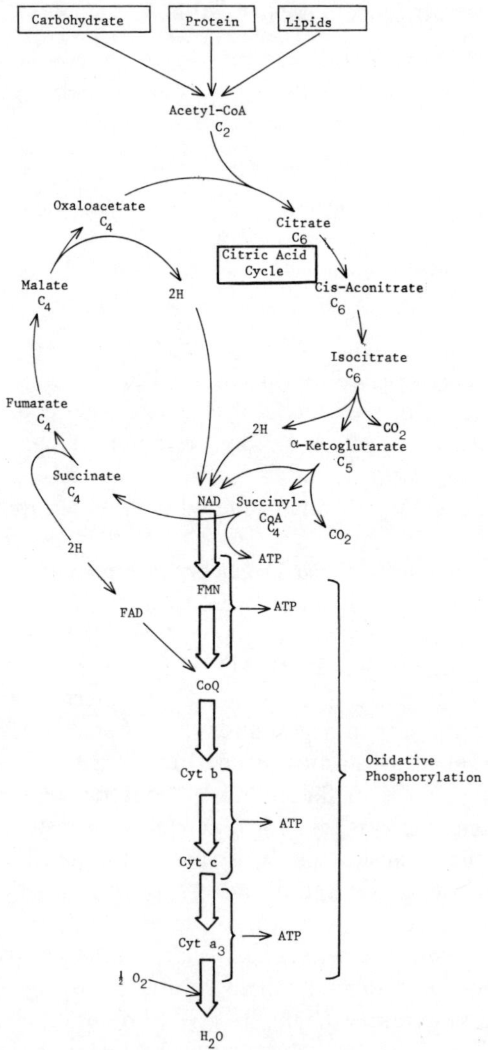

Fig. 17. Role of NAD, FMN and FAD in the complete oxidation of acctyl-CoA and in the generation of ATP in cells. Coenzyme Q (CoQ) is the collecting point in the respiratory chain for reducing equivalents derived from other substrates, e.g. succinyl-CoA, that are linked directly to it through FAD-linked dehydrogenases. Electrons flow from CoQ through the series of cytochromes indicated to molecular oxygen. In the process, oxidation of substrate through NAD- and FAD-linked dehydrogenases yields 3 mol and 2 mol of ATP, respectively, per $\frac{1}{2}$ mol of O_2 consumed. These reactions are known as oxidative phosphorylation at the respiratory chain level. (Modified from Ref. 3.)

The major clinical symptoms of niacin deficiency are dermatitis, diarrhoea and dementia. In humans, skin changes are characteristic symptoms, known as pellagra (meaning rough skin), and occur in areas exposed to sunlight. Depending on the degree of deficiency, the observable changes associated with the gastrointestinal tract include loss of appetite, dizziness, vomiting, constipation and diarrhoea. Nervous disorders include insomnia, fatigue, vertigo and headache. Depression and confusion are evident in subjects with severe deficiency.

8.3. Human Requirement and Food Sources

Requirements for niacin are influenced by many variables including dietary content of essential amino acids, vitamins (B_1, B_6 and biotin), digestive disorders, feverish illnesses, periods of growth, hard physical work, pregnancy, lactation and ingestion of large quantities of fluids. Decreased intake of essential amino acids other than tryptophan increases the availability of tryptophan for niacin synthesis, whereas decreased vitamin intake decreases this process. Other variables listed increase the niacin requirements of the individual. To prevent pellagra, a minimum of 4·4 mg of niacin per 1000 cal is required, provided that there is a minimum intake of 2000 cal. Liver and meat of hoofed animals are nutritionally important sources of this vitamin. It is also present in maize and other cereals in a form not utilized by man.

8.4. Niacin in Milk

The concentrations of niacin in mature bovine and human milks are typically about 900 and 1800 μg litre^{-1}, respectively. Unlike plant material, almost all the niacin potency of milk is in the nicotinamide form.[109] Similar concentrations of niacin are observed in bovine colostrum and mature milk[89] but the concentration in human colostrum is approximately 30–60% of that in mature human milk.[70] There are no significant nutritional or seasonal influences on the concentration of niacin in milk.[82,110] However, there is some evidence that supplementation of rations of dairy heifers and cows in early lactation may, in certain circumstances, increase milk production and the concentration of fat and protein in their milks.[111–113]

It is estimated that 60 mg of tryptophan in the diet is metabolically equivalent to 1 mg of niacin in the body.[8] Bovine and human milks contain about 90 and 140 mg of tryptophan per gram of total nitrogen, respectively,[114] a concentration that is sufficient to protect against niacin deficiency, i.e. pellagra, in humans in Western countries.

9. PANTOTHENIC ACID

9.1. Chemical Structure and Physical Properties

Pantothenic acid (Fig. 18) consists of a molecule of pantoic acid bonded to β-alanine by an acid amide linkage. Free pantothenic acid is an unstable, extremely hygroscopic oil which consequently is unsuitable for commercial application. It is used mainly in the form of its calcium and sodium salts which are water-soluble white powders. In the absence of moisture, these salts are fairly stable to atmospheric oxygen and light when cold-stored. Aqueous solutions are thermolabile and undergo hydrolytic cleavage, especially if acidic or alkaline. The alcohol, pantothenol, possesses full vitamin activity and has also assumed great importance. Aqueous solutions of D-pantothenol, especially if acidic, are significantly more stable. Pantothenic acid occurs rarely in nature but rather occurs in its cofactor form, i.e. as a component of coenzyme A (Fig. 18). Pantothenic acid is optically active and only the D form possesses vitamin activity. Pantothenate is determined in foods and feedstuffs by microbiological assay.

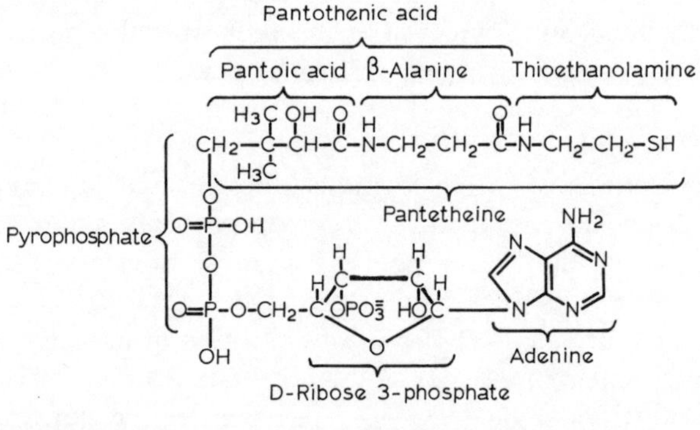

FIG. 18. Structural formulae of pantothenic acid and acetyl-CoA.

9.2. Physiological Function and Deficiency Symptoms

Pantothenate participates in metabolism as the cofactor, coenzyme A. The latter has a nucleotide structure incorporating adenine and D-ribose-3-phosphate, attached to pantetheine through a pyrophosphate bridge (Fig. 18). Pantetheine consists of a molecule of pantothenic acid bonded to thioethanolamine through an acid amide linkage. Coenzyme A occupies a central role in intermediary metabolism, forming energy-rich thioesters with weakly reactive carboxylic acids. Metabolites are activated by combining with coenzyme A at the sulphydryl (SH) group of pantetheine, thus forming a high-energy sulphur bond. The structure of the free coenzyme is usually abbreviated to CoA.SH whereby only the reactive SH group is indicated. Acetyl-CoA is the most important acyl-CoA derivative appearing in intermediary metabolism. It plays a pivotal role in the degradation of fats, carbohydrates and a variety of amino acids (Fig. 19). As discussed previously (Sections 7 and 8), the acetyl moiety of acetyl-CoA is completely oxidized to carbon dioxide and water in the citric acid cycle, resulting in the generation of ATP.

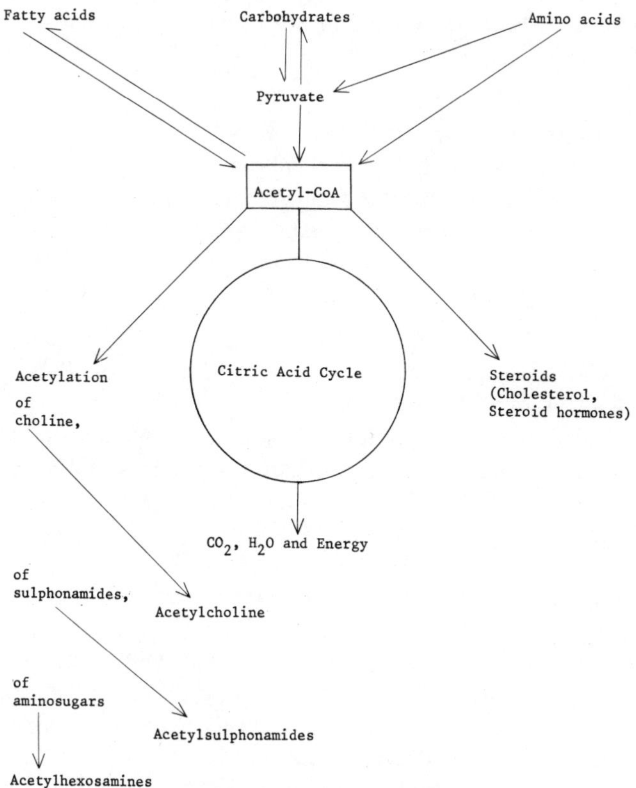

FIG. 19. The central role of acetyl-CoA in metabolism. (Reproduced with permission from Ref. 5.)

Acetyl-CoA (Fig. 19) is the basic building unit from which long-chain fatty acids, phosphatides, cholesterol, steroid hormones and bile acids are synthesized. Acetyl-CoA acts as a donor of acetyl groups to acceptors with, for example, resultant formation of the neurotransmitter acetylcholine at nerve endings, which is of basic importance. Acetyl-CoA acetylates amino sugars, which are constituents of various mucopolysaccharides of cartilage and other skeletal substances. Acetyl-CoA plays an important role in detoxification reactions, e.g. the acetylation of sulphonamides and related drugs. CoA-SH activates succinate to succinyl-CoA, an intermediate of the citric acid cycle and an essential precursor for porphyrin synthesis in blood formation. Deficiency symptoms observed in human volunteers include reduced growth or weight loss, skin lesions, disorders of the nervous system including 'burning feet' syndrome, gastrointestinal disturbances, inhibition of antibody formation and impaired adrenal function.

9.3. Human Requirement and Food Sources

A typical mixed diet containing 2500 cal provides 10 mg of pantothenate, which apparently represents the daily requirement. Pantothenate occurs widely in nature and especially good sources include liver, kidney, muscle, brain, egg yolk, yeast, cereals and some green plants, especially legumes. Consequently, clear-cut deficiency symptoms are rarely observed in man.

9.4. Pantothenic Acid in Milk

Typical concentrations of pantothenic acid in bovine and human milks are 3500 and 2500 µg litre^{-1}, respectively. Most pantothenic acid in human milk is free, but 10–15% is bound and must be released by enzyme treatment prior to assay.[115] The concentration of pantothenic acid in bovine and human colostra is lower than that in the mature milks; it reaches a maximum within a few days of parturition and subsequently decreases again.[69,89] Song et al.[115] observed no difference in the pantothenic acid concentration in human milk samples 2 and 12 weeks after parturition; they also observed a significant correlation between dietary intake, circulating maternal plasma concentration and urinary excretion of pantothenic acid.

Pantothenic acid is considered to be heat-stable, although Gorner and Uherova[105] reported losses of 30–35% in UHT-treated cow's milk when stored at room temperature for 6 weeks.

10. VITAMIN B_6 (PYRIDOXINE)

10.1. Chemical Structure and Physical Properties

The substituted pyridine derivatives pyridoxine (pyridoxol, pyridoxal and pyridoxamine) exhibit vitamin B_6 activity (Fig. 20). These compounds differ from each other only with regard to the functional group at the 4-position and are together often considered as pyridoxine. All form salts but the commercial form is almost exclusively the hydrochloride of

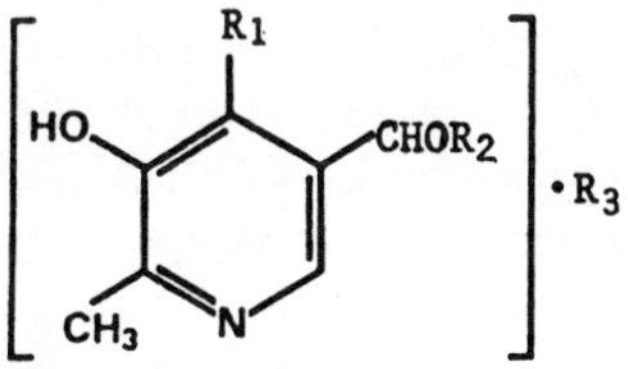

Pyridoxine: $R_1 = CH_2OH$; $R_2 = H$
Pyridoxal: $R_1 = CHO$; $R_2 = H$

Pyridoxal-5'-phosphate: $R_1 = CHO$; $R_2 = -\overset{\overset{O}{\|}}{\underset{\underset{OH}{|}}{P}}-OH$

Pyridoxamine-5'-phosphate: $R_1 = CH_2NH_2$; $R_2 = -\overset{\overset{O}{\|}}{\underset{\underset{OH}{|}}{P}}-OH$

Pyridoxine hydrochloride: $R_1 = CH_2OH$; $R_2 = CH_2OH$; $R_3 = HCl$

FIG. 20. Physicochemical properties of pyridoxine.

the alcohol, i.e. pyridoxine hydrochloride. Pyridoxine hydrochloride is a white crystalline powder which is readily soluble in water and sparingly soluble in ethanol. Absorption maxima in acid, neutral and alkaline aqueous solutions occur at 291, 254 and 324, and 245 and 309 nm, respectively. It is stable to heat and oxygen, but is degraded by light in alkaline or neutral solutions, and to a lesser extent in acid solution. The vitamin B_6 activity of non-enriched foods and tissues is usually determined by microbiological assay, although a high performance liquid chromatographic method has recently been described for milk.[116] With either of these methods it is possible to quantify the contribution of the three forms of pyridoxine to vitamin B_6 activity in milk.

10.2. Physiological Function and Deficiency Symptoms

As pyridoxal-5'-phosphate, vitamin B_6 participates as a cofactor for apoenzymes that catalyse transamination, decarboxylation, deamination, desulphydration and the cleavage or synthesis of amino acids. The aminotransferases represent an important link between the metabolism of amino acids, carbohydrates and fatty acids and energy generation in the citric acid cycle. Pyridoxine functions in the biosynthesis of porphyrins, specifically in the decarboxylation of α-amino-β-ketoadipate to laevulinic acid. The decarboxylases also convert amino acids into biogenic amines including histamine, hydroxytyramine, serotonin, γ-aminobutyric acid, ethanolamine and taurine. Some of these substances have important physiological roles as regulators of blood vessel diameter, neurohormonal agents, and essential components of phospholipids and bile acids. Because of these important roles, the enzymes for their production are present in most tissues. Deamination and desulphydration reactions are involved in amino acid catabolism and anabolism especially, and consequently are localized mainly in liver.

Pyridoxal phosphate is required as a cofactor for kynureninase in the normal metabolism of tryptophan to niacin in the liver. Xanthurenic acid, especially after a test dose of tryptophan, shows up as an abnormal urinary metabolite and is used as the basis of a test to diagnose vitamin B_6 deficiency. Normal amino acid metabolism is essential for reactions in the body whereby harmful chemicals are detoxified and eliminated from the body. Vitamin B_6 is also required for coenzyme A synthesis and for homeostasis. Consequently, reduced coenzyme A levels and altered fatty acid metabolism occur in vitamin B_6 deficiency. Diets high in fat simultaneously increase the body's requirement for vitamin B_2 and decrease the pyridoxal-5'-phosphate content of the liver, because its formation by pyridoxal phosphate oxidase is a vitamin B_2 dependent step. Vitamin B_6 may also play a direct role in the absorption of amino acids and their concentrations in cells. The widespread involvement of vitamin B_6 in protein metabolism explains the interrelationship observed between increased protein intakes and increased vitamin B_6 requirements.

Severe vitamin B_6 deficiency is characterized by skin changes in the areas of the nose, eyes and mouth, and by nervous disorders including peripheral neuritis and epileptiform convulsions; the latter occur especially in actively growing children. In mild vitamin B_6 deficiency, a number of non-specific symptoms occur resembling those associated with a deficiency of other B vitamins.

10.3. Human Requirement and Food Sources

Human requirements depend on the level of protein intake and are estimated at 1–2 mg per day. Useful sources of vitamin B_6 in human nutrition are red meats, liver, kidney, brain, cod liver and roe, egg yolk, yeast, cereals, green vegetables, milk and cheese.

10.4. Vitamin B_6 in Milk

The concentration of vitamin B_6 in bovine milk (500 µg litre^{-1}) is ~5 times that observed in human milk. The relative distribution of vitamin B_6 activity in bovine and human milks between pyridoxal, pyridoxamine, pyridoxine, pyridoxamine phosphate and pyridoxal phosphate may also be different. About $67 \pm 18\%$ of the B_6 activity of the milks of 12 women consuming a daily supplement of 0·4–0·5 mg of pyridoxamine was due to pyridoxal, $27 \pm 16\%$ to pyridoxal phosphate, and traces (usually) to pyridoxamine.[116] In contrast, the vitamin B_6 activity of raw cow's milk is associated with pyridoxal (80%) and pyridoxamine (20%),[117] and pyridoxamine phosphate (traces).[118]

As with niacin, riboflavin and thiamine, the concentration of vitamin B_6 in human colostrum is less than that in mature human milk,[83] but it progressively increases during the weeks following parturition.[119] Nutrition and breed have no effect on the concentration of vitamin B_6 in bovine milk, but elevated concentrations have been reported in colostrum.[89,120] Administration of vitamin B_6 supplements to poorly-nourished women resulted in a two-fold increase in its concentration in milk,[121] but little change was observed in similar studies in well-nourished women.[110]

Vitamin B_6 in milk is relatively stable to normal commercial processing, including UHT treatment,[107] HTST treatment and normal storage conditions,[106] and to light; ~21% destruction occurred after 8 hours exposure to daylight.[122] During in-bottle sterilization of milk, pyridoxal is converted into pyridoxamine, which, during storage, can complex with sulphur compounds to form bis-4-pyridoxy disulphide which has a low biological potency.[123]

11. VITAMIN B_{12} (CYANOCOBALAMIN)

11.1. Chemical Structure and Physical Properties

The compounds active as vitamin B_{12} have complicated chemical structures; the best-known and most important is cyanocobalamin (Fig. 21)

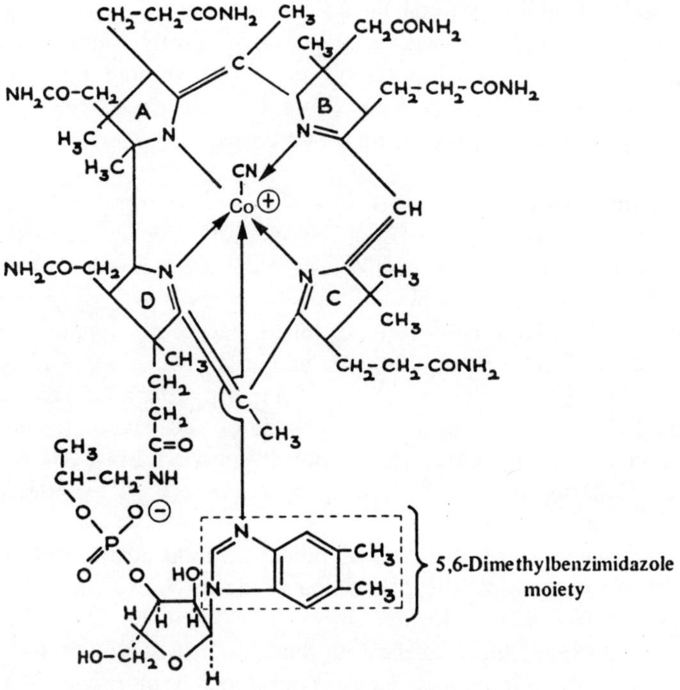

FIG. 21. Chemical structure of cyanocobalamin.

which differs only slightly from the others. Cyanocobalamin is a dark red, crystalline powder which is slightly soluble in water and soluble in ethanol. Aqueous solutions exhibit absorption maxima at about 278, 361 and 550 nm. Crystalline cyanocobalamin and its neutral to weakly acid solutions are relatively stable to air and heat, but are attacked by light and ultraviolet radiation. Vitamin B_{12} has poor stability to alkalis, strong acids and reducing agents. It is quantitatively measured in complex systems by microbiological assay, and more recently by the isotopic dilution principle. Vitamin B_{12} occurs only in animal products and in metabolites of microorganisms. The synthesis of this vitamin in the alimentary tract is of considerable importance for animals because, if sufficient cobalt is available, ruminants are independent of external sources of vitamin B_{12}.

Cyanocobalamin consists of four reduced and extensively substituted pyrrole rings surrounding a single cobalt atom. This 'corrin' ring system is similar to the porphyrins but differs in that two of the pyrrole rings (A and D) are joined directly rather than through a single methylidyne carbon. The basic corrin structure of vitamin B_{12} is, like the porphyrins, made from δ-aminolaevulinic acid in a vitamin B_6 dependent process. Below the corrin ring system is a 5,6-dimethylbenzimidazole riboside connected at one end to the central cobalt atom and at the other end from the ribose moiety through phosphate and aminopropanol to a side chain on ring D of the tetrapyrrole nucleus.

Cyanide appears in the vitamin molecule coordinately bonded to cobalt, mainly as an artifact of vitamin B_{12} isolation from natural sources. Removal of cyanide yields cobalamin, and its substitution by a hydroxy-, methyl- or 5'-deoxy-adenosyl nucleotide (Fig. 22) yields their respective cobalamin coenzymes, which occur in mammalian tissues.

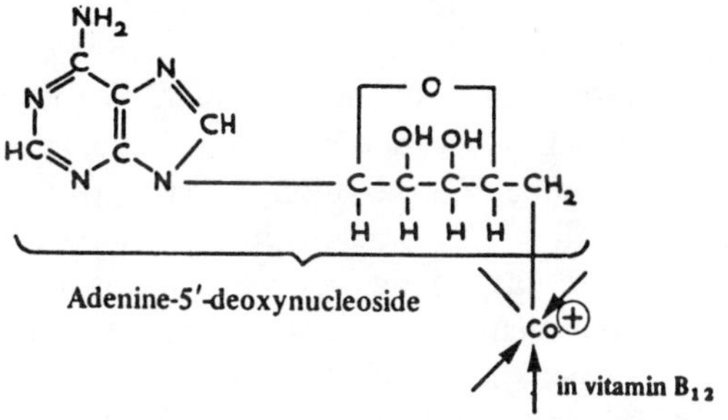

FIG. 22. Formation of cobamide coenzyme by attachment of adenine-5'-deoxynucleotide to vitamin B_{12} through C-5' to cobalt.

11.2. Physiological Function and Deficiency Symptoms

Metabolic reactions in which vitamin B_{12} participates as a cofactor include the conversion of methylmalonyl-CoA into succinyl-CoA by methylmalonyl-CoA isomerase, reciprocal oxidation and reduction of active formaldehyde and active formate by hydroxymethyltetrahydrofolic acid dehydrogenase, and methylation of homocysteine to methionine by N^5-methyltetrahydrofolic acid: homocysteine methyltransferase. A deficiency of vitamin B_{12} induces a loss in activity of methylmalonyl-CoA isomerase, an enzyme that facilitates the transformation of propionate to succinate, an intermediate of the citric acid cycle. At an early stage of vitamin B_{12} deficiency, a sharp increase occurs in urinary methylmalonic acid which provides a diagnostic test.

Vitamin B_{12} probably acts as a cofactor for hydroxymethyltetrahydrofolic acid dehydrogenase which acts in the reduction of one-carbon compounds of the oxidation state of formate and formaldehyde. In this critical step it participates, with folic acid, in the biosynthesis of labile methyl groups. The latter are essential for the synthesis of purines, pyrimidines and consequently DNA and RNA. Vitamin B_{12} is therefore essential for all cells of the body, and in its absence cell division and growth are impaired. As an example, inadequate amounts of vitamin B_{12} result in failure of cell nuclei to mature and in failure of cells to divide. The occurrence of macrocytosis, large immature red blood cells, acts as a sensitive indicator of vitamin B_{12} deficiency in humans.

Labile methyl group metabolism involving vitamin B_{12} and folate is also important for the biosynthesis, in the body, of methionine from homocysteine and choline from ethanolamine. Methionine is essential for protein synthesis and acts as a methyl group donor for the biosynthesis of lipotrophically active choline, which prevents the development of fatty livers in animals. It is also important for the formation of creatine, which serves, after conversion into creatine phosphate, for energy storage in muscle tissue. Vitamin B_{12} also participates in intermediary metabolism in maintaining glutathione and the sulphydryl groups of enzymes in the reduced state. The fall in glyceraldehyde 3-phosphate dehydrogenase activity, which requires glutathione as a coenzyme, is possibly the reason for the observed impairment in carbohydrate metabolism in vitamin B_{12} deficiency. The influence of vitamin B_{12} on lipid metabolism probably also follows from its action on thiols. Vitamin B_{12} may also be required for storage of folic acid in the liver.

Absorption of vitamin B_{12} from the digestive tract of man occurs specifically in the ileum and is dependent on an 'intrinsic factor', a constituent of normal gastric juice, for its occurrence. Vitamin B_{12} deficiency in humans is characterized by a macrocytic anaemia or by characteristic lesions of the central nervous system, or both. The human daily requirement for vitamin B_{12} is difficult to estimate because it is synthesized by the intestinal microflora, but it appears to be about 2 µg. In the absence of 'intrinsic factor', the vitamin is not absorbed and if unchecked will result in pernicious anaemia. In such patients, 0·5 µg per day administered intraperitoneally results in complete neurological and haematological remission. Important sources of vitamin B_{12} in human nutrition are animal tissues including liver, kidney and egg yolk.

11.3. Vitamin B_{12} in Milk

Average concentrations of vitamin B_{12} in bovine and human milks are

4·5 and 0·4 µg litre^{-1}, respectively. Wide variations exist in reported vitamin B_{12} concentrations in human milk. A survey by the UK Department of Health and Social Security[79] showed concentrations less than 0·1 µg litre^{-1} in human milk, while Sandberg et al.[124] reported that the concentration of vitamin B_{12} in milks from 19 women ranged from 0·33 to 3·20 µg litre^{-1}, with a mean concentration of 0·97 µg litre^{-1}. The latter authors identified methodological problems, including failure to destroy binding proteins prior to vitamin B_{12} assay, as reasons for discrepancies between reported vitamin B_{12} concentrations in milk. Vitamin B_{12} occurs in milk in several chemical forms; methylcobalamin is the predominant form in human milks, followed by adenosylcobalamin and small amounts of hydroxy- and cyano-cobalamins.[124] Hydroxycobalamin predominates in cow's milk with minor amounts of methyl- and adenosylcobalamins.[125] In milk, vitamin B_{12} exists primarily bound to R-binder glycoprotein (20·5 µg B_{12} per mg of protein) and must be released from this complex by pancreatic digestion prior to absorption from the gastrointestinal tract. A considerable excess of R-binder protein exists in human and sow's milk whereas little or no excess exists in cow's milk.[114] Suggested functions of the R-proteins in milk include (i) partition of B_{12} from blood into the mammary gland, (ii) inhibition of bacterial growth in milk, i.e. those microorganisms that require vitamin B_{12} for multiplication, and (iii) favourable development of the microfloral composition of the infant's large intestine.[126] In ultrafiltration experiments, Kim et al.[127] observed that 5% of vitamin B_{12} in cow's milk was present in the unbound form. Human and bovine colostra have 6–10 times the vitamin B_{12} concentration of the corresponding mature milks. Ford et al.[119] observed a gradual decline in the concentration of vitamin B_{12} in milk from women with full-term and pre-term infants. Oral vitamin B_{12} supplementation of poorly- and well-nourished women did not increase its concentration in human milk,[110,128] and it was not decreased by vegetarianism.[129] Administration of cyanocobalamin supplements of 5–100 µg per day to 10 women did not significantly increase the concentration of vitamin B_{12} in human milk.[124] The lower concentration observed in human milk does not increase the risk of vitamin B_{12} deficiency in the neonate if the mother is well nourished during pregnancy, as tissue stores acquired *in utero* provide the vitamin B_{12} requirements of the neonate during the early months of life.[130]

Dietary supplementation of cows ration with up to 3·3 mg of vitamin B_{12} daily did not influence its concentration in milk.[131] Because vitamin

B_{12} contains cobalt, ruminants with normally functioning rumens can synthesize it, provided that they are supplied with adequate cobalt in the diet, i.e. 0·11 mg per kg of dry matter intake.[132] Smith and Marston[133] estimate that 40% of rumen cobalt is incorporated into vitamin B_{12}. Therefore dietary cobalt is the major determinant of vitamin B_{12} concentration in bovine milk and its synthesis in the rumen. Rumen synthesis varies from 0·1 µg per gram of digestible dry matter intake in cobalt deficiency to 4 µg per gram when cobalt intake is adequate.[134] Cobalt supplementation of the diet increases the vitamin B_{12} concentration of milk only if the cobalt content of the diet is suboptimal.[135,136]

Under normal commercial and domestic conditions of milk processing and consumption the concentration of vitamin B_{12} in milk is not significantly impaired.[106]

12. FOLIC ACID (PTEROYLGLUTAMIC ACID)

12.1. Chemical Structure and Physical Properties

Compounds with folic acid activity contain a pteridine nucleus attached to *p*-aminobenzoic acid and one or more glutamic acid residues (Fig. 23). The latter, polyglutamates, are conjugates of folic acid with several glutamic acid residues and represent the form in which it occurs most commonly in nature. Commercially synthesized folic acid (pteroylmonoglutamate) has one glutamic acid residue attached, the glutamate moiety being essential for folic acid activity. Folic acid is an orange to yellow crystalline powder with a characteristic absorption spectrum dependent on the pH of the solution. In 0·1 N NaOH, the maxima occur at 256, 283 and 365 nm. Crystalline folic acid is fairly stable to air and heat, but it is degraded by light and ultraviolet radiation. Neutral solutions are relatively stable whereas acids, alkalis, and oxidizing and reducing agents have destructive effects. Folic acid is determined in biological tissues, food and feedstuffs by a non-specific microbiological assay, using a variety of microorganisms.

12.2. Physiological Function and Deficiency Symptoms

Folic acid participates as a coenzyme in biochemical reactions involving the transfer and utilization of one-carbon compounds. Before it can act as a cofactor in one-carbon metabolism, folic acid must be reduced to 7,8-dihydrofolic acid and then to 5,6,7,8-tetrahydrofolic acid by folic acid

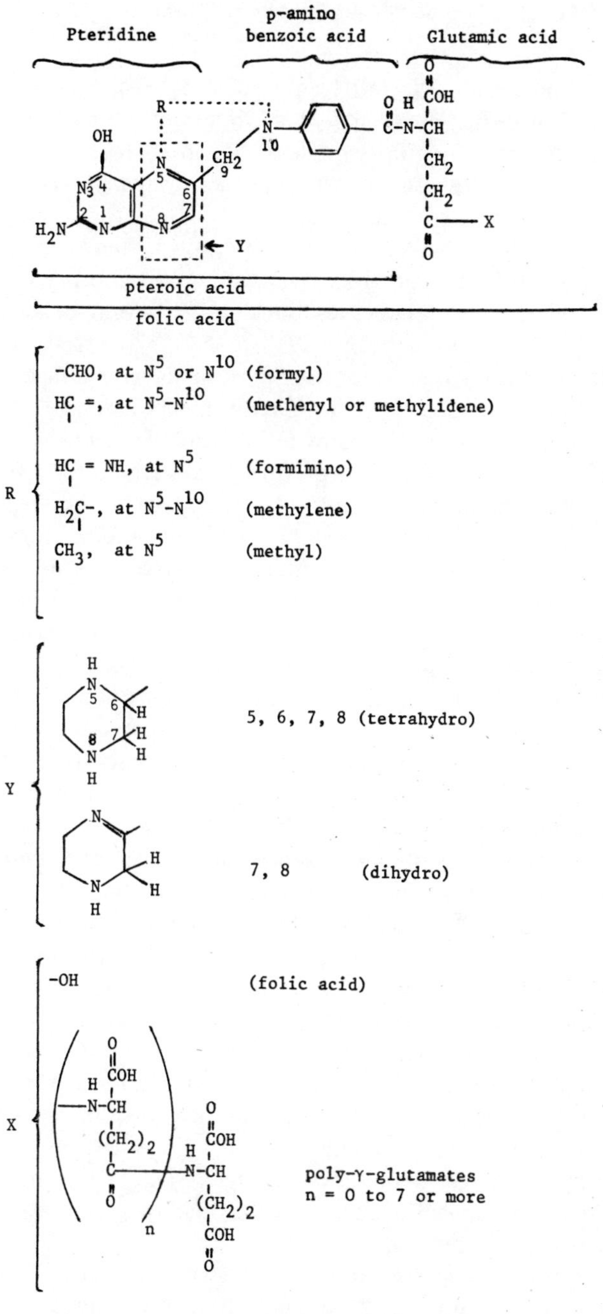

-CHO, at N^5 or N^{10} (formyl)

HC =, at N^5–N^{10} (methenyl or methylidene)

HC = NH, at N^5 (formimino)

H_2C–, at N^5–N^{10} (methylene)

CH_3, at N^5 (methyl)

R

5, 6, 7, 8 (tetrahydro)

7, 8 (dihydro)

Y

-OH (folic acid)

X

poly-γ-glutamates
n = 0 to 7 or more

FIG. 23. Chemical structure of folic acid and its derivatives.

reductases that require NADPH as a cofactor. Folic acid antagonists, such as methotrexate, act as potent inhibitors of 5,6,7,8-tetrahydrofolic acid formation from 7,8-dihydrofolic acid. The one-carbon moiety carried on tetrahydrofolic acid may be the formyl (–CHO), formate (–COOH), methyl (–CH$_3$), hydroxymethyl (–CH$_2$OH) or formimino group (–CH=NH–), all of which are metabolically interconvertible. N^{10}-Formyltetrahydrofolic acid (Fig. 23) is the active form of folic acid coenzymes in most metabolic reactions. N^5,N^{10}-Methylenetetrahydrofolic acid can be oxidized by NADP$^+$ to N^5,N^{10}-methenyltetrahydrofolic acid, and both act as major donors of the formyl (single) carbon in many synthetic reactions. N^5,N^{10}-Methylenetetrahydrofolic acid can also be converted into N^5-methyltetrahydrofolic acid by an NAD-dependent reductase. Its methyl group is transferable to cobalamin to form methylcobalamin (see vitamin B$_{12}$ coenzyme), an important donor of methyl groups, as in the methylation of homocysteine to methionine. There is reason to believe that vitamin B$_{12}$ participates in the reciprocal transformation of active formaldehyde and active formic acid by oxidation or reduction through the enzyme hydroxymethyltetrahydrofolic acid dehydrogenase. In short, the above reaction sequence, in association with others, enables the body to prepare the one-carbon compounds for a series of biosynthetic reactions resulting in the synthesis of cellular constituents required for important biological processes (Fig. 24). These include the formation of purine bases, thymine, histidine, serine (from glycine), methionine, choline and methylnicotinamide.

The participation of folic acid coenzymes in reactions leading to the synthesis of the purine bases adenine and guanine, as well as the pyrimidine thymine, emphasizes its fundamental role in cell growth and reproduction. Folic acid deficiency therefore leads to serious disturbance of nucleic acid biosynthesis, which is essential for normal cell formation and function. The megaloblastic anaemias observed in folic acid deficiency are a consequence of this role of folate in nucleic acid synthesis. In folic acid deficiency, the formiminoglutamate formed as an intermediate in the degradation of histidine can no longer be transformed by a transferase enzyme into glutamate and formiminotetrahydrofolic acid, and is therefore excreted in the urine. Excretion of this metabolite is suitable as a biochemical indicator of folic acid deficiency, as it appears at an early stage of deficiency.

Folic acid deficiency is considered to be the most common hypovitaminosis of man. This is especially true in pregnancy, where folic acid deficiency is regarded as the most frequent cause of megaloblastic

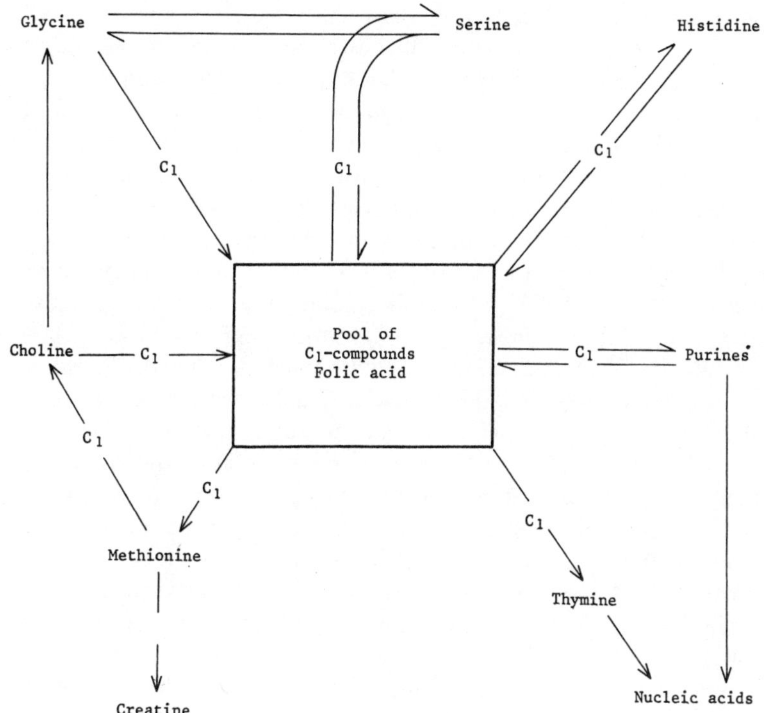

FIG. 24. Role of folic acid in the metabolism of C_1 compounds. (Reproduced with permission from Ref. 5.)

anaemia. Other conditions predisposing to folic acid deficiency include sulphonamide or antibiotic interference with intestinal synthesis, alcoholism, malabsorption syndromes, zinc deficiency, haemolytic anaemias and the anaemias occurring in infancy, pregnancy or malignancies. The recommended dietary allowance for folic acid is 400 µg per day for adults; this allowance is increased in pregnancy and lactation.

12.3. Human Requirement and Food Sources

Folates are present in a wide variety of plant and animal tissues, mainly as polyglutamates in reduced methyl or formyl forms (Fig. 23). Dietary folate intake varies greatly among various individuals depending on economic status and dietary habits. The richest dietary sources are yeast, liver, kidney and green vegetables, with moderate amounts in dairy foods, meat and fish.

12.4. Folic Acid in Milk

Typical concentrations of folate in bovine and human milks are 50–60 and $40\,\mu g$ litre^{-1}, respectively. The findings of a number of authors[83,137,138] indicate that the concentration of folates is lowest in human colostrum, increases during the early months of lactation, reaches a plateau,[83,138] and decreases again towards the end of lactation.[83] In contrast, the highest concentrations of folate in bovine lactation are found in colostrum. Its concentration in bovine colostrum decreases rapidly initially and then more slowly.[139,140] In a study of folate concentrations in HTST bulk milk, Scott *et al.*[14] observed seasonal differences with peak and trough concentrations of 75 and $45\,\mu g$ litre^{-1}, thus implying a nutritional and/or environmental influence. Lardinois *et al.*[141] observed no correlation between folate concentration in the diet and ration composition. This does not proscribe a nutritional influence as folate is synthesized by the rumen microflora and therefore dietary constituents that affect rumen function may alter its rate of synthesis and secretion to blood and milk. Folate in human milk is mainly in the form of pteroylmonoglutamate, but a little folyl-di- and -tri-glutamate are also present.[137,138] Unlike folyl polyglutamates, these latter forms of folate do not require to be hydrolysed by γ-carboxyglutamyl peptidase (conjugase) prior to microbiological assay or to absorption to the bloodstream.[142] In bovine milk, 5-methyltetrahydrofolic acid is the principal chemical form.[143] In bovine and human milks, folate is bound to specific binding proteins whose binding capacity is greatly in excess of the concentration of folate in milk.[114] Like vitamin B_{12}-binding protein, cow's milk folate binding protein is a glycoprotein. It occurs in the whey fraction and has a molecular weight of 35 000.[143] Human milk folate-binding protein was characterized by Waxman and Schreber.[144] *In vivo*, both proteins are resistant to digestion.[145]

13. BIOTIN

13.1. Chemical Structure and Physical Properties

Biotin (hexahydro-2-oxo-1-thieno-3,4-imidazole-4-valeric acid) contains three asymmetric carbon atoms, and consequently eight different stereoisomers are possible, of which only the dextrorotatory form *d*-biotin (Fig. 25) occurs in nature and possesses full vitamin activity. Biotin occurs in the free state in vegetables, fruit, milk and rice bran, and in a combined form, biocytin, in animal tissues, plants, seeds and yeast.

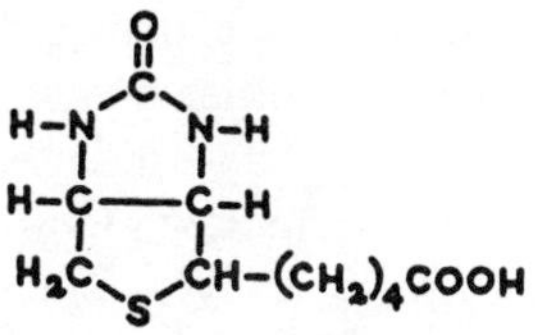

FIG. 25. Chemical structure of biotin (hexahydro-2-oxo-1-thieno-3,4-imidazole-4-valeric acid).

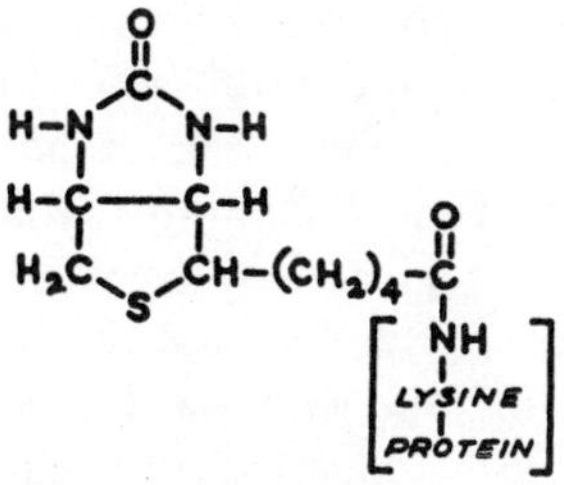

FIG. 26. Attachment of biotin coenzyme to apoenzyme through the ε nitrogen of lysine.

Biocytin is an ε-N-biotinyl lysine conjugate (Fig. 26) produced during biotin isolation and may represent a degradation product of the physiologically important form of biotin which occurs tightly bound to the apoenzyme carboxylase. Biotin is a white crystalline powder which is sparingly soluble in water but soluble in dilute alkali. It has an $[\alpha]^{20}$ of $+90$–$94°$ in 0.1 N NaOH (c 1.0). When dry, crystalline d-biotin is fairly stable to atmospheric oxygen, daylight and heat, but it is gradually destroyed by ultraviolet radiation. In aqueous solutions, biotin is relatively stable under weakly acidic or weakly alkaline conditions, but biological activity is destroyed by heating in strongly acidic or alkaline solutions. Biotin occurs at very low concentrations in foods, and consequently microbiological methods must be used for its assay.

13.2. Physiological Function and Deficiency Symptoms

As a tightly bound cofactor of carboxylase (Fig. 26), biotin facilitates carboxylation reactions in metabolism. In this complex, biotin is capable of taking up carbon dioxide with the formation of an *active* CO_2–biotin enzyme complex. *Active* carbon dioxide can then be transferred to suitable substrates, regenerating the free biotin–enzyme complex in the process. Such carboxylation reactions are involved in the degradation of the amino acids leucine and isoleucine, and in the formation of malonyl-CoA from acetyl-CoA (see pantothenic acid, Section 9). The latter reaction is basic to *de novo* fatty acid synthesis from acetyl-CoA in the body (Fig. 19). In this process, malonyl-CoA combines with another molecule as acetyl-CoA to form acetyl-malonyl-CoA. The latter is decarboxylated, reduced and dehydrated resulting in butyryl-CoA formation. The carbon chain is then elongated by a stepwise series of repetitive reactions involving malonyl-CoA, in the first of which butyryl-CoA is linked to malonyl-CoA and the reaction sequence described for butyryl-CoA formation is repeated with the initial formation of a six-carbon activated fatty acid and eventually the formation of palmitic and shorter-chain acids. The biotin-dependent carboxylation of acetyl-CoA is thus a key cellular reaction in the *de novo* synthesis of fatty acids. Biotin plays an important part in gluconeogenesis as a cofactor for pyruvate carboxylase which catalyses the reversible carboxylation of pyruvate to oxaloacetate—a connecting link with the citric acid cycle. In this process, body fat and protein are converted into carbohydrate to maintain normal blood glucose concentrations for brain and other key tissues.

Reduced protein synthesis associated with biotin deficiency reflects inadequate formation of dicarboxylic acids (oxaloacetate, succinate), leading to reduced incorporation of amino acids into protein, rather than a direct role of biotin in protein synthesis. A number of enzyme systems are reported to be influenced by biotin, including succinic dehydrogenase and decarboxylase and the deaminases of the amino acids aspartate, serine and threonine.

Biotin deficiency in man is always characterized by skin changes, and non-specific symptoms include fatigue, anorexia, nausea, muscular pain, hyperaesthesia, paraesthesia and a fall in haemoglobin concentration. Biotin deficiency appears spontaneously in the suckling infant and young child (seborrhoeic dermatitis), and also in various animals. Ingestion of biotin-binding proteins, including avidin from raw egg white and stravidin and streptavidin from certain *Streptomyces*, and drugs such as the sulphonamides, which interfere with the biosynthesis of biotin by the intestinal bacteria, potentiate the development of biotin deficiency.

13.3. Human Requirement and Food Sources

Establishment of human requirement is complicated by intestinal synthesis. In careful balance studies in man, urinary excretion often exceeds dietary intake. A daily intake of 150–300 μg of *d*-biotin prevents deficiency symptoms. It is doubtful whether any but the most severely deficient diets would result in biotin deficiency in adult man as the vitamin is widely distributed in natural foods. Egg yolk, kidney, liver, tomatoes and yeast are excellent sources.

13.4. Biotin in Milk

Typical concentrations of biotin in bovine and human milks are 35 and 7 µg litre^{-1}, respectively. As with thiamine, riboflavin, B$_6$, folic and pantothenic acid, but unlike vitamin B$_{12}$, the biotin concentrations in human and bovine colostra are lower than those in their respective mature milks. Post-parturition, the concentration of biotin rises gradually to a maximum in human milk[30,119] and within a few days in cow's milk.[89,146] Biotin supplementation of women of poor nutritional status effected an increase from 1·5 to 5 µg litre^{-1},[121] but no such effect was observed in women with an initial concentration of 8 µg litre^{-1} of milk.[147] Based on earlier reports,[148,149] it was concluded that a direct nutritional influence on the biotin content of bovine milk was questionable.[21] More recently, Roberts and Baggott,[150] in a supplementation experiment with a herd of 100 cows with a history of severe lameness, administered 20 mg of biotin per cow per day to half the cows and found significantly higher biotin concentrations in the blood and milk of supplemented animals. Biotin supplementation was not observed to affect milk yield or the concentrations of fat and protein in the milk, but a marked reduction was observed in the number of cows suffering from lameness due to soft-horn tissue.

14. VITAMIN C (ASCORBIC ACID)

14.1. Chemical Structure and Physical Properties

Ascorbic acid is the most important of the various compounds that possess vitamin C activity. It occurs in all living tissues as an important redox compound in cell metabolism. Its chemical structure (Fig. 27) resembles that of a monosaccharide, but it possesses an ene-diol group between carbons 2 and 3 which is readily and reversibly oxidized to a diketo group with the resultant formation of dehydroascorbic acid. The ene-diol group may be involved in its physiological function, for example, in hydrogen transfer systems as in the oxidation of tyrosine. In addition to L-ascorbic acid, sodium ascorbate is frequently used in commercial practice. Both are white to yellow-tinged, readily soluble in water, crystalline powders exhibiting pH-dependent absorption maxima.

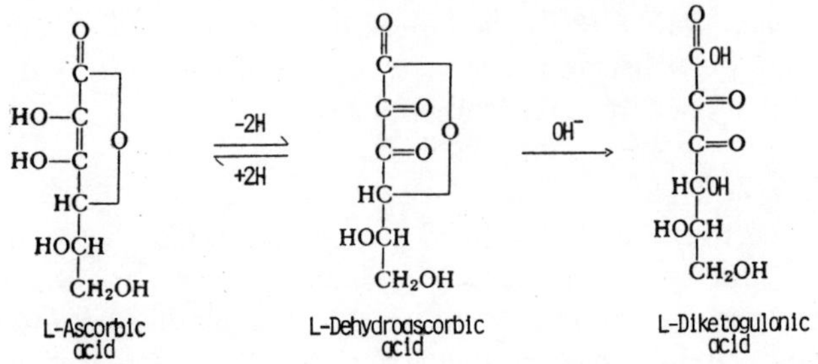

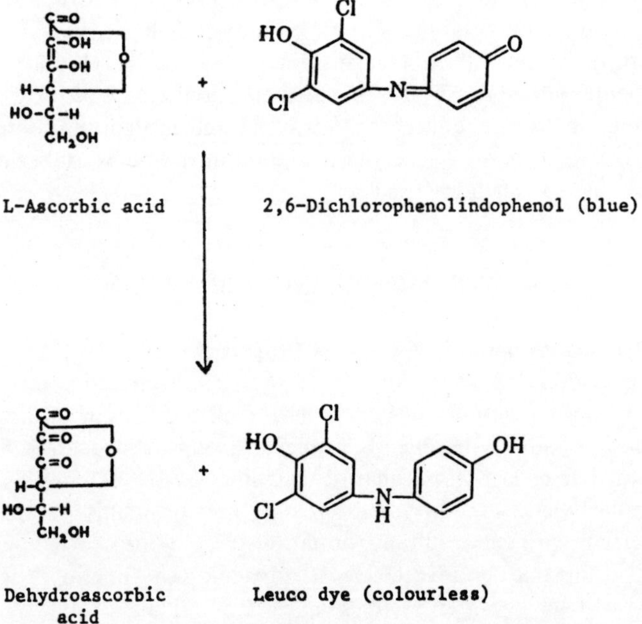

FIG. 27. Oxidation of ascorbic acid (vitamin C).

In ultraviolet light, ascorbic acid in strongly acidic solutions has an absorption maximum at ~ 245 nm, which shifts at neutrality to 265 nm and at pH 14 to 300 nm. It is optically active: ascorbic acid, $[\alpha]_D^{20} = -22 - 23°$ (c 2, in water); sodium ascorbate, $[\alpha]_D^{20} = +103–106°$ (c 5, in water). Crystalline ascorbic acid is stable in air in the absence of water, whereas

FIG. 28. Oxidation of L-ascorbic acid by 2,6-dichlorophenolindophenol as used in the titrimetric determination of vitamin C.

sodium ascorbate tends to turn yellow. Aqueous solutions are attacked by atmospheric oxygen and other oxidizing agents; dehydroascorbic acid is the first oxidation product which is subsequently irreversibly oxidized to diketogulonic acid (Fig. 27). Alkalis and traces of heavy metal ions, including copper and iron, act as catalysts for its destruction. Most of the assays for vitamin C in foods are based on the reductive properties of the carbonyl ene-diol group, which is oxidized, yielding dehydroascorbic acid. The end-point determination is by visual indicator titration, using 2,6-dichlorophenolindophenol (Fig. 28), photometry and potentiometry.

14.2. Physiological Function and Deficiency Symptoms

In the body, vitamin C is essential for the formation of intercellular substances of skeletal tissues, including connective tissues, bones, cartilage and dentine. The impaired formation of these substances in vitamin C deficiency is caused by impairment of the hydroxylation of proline to hydroxyproline and lysine to hydroxylysine, which are important structural components of collagen fibres.

L-Ascorbic acid and folic acid interact in metabolism, as folic acid requirements are increased in ascorbic acid deficiency. Man, primates, guinea pigs and fish (so far investigated) depend on their food supply for ascorbic acid, because their cells lack the enzyme L-gulonolactone oxidase required for ascorbic acid synthesis from D-glucose and D-galactose.

Unlike the other vitamins, severe deficiency symptoms are observed mainly in humans and are characterized as scurvy in adults and as Moller Bärlow disease in children. Mild deficiency symptoms include weakness, fatigue, dyspnoea, aching bones, hyperaesthesia and pain in the lingual and buccal mucosa, follicular hyperkeratosis, bleeding of gums and haemorrhagic diathesis.

Groups susceptible to scurvy in the population include 6–12 month old infants fed processed milk formulae and whose diets are not supplemented with fresh fruits and vegetables (infantile scurvy), elderly bachelors and widowers living alone and preparing their own food (bachelor scurvy), and food faddists, especially those living exclusively on macrobiotic diets.

14.3. Human Requirement and Food Sources

The adult recommended daily dietary allowance for ascorbic acid is 60 mg. This allowance is considerably increased by physiological and pathological conditions including pregnancy and lactation and stress induced by burns, surgery or infection. Arthritic patients who are consuming aspirin regularly may require additional vitamin C as it has been observed *in vitro* to block the uptake of vitamin C by blood platelets. Vitamin C is metabolized to oxalate in the body, and excess intakes may result in the formation of calcium oxalate stones in the kidney and urinary bladder.

Important sources of vitamin C are fresh fruits, especially citrus fruits, blackcurrants, rose-hips, sea buck-thorn, paprika, tomatoes and vegetables including cabbage, potatoes and lettuce. The only animal organs that contain significant amounts of ascorbic acid are adrenals, liver and pituitary. The quantity of an individual food consumed determines its significance as a source of vitamin C in the diet, rather than its concentration of vitamin C. The infant is usually well supplied with vitamin C at birth.

14.4. Vitamin C Activity in Milk

The concentration of vitamin C in human milk is approximately twice that in cow's milk, typical values being about 45 and 20 mg litre^{-1}, respectively. Scott *et al.*[14] reported average vitamin C concentrations of 14·5 mg litre^{-1} in bulk HTST milk, and no seasonal variation was evident. Vitamin C in freshly secreted cow's milk is predominantly in the L-ascorbate form, but in the presence of oxygen it is rapidly oxidized to dehydroascorbate.

The destruction of vitamin C in milk can be substantial and, like folic acid, the extent of destruction is determined mainly by the dissolved oxygen content of milk.[151] This process is greatly accelerated by the photochemical destruction of riboflavin with the resultant formation of lumichrome and lumiflavin, both of which effect the catalytic destruction of vitamin C. Scott *et al.*[106] reported that the total vitamin C concentration in milk was reduced from 13·8 to 3·9 mg litre^{-1} after 6 hours while the ascorbic acid concentration was lowered from 10·8 to 0·3 mg litre^{-1} after 1 hour in milk stored in glass bottles and exposed to sunlight. Further losses occurred during subsequent storage of these samples under refrigerated conditions, whereas only small losses occurred in refrigerated milk that was not previously exposed to sunlight. Haddad and Loewenstein[76] reported that ~30% of vitamin C in fresh raw milk was destroyed after storage at −40°C for 2 weeks. Oxidation of vitamin C occurs by a two-stage process through a reversible first step to dehydroascorbate and an irreversible second step to diketogulonate (Fig. 27) which possesses no biological activity. Losses of vitamin C activity during HTST and UHT treatment of milks average 20%,[76,86,152] but losses incurred during in-bottle sterilization of milk are higher at 40–79%.[85,152]

The vitamin C concentration of human colostrum is only slightly elevated compared with mature milk,[69,114] whereas high concentrations occur in bovine colostrum, but these fall within a few days.[98,153] Seasonal variations are reported for human[31] but not for bovine milks.[14,21] Citrus fruit supplementation of the diets of Hungarian women observed to have low serum vitamin C concentrations doubled the vitamin C concentration of their milks.[27] This effect was not observed in North American women whose diet was supplemented with 130% of the RDA for vitamin C.[110] There is evidence that the mammary glands of human and bovine species have some vitamin C biosynthetic capacity.[79,154,155]

REFERENCES

1. WAGNER, F. and FOLKERS, K., *Vitamins and Coenzymes*, 1964, Interscience Publishers, New York.
2. DYKE, S. F., *The Chemistry of Vitamins*, 1965, Interscience Publishers, New York.
3. HARPER, H. A., RODWELL, V. W. and MAYES, P. A., *Review of Physiological Chemistry*, 1979, Lange Medical Publications, Los Altos, California.
4. GOODHART, R. S. and SHILS, M. E., *Modern Nutrition in Health and Disease*, 5th edn, 1973, Lea and Febiger, Philadelphia.
5. F. Hoffmann-La Roche & Co., *Vitamin Compendium*, 1976, Basle, Switzerland.
6. Department of Health and Social Security, *Recommended Intakes of Nutrients for the United Kingdom*, Report No. 120, 1969, HMSO, London.
7. FAO/WHO, *Requirements of Vitamin A, Thiamine, Riboflavin and Niacin*, Report No. 41, 1967, WHO Technical Report Series No. 362.
8. NRC, National Academy of Sciences, Committee on Food and Nutrition, *Recommended Daily Dietary Allowances*, 1980.
9. PARRISH, D. B., WISE, G. H. and HUGHES, J. S., *J. Biol. Chem.*, 1947, **167**, 673.
10. PARRISH, D. B., WISE, G. H. and HUGHES, J. S., *J. Dairy Sci.*, 1947, **30**, 849.
11. KRUKOVSKY, V. N. and WHITING, F., *J. Dairy Sci.*, 1948, **31**, 666.
12. WAGNER, K. H., *Milchwissenschaft*, 1952, **7**, 250.
13. THOMPSON, S. Y., HENRY, K. M. and KON, S. K., *J. Dairy Res.*, 1964, **31**, 1.
14. SCOTT, K. J., BISHOP, D. R., ZECHALKO, A. and EDWARDS-WEBB, J. D., *J. Dairy Res.*, 1984, **51**, 37.
15. HARTMAN, A. M. and DRYDEN, L. P., Vitamins in Milk and Milk Products, In: *Fundamentals of Dairy Chemistry*, Webb, B. H., Johnson, A. H. and Alford, J. A. (eds), 1984, Avi Publishing Co., Westport, CT, pp. 325–401.

16. RENNER, E., *Milk and Dairy Products in Human Nutrition*, 1983, Volkswirtschafticher Verlag, München.
17. HAMBRAEUS, L., *Nutr. Abst. Rev.*, 1984, **54**, 219.
18. SHEPHERD, J. B., WEISMAN, H. G., ELY, R. E., MELIN, G. C., SWEETMAN, W. J., GORDON, C. H., SCHOENLEBER, L. G., WAGNER, R. E., CAMPBELL, L. E., ROANE, G. D. and HOSTERMAN, W. H., US Department of Agriculture Technical Bulletin No. 1079, 1954.
19. DIJKSTRA, N. D., *Versl. Landbouwk Onderzoek.*, 1957, **63**, 10.
20. HAUGE, S. M., WESTFALL, R. J., WILBUR, J. W. and HILTON, J. H., *J. Dairy Sci.*, 1944, **27**, 63.
21. KIRCHGESSNER, M., FRIESECKE, H. and KOCH, G., Nutrition and the Vitamin Content of Milk, In: *Nutrition and the Composition of Milk*, 1967, Crosby Lockwood, London, pp. 164–208.
22. HANSEN, R. G., PHILLIPS, P. H. and SMITH, V. R., *J. Dairy Sci.*, 1946, **29**, 809.
23. HENRY, K. M., HOUSTON, J. and KON, S. K., *J. Dairy Res.*, 1940, **11**, 1.
24. PARRISH, D. B., WISE, G. H., ATKESON, F. W. and HUGHES, J. S., *J. Dairy Sci.*, 1949, **32**, 209.
25. HENNAUX, L., ANTOINE, A. and LECOMTE, R., *Gembloux*, 1951, **19**, 251.
26. GABRE-MEDHIN, M., *Am. J. Clin. Nutr.*, 1976, **29**, 441.
27. TARJAN, R., KRAMER, M., SZOKE, K. and LINDER, K., *Nutr. Diet.*, 1963, **5**, 12.
28. TARJAN, R., KRAMER, M., SZOKE, K., LINDER, K., SZARVAS, T. and DWORSCHACK, E., *Nutr. Diet.*, 1965, **7**, 136.
29. MACY, I. G., *Am. J. Dis. Child.*, 1949, **78**, 589.
30. MACY, I. G. and KELLY, H. J., Human Milk and Cow's Milk in Infant Nutrition, In: *The Mammary Gland and Its Secretion*, Vol. 2, Kon, S. K. and Cowie, A. T. (eds), 1961, Academic Press, New York, pp. 265–304.
31. KON, S. K. and MAWSON, E. H., Medical Research Council Special Report No. 269, 1950, HMSO, London.
32. REEVE, L. E., CHESNEY, R. W. and DE LUCA, H. F., *Am. J. Clin. Nutr.*, 1982, **36**, 122.
33. REEVE, L. E., JORGENSEN, N. A. and DE LUCA, H. F., *J. Nutr.*, 1982, **112**, 667.
34. JACKSON, P., SHELTON, C. J. and FRIER, P. J., *Analyst*, 1982, **107**, 1363.
35. HOLLIS, B. W., *Anal. Biochem.*, 1983, **131**, 211.
36. WEISMAN, Y., BAWNIK, J. C., EISENBERG, Z. and SPIRER, Z., *J. Pediat.*, 1982, **100**, 745.
37. LEERBECK, E. and SØNDERGAARD, H., *Brit. J. Nutr.*, 1980, **44**, 7.
38. TANAKA, Y. and DE LUCA, H. F., *Endocrinology*, 1973, **92**, 417.
39. HESS, A. F., LEWIS, J. M., MACLEOD, F. L. and THOMAS, B. H., *J. Am. Med. Assoc.*, 1931, **97**, 370.
40. HESS, A. F., LIGHT, R. F., FREY, C. N. and GROSS, J., *J. Biol. Chem.*, 1932, **97**, 369.
41. KRAUSS, W. E., BETHBE, R. M. and MONROE, C. F., *J. Nutr.*, 1932, **5**, 467.
42. RUSSELL, W. C., WILCOX, D. E., WADDELL, J. and WILSON, L. T., *J. Dairy Sci.*, 1934, **17**, 445.

43. LIGHT, R. F., WILSON, L. T. and FREY, C. N., *J. Nutr.*, 1937, **14**, 453.
44. WALLIS, G. C., *J. Dairy Sci.*, 1938, **21**, 11.
45. FREY, C. N., LIGHT, R. F. and WILSON, L. T., *J. Dairy Sci.*, 1936, **19**, 94.
46. GUNTHER, K. and LENKEIT, W., *Mitteilung. Ztschr. Tierphysiol. Tierernährung Futtermittelk.*, 1962, **17**, 275.
47. ROCHE, *Vitamins and Carotenoids for Livestock 1982–1983*, 1982, Roche Products Ltd., Chemical Division, 318 High Street North, Dunstable, Bedfordshire LU6 1BG, England.
48. LENKEIT, W., BRUNE, H. and GUNTHER, K., *Mitteilung. Ztschr. Tierphysiol. Tierernährung Futtermittelk.*, 1960, **15**, 155.
49. THOMPSON, J. N. and HIRIROYLAN, M., *J. Dairy Sci.*, 1983, **66**, 1638.
50. HENRY, K. M., HOUSTON, J. and KON, S. K., *Biochem. J.*, 1937, **31**, 2199.
51. EATON, H. D., SPIELMAN, A. A., LOOSLI, J. K., THOMAS, J. W., NORTON, C. C. and TURK, K. L., *J. Dairy Sci.*, 1947, **30**, 787.
52. SAHASHI, Y., SUZUKI, T., HIGAKI, M. and ASANO, T., *J. Vitaminol.*, 1969, **15**, 78.
53. LE BOULCH, N., GULAT-MARNAY, C. and RAOUL, Y., *Intern. J. Vit. Res.*, 1974, **44**, 167.
54. LAKDAWALA, D. R. and WIDDOWSON, E. W., *Lancet*, 1977, **i**, 167.
55. ANTILA, P., ANTILA, V. and KUUJO, S., *Finn. J. Dairy Sci.*, 1979, **37**, 1.
56. HOLLIS, B. W., ROOS, B. A., DRAPER, H. H. and LAMBERT, P. W., *J. Nutr.*, 1981, **111**, 384.
57. REEVE, L. E., DE LUCA, H. F. and SCHNOES, H. K., *J. Biol. Chem.*, 1981, **256**, 823.
58. DE LUCA, H. F., *Nutr. Rev.*, 1980, **38**, 169.
59. BROWN, F., *Biochem. J.*, 1952, **51**, 237.
60. KOBAYASHI, H., KANNO, C., YAMAUCHI, K. and TSUGO, T., *Biochim. Biophys. Acta*, 1975, **380**, 282.
61. LETH, T. and SØNDERGAARD, H., *J. Nutr.*, 1977, **107**, 2236.
62. KRAMER, M., SZOKA, K., LINDER, K. and TARJAN, R., *Nutr. Diet*, 1965, **7**, 71.
63. BROWN, F., *J. Sci. Fd. Agr.*, 1953, **4**, 161.
64. VAN DER KAAY, F. C., TEUNISSEN, G. H. B., EMMERIE, A. and VAN ECKELEN, M., *Internat. Ztschr. Vitaminforsch.*, 1949, **21**, 140.
65. WHITING, F., LOOSLI, J. K., KRUKOVSKY, V. N. and TURK, K. L., *J. Dairy Sci.*, 1949, **32**, 133.
66. KRUKOVSKY, V. N., LOOSLI, J. K. and THEOKAS, D. A., *J. Dairy Sci.*, 1949, **32**, 700.
67. KRUKOVSKY, V. N., LOOSLI, J. K. and WHITING, F., *J. Dairy Sci.*, 1949, **32**, 196.
68. HVIDSTEN, H., MEHIUM, J. and SIMONSON, H., *Meieriposten*, 1952, **41**, 186.
69. BLANC, B., *World Rev. Nutr. Diet.*, 1981, **36**, 1.
70. NICHOLS, B. L. and NICHOLS, V. N., *Nutr. Abstr. Rev.*, 1983, **53**, 259.
71. KEENAN, W. J., JEWETT, T. and GLULCH, H. I., *Am. J. Dis. Child.*, 1971, **121**, 271.
72. GYORGY, P., *Am. J. Clin. Nutr.*, 1971, **24**, 970.

73. HAROON, Y., SHEARCE, M. J., SEEMA, R., GUNN, W. G., MCENERY, G. and BARKHAN, P., *J. Nutr.*, 1982, **112**, 1105.

74. Department of Health and Social Security, *The Composition of Mature Human Milk*, 1970, HMSO, London.

75. MCELROY, L. W. and GOSS, H., *J. Nutr.*, 1940, **20**, 541.

76. HADDAD, G. S. and LOEWENSTEIN, M., *J. Dairy Sci.*, 1983, **66**, 1601.

77. LUECKE, B. W., DUNCAN, C. W., ELY, R. E., JONES, E., GREENE, A. J., and TULL, C., *Arch. Biochem.*, 1947, **13**, 277.

78. BELAVADY, B., *Indian J. Med. Res.*, 1969, **57**, 63.

79. JELLIFFE, D. B. and JELLIFFE, E. F. P., *Human Milk in the Modern World*, 1978, Oxford University Press, Oxford.

80. BARTLETT, S., HENRY, K. M. and KON, S. K., *J. Dairy Res.*, 1940, **11**, 22.

81. HOLMES, A. D., JONES, C. P. and WERTZ, A. W., *Am. J. Dis. Child.*, 1944, **67**, 376.

82. MARSH, D. C., PEARSON, P. B. and RUPEL, I. W., *J. Dairy Sci.*, 1947, **30**, 867.

83. CAUSERET, J., *Ann. Nutr. Aliment.*, 1977, **25**, A313.

84. PACKARD, V. S., Vitamins, In: *Human Milk and Infant Formula*, 1982, Academic Press, New York, pp. 29–49.

85. VAN ECKLEN, M. and HEIJNE, J. J. I. G., Nutritive Value of Sterilized Milk, In: *Milk Sterilization*, 1965, Food and Agricultural Organization of the United Nations, Rome, pp. 33–41.

86. FORD, J. E., PORTER, J. W. G., THOMPSON, S. W., TOOTHILL, J. and EDWARDS-WEBB, J., *J. Dairy Res.*, 1969, **36**, 447.

87. GREGORY, M. E. and BURTON, H., *J. Dairy Res.*, 1965, **32**, 13.

88. Ministry of Agriculture, Fisheries and Food, *Household Food Consumption and Expenditure*, 1980 Annual Report of National Food Supply Committee, 1982, HMSO, London.

89. GREGORY, M. E., FORD, J. E. and KON, S. K., *J. Dairy Res.*, 1958, **25**, 447.

90. ANAGAMA, Y. and KUSUGA, Y., *Res. Bull. Fac. Agric. Gifu Univ.* (Japan), 1970, **29**, 247.

91. ANAGAMA, Y. and KUSUGA, Y., *Res. Bull. Fac. Agric. Gifu Univ.* (Japan), 1973, **34**, 387.

92. FUNAI, Y., *J. Exp. Med.*, 1955, **2**, 194.

93. MODI, V. V. and OWEN, E. L., *Nature*, 1956, **178**, 1120.

94. NAGASAWA, T., KUZUYA, Y. and SHIGETA, N., *Jap. J. Zootech. Sci.*, 1961, **32**, 235.

95. NAGASAWA, T., KUZUYA, Y. and SHIGETA, N., *Jap. J. Zootech. Sci.*, 1961, **32**, 240.

96. SWOPE, F. C., BRUNNER, J. R. and VADEHRA, D. V., *J. Dairy Sci.*, 1965, **48**, 1707.

97. SUTTON, J. S., WARNER, R. G. and KAESER, H. E., *J. Dairy Sci.*, 1947, **30**, 927.

98. VAN LANDNGHAM, A. H., FLANDERS, L. A. and ACKERMAN, R. A., *J. Dairy Sci.*, 1949, **32**, 720.

99. HUNT, C. H., BURROUGHS, E. W., BETHKE, R. M., SCHALK, A. F. and GERLAUGH, P., *J. Nutr.*, 1943, **25**, 207.

100. HUNT, C. H., KICK, C. H., BURROUGHS, E. W., BEIHKE, E. M., SCHALK, A. F. and GERLAUGH, P., *J. Nutr.*, 1941, **21**, 85.
101. CONRAD, H. R. and HIBBS, J. W., *J. Dairy Sci.*, 1954, **37**, 512.
102. POPPE, S., *Arch. Tierernährung*, 1958, **8**, 9.
103. POPPE, S., *Arch. Tierernährung*, 1958, **8**, 99.
104. SINGH, R. and MERILAN, C. P., *Univ. Miss. Agric. Exp. Stat. Res. Bull.* No. 688, 1959.
105. GORNER, F. and UHEROVA, R., *Nährung*, 1980, **24**, 373.
106. SCOTT, K. J., BISHOP, D. R., ZECHALKO, A. and EDWARDS-WEBB, J. D., *J. Dairy Res.*, 1984, **51**, 51.
107. BENDER, A. E., *Food Processing and Nutrition*, 1978, Academic Press, London.
108. *Dairy Council Digest*, 1977, **48**, 13.
109. KREHL, W. A., DELA HUERGA, J., ELVEHJEM, C. A. and HART, E. B., *J. Biol. Chem.*, 1946, **166**, 53.
110. THOMAS, M. R., KAWAMOTO, J. and SNEED, S. M., *Am. J. Clin. Nutr.*, 1979, **32**, 1679.
111. KUNG, L., GUBERT, K. and HUBER, J. T., *J. Dairy Sci.*, 1980, **63**, 2020.
112. HARMEYER, J. and GRABE, C., *Deutsche Tierarzk Wschr.*, 1981, **88**, 401.
113. RIDDEL, D. O., BARTLEY, E. E. and DAYTON, A. D., *J. Dairy Sci.*, 1981, **64**, 782.
114. GURR, M. I., *J. Dairy Res.*, 1981, **48**, 519.
115. SONG, W. O., CHAN, G. M., WYSE, B. W. and HANSEN, R. G., *Am. J. Clin. Nutr.*, 1984, **40**, 317.
116. VANDERSLICE, J. T., BROWNLESS, S. G., MAIRE, C. E., REYNOLDS, R. D. and POLANSKY, M., *Am. J. Clin. Nutr.*, 1983, **37**, 867.
117. GREGORY, M. E., *J. Dairy Res.*, 1959, **26**, 203.
118. GREGORY, M. E. and MABBITT, L. A., *J. Dairy Res.*, 1961, **28**, 293.
119. FORD, J. E., ZECHALKO, A., MURPHY, J. and BROOKE, O. G., *Arch. Dis. Child.*, 1984, **58**, 367.
120. KARLIN, R. and PORTAFAIX, D., *C. R. Soc. Biol.*, 1960, **154**, 623.
121. BELAVADY, B. and GOPOLAN, C., *Indian J. Med. Res.*, 1960, **48**, 518.
122. DEBRIT, F. B., *Internat. Zeitschr. Vitaminforsch.*, 1952, **24**, 331.
123. BERNHART, F. W., D'AMATO, E. and TOMARELLI, R. M., *Arch. Biochem. Biophys.*, 1960, **88**, 267.
124. SANDBERG, D. P., BERGLEY, J. A. and HALL, C. A., *Am. J. Clin. Nutr.*, 1981, **34**, 1717.
125. FARQUHARSON, J. and ADAMS, J. F., *Brit. J. Nutr.*, 1976, **36**, 127.
126. FORD, J. E., *Brit. J. Nutr.*, 1974, **31**, 243.
127. KIM, Y. P., GIZIS, E., BRUNNER, J. R. and SCHWEIGERT, B. S., *J. Nutr.*, 1965, **86**, 394.
128. DEODHAR, A. D., RAJALAKSKMI, R. and RAMAKRISHNAN, C. V., *Acta Paediatr. Scand.*, 1964, **53**, 42.
129. JATHAR, V. S., KAMATH, S. A., PARIKH, M. N., REGE, D. V. and SATOSKAR, R. S., *Arch. Dis. Child.*, 1970, **45**, 236.
130. HALL, C. A., *Gastroenterology*, 1973, **65**, 684.
131. RUSOFF, F. A. and HAQ, M. O., *J. Dairy Sci.*, 1954, **37**, 677.

132. Agricultural Research Council, *The Nutrient Requirements of Ruminant Livestock*, Commonwealth Agricultural Bureau, 1980.
133. SMITH, R. M. and MARSTON, H. R., *Brit. J. Nutr.*, 1970, **24**, 879.
134. MACPHERSON, A., Dietary Vitamin B12 and Cobalt for Ruminants, In: *Recent Research on the Vitamin Requirements of Ruminants* (Roche Vitamin Symposium), Nov. 1982, London.
135. KORSHAKOV, P. N., *Nutr. Abstr. Rev.*, 1954, **25**, 350.
136. DAVIDOV, R. A. and KRUGLOVA, A., *Nutr. Abstr. Rev.*, 1958, **29**, 1186.
137. COOPERMAN, J. M., DWECK, H. S., NEWMAN, L. J., GARBARINO, C. and LOPEZ, R., *Am. J. Clin. Nutr.*, 1982, **36**, 576.
138. EK, J., *Am. J. Clin. Nutr.*, 1983, **38**, 929.
139. COLLINS, R. A., HARPER, A. E., SCHREIBER, M. and ELVEHJEM, C. A., *J. Nutr.*, 1951, **43**, 313.
140. COLLINS, R. A., BOLDT, R. E., ELVEHJEM, C. A., HART, E. B. and BOMSTEIN, R. A., *J. Dairy Sci.*, 1953, **36**, 24.
141. LARDINOIS, C. C., Mills, R. L., ELVEHJEM, C. A. and HART, E. B., *J. Dairy Sci.*, 1944, **27**, 579.
142. TAMURA, T., SHIN, Y., WILLIAMS, M. A. and STOCKSTAD, E. L. R., *Anal. Biochem.*, 1972, **49**, 517.
143. SALTER, D. N., SCOTT, K., SLADE, H. and ANDREWS, P., *Biochem. J.*, 1981, **193**, 469.
144. WAXMAN, S. and SCHREBER, C., *Biochemistry*, 1975, **14**, 5422.
145. SALTER, D. N. and MOWLEW, A., *Brit. J. Nutr.*, 1983, **50**, 589.
146. LAWRENCE, J. M., HERRINGTON, B. L. and MAYNARD, L. A., *J. Nutr.*, 1946, **32**, 73.
147. PRATT, J. P., HAMIL, B. M. and MOYER, E. Z., *J. Nutr.*, 1951, **44**, 141.
148. STEFANIAK, J. J. and PETERSON, W. H., *J. Dairy Sci.*, 1949, **29**, 783.
149. DAVIDOV, R. A. and KRUGLOVA, A., *Nutr. Abstr. Rev.*, 1959, **29**, 829.
150. ROBERTS, G. J. and BAGGOTT, D. G., Biotin in Ruminant Nutrition, In: *Recent Research on the Vitamin Requirements of Ruminants* (Roche Vitamin Symposium), Nov. 1982, London.
151. FORD, J. E., *J. Dairy Res.*, 1967, **34**, 239.
152. MOTTAR, J. and NAUDTS, M., *Lait*, 1979, **59** (588), 476.
153. KON, S. K. and WATSON, M. B., *Biochem. J.*, 1937, **31**, 223.
154. VON WENDT, G., *Skand. Arch. Physiol.*, 1938, **80**, 398.
155. REINEKE, E. P., GARRISON, E. R. and TURNER, C. W., *J. Dairy Sci.*, 1941, **24**, 41.

INDEX